DWARFED
FRUIT TREES

DWARFED
FRUIT TREES

for ORCHARD, GARDEN, *and* HOME
. . . . . *with Special Reference to*
the Control of Tree Size and Fruiting
in Commercial Fruit Production

BY HAROLD BRADFORD TUKEY

COMSTOCK PUBLISHING ASSOCIATES a division of
CORNELL UNIVERSITY PRESS | Ithaca and London

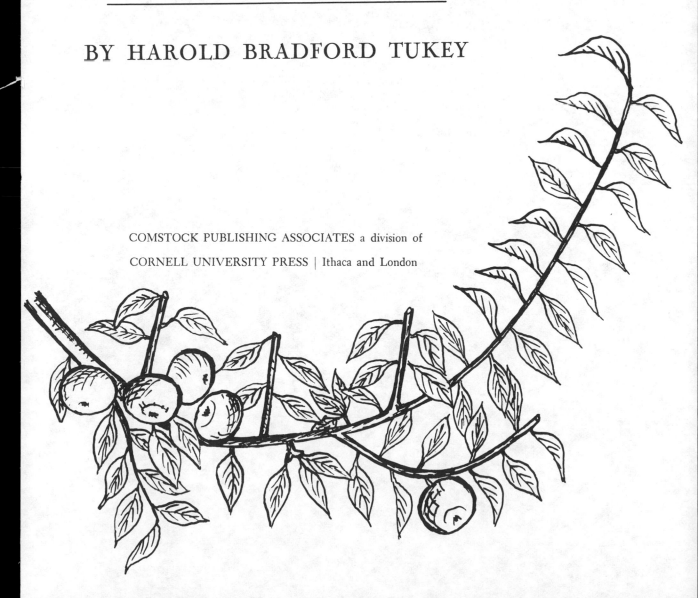

First published in 1964 by The Macmillan Company, New York. Reissued 1978 by Cornell University Press. Published in the United Kingdom by Cornell University Press Ltd., London.

Third printing, 1983

International Standard Book Number 0–8014–1126–2
Library of Congress Catalog Card Number 77–12289
Designed by Jack Meserole
Printed in the United States of America
Librarians: Library of Congress cataloging information appears on the last page of the book.

ACKNOWLEDGMENTS

The author is indebted to Dr. R. F. Carlson, Michigan State University, and Professor George Slate, New York State Agricultural Experiment Station, for reading the entire manuscript and for making many valuable suggestions. Colleagues at the East Malling Research Station have been most helpful with illustrations, material, and suggestions, including Dr. H. M. S. Montgomery, T. N. Hoblyn, Barbara Mosse, R. J. Garner, and especially A. N. Preston, who reviewed the chapters dealing with pruning, apple rootstocks, and commercial fruit production. Dr. P. Rémy, Villefranche-sur-Saône, France, reviewed the chapter on pear rootstocks; Dr. W. T. Chang, formerly of the University of Nanking, China, reviewed the chapter on bonsai; Dr. W. P. Bitters, University of California, Riverside, reviewed the material on citrus; Raymond Bush, Surrey, England, and Shinobu Nozaki, Tokyo, Japan, permitted free use of their published material. Permission was granted to use material and illustrations by Établissement Georges Truffaut, Paris, France; Faber & Faber, Ltd., London, England; Cassell & Company, Ltd., London, England; John Wiley & Sons, New York City, New York; Penguin Books Limited, Middlesex, England; and the Royal Horticultural Society, London, England.

Photographs and drawings were supplied by the East Malling Research Station, Kent, England; Dr. George Avery, Brooklyn Botanic Garden, Brooklyn, New York; Dr. John Harris, Saanichton, British Columbia; Dr. R. H. Sharpe, University of Florida, Gainesville, Florida; Dr. R. J. Seibert, Longwood Gardens, Kennett Square, Pennsylvania;

E. J. Rasmussen, University of New Hampshire, Durham, New Hampshire; Dr. John R. Magness, United States Department of Agriculture, Beltsville, Maryland; Dr. David Armstrong, Armstrong Nurseries, Ontario, California; Croux Fils, Châtenay-Malabry, France; R. T. Meister and the American Fruit Grower, Willoughby, Ohio; Dr. W. P. Bitters, University of California, Riverside, California; Cal Bosch, Goodfruit Grower, Yakima, Washington; Gordon MacLean, Abingdon, England; F. L. O'Rourke, Dr. John Bukovac, and Dr. S. H. Wittwer, Michigan State University, East Lansing, Michigan; and the Four Winds Nurseries, San Jose, California.

My wife, Dr. Ruth S. Tukey, added many valuable suggestions to her constant encouragement. My daughter, Dr. Ann Tukey, assisted with French translations. Cecil Scott, and the editorial and publishing staff of The Macmillan Company were most understanding and helpful. Ted Cichoz, Institute of Pomology, Skierniewice, Poland, made most of the drawings; my daughter, Mrs. Lois Tukey Baker, made the others. Mrs. M. S. Barrett, Michigan State University, performed all the typing and did much of the detail work.

H. B. TUKEY

East Lansing, Michigan

PREFACE

The small, compact, dwarfed, or size-controlled fruit tree seems at once the natural and obvious answer to many problems with fruit trees. To the landscape gardener it suggests the possibilities of fruit trees used as ornamentals or in a more complete garden. To the amateur horticulturist it suggests growing a variety of choice fruits. To the suburban dweller, to the man with a few acres of land, and to the subsistence farmer it suggests the possibilities of growing a sufficient quantity of fruit to satisfy his own and local needs. To the commercial orchardist it suggests early fruiting, a more rapid turnover in varieties to meet changing market requirements, lower cost of production, higher proportion of high-grade fruit, and easier thinning, pruning, spraying, and harvesting. Finally, there is the elderly gentleman who writes: "I can no longer climb a ladder, yet I want to grow fruit. I must do my orcharding with my feet on the ground, hence my interest in dwarf fruit trees."

It is now between sixty and eighty years since the last surge of interest in dwarfed fruit trees in America. Much has occurred in that time to alter the entire viewpoint. New rootstocks have made their appearance; improvements in spraying and dusting have made the control of insects and diseases a much more dependable and far less laborious undertaking; markets are demanding higher-quality fruits; and interest in plants, in gardening, and in country life is increasing. Added to this, the scientist has contributed much that is fundamental and helpful in understanding and carrying out some of the practices of dwarf fruit tree culture. Accordingly, another discussion of dwarfed fruit trees seems timely and worthwhile.

This book attempts to draw together under one cover a wide range of subject matter dealing with dwarfed fruit trees, with major emphasis on the apple; as, the historical background of dwarfed fruit trees, features and limitations, the structure and physiology of fruit trees, how a tree is dwarfed, and stock and scion relations; the nature and identity of dwarfing rootstocks, propagation of rootstocks, and nursery management; planting, pruning, training, nutrient requirements, trellising, pollination, fruit set, and thinning; yields, fruit quality, and costs; trees trained to artistic and novel forms; trees in pots and under glass; bonsai and dwarfed trees as ornamentals. Some of the discussion is necessarily similar to that found in other books dealing with fruit culture. It has seemed worthwhile, nevertheless, to place a number of these items together where they are easily accessible without requiring the reader to resort too frequently to specialized publications.

While the approach has been largely from the practical viewpoint and has been flavored in good measure with the personal experiences of the author, an attempt has been made also to include wherever possible some of the more recent findings of science, especially those of a more fundamental nature that tend to explain and rationalize practices formerly accepted but little understood.

H. B. TUKEY

East Lansing, Michigan

CONTENTS

Part Four

PROPAGATION OF DWARFING ROOTSTOCKS AND DWARFED FRUIT TREES

Part Five

ESTABLISHMENT AND MANAGEMENT OF THE ORCHARD

Part Six

ORCHARD PERFORMANCE OF DWARFED FRUIT TREES

DWARFED FRUIT TREES AS SPECIAL INTERESTS: IN POTS, UNDER GLASS, AS BONSAI, AND AS ORNAMENTALS

Part Seven

Part One

THE GENERAL BACKGROUND OF DWARFED FRUIT TREES

1

WHAT IS A DWARFED FRUIT TREE?

THE word "dwarf" immediately calls to mind the diminutive manlike figures of mythology and folklore—frequently misshapen, ugly, bearded, and gray, living to great ages in subterranean caverns and rocks. They were expert with minerals and metals and forged wonderful weapons and treasures for the gods. They were given to mischievous and malicious pranks in their dealings with men. In the Norse country, an echo is often called "dwarf language." "Dwarf," then, implies something more than merely small stature. It carries with it an image of additional distinctive qualities and characteristics.

In somewhat similar vein, a dwarf plant or a dwarf fruit tree not only does not attain the stature typical of the group, species, or kind to which it belongs; it is decidedly smaller or even miniature. Further, it is special and distinctive, often differing in shape, in habit of growth and fruiting, in time of blossoming, in time of fruit maturity, and in adaptability to various environmental conditions. Such plants are often referred to as "true dwarfs," as though to emphasize the exactness of the term (Fig. 1).

The small stature of true dwarf plants, however, does not necessarily imply that they are sickly or weak-growing. On the contrary, most of them are healthy and vigorous, though not so continuously strong-growing as the standard.

Speaking of plants in general, there are two general types of dwarfs; namely, (1) those that are dwarfed by nature or genetic makeup, and

3

Fig. 1. **True dwarf or very dwarf Baldwin apple tree propagated on the very dwarfing EM IX rootstock (Baldwin/EM IX): third season in the orchard; it is the same age as the girl in the photograph. The tree bore some fruit the second year planted and will never become taller than a man can reach.** *Courtesy of Jackson and Perkins Company*

(2) those that are made dwarf by environmental factors or artificial means. Dwarf corn (*Zea mays*), dwarf bush beans (*Phaseolus vulgaris*), and dwarf maple (*Acer glabrum*) are examples of natural dwarfs. Examples of environmental or artificial dwarfs are plants that are dwarfed

by high mountain altitude, by Japanese bonsai training, and by grafting onto rootstocks that restrict size decidedly.

Additional Meanings of the Word "Dwarf"

But while this use of the word "dwarf" is the original and correct connotation, it is not the only one. As is true with all specializations, usage has developed other terms and meanings particularly suited to the situation. Thus, in the case of dwarf fruit trees, the semantic emphasis is placed upon the dwarfing rootstock that brings about the dwarfing of the scion variety propagated upon it. Almost exclusively, dwarf fruit trees are budded or grafted plants. In the nursery trade and in fruit circles, dwarf fruit trees of cultivated varieties of apple (*Malus domestica*, pear (*Pyrus communis*,), cherry (*Prunus avium* and *P. cerasus*), peach (*P. persica*), plum (*P. spp.*), and apricot (*P. amygdalus*) are understood to be propagated on recognized dwarfing rootstocks (Fig. 2).

Unfortunately, by going one step further, the word "dwarf" is sometimes used to refer to any fruit tree budded or grafted onto a dwarfing rootstock or interstock, which restricts size only slightly. In fact, some trees have been called "dwarf" that are scarcely smaller at all than the group, variety, or species to which they belong or with which they are compared. Fruit growers not uncommonly lapse into this terminology. And while it is clear to them that they are referring to fruit trees that are merely not full size or standard size, the practice often leads to confusion. It may not be wrong, because after all it is usage that determines the meaning of words; but it is not sufficiently specific and it is poor usage.

Suggested Terminology

A way of meeting the problem—and this is frequently done—is to use a modifying adjective in connection with the word "dwarf." Thus, such expressions have arisen as "full dwarf," "very dwarf," "semidwarf," and "slightly dwarf," as the case may be. These are descriptive terms, and carry a more precise meaning.

Another procedure—and this is perhaps better—is to employ the word "dwarfed" rather than "dwarf" as the all-inclusive term. Thus, some plants are only slightly dwarfed, whereas others are decidedly dwarfed. Yet they are all smaller than the standard to which they are compared and may all be properly termed "dwarfed plants." In other words, all "true dwarf" fruit trees are "dwarfed" fruit trees, but not all "dwarfed" fruit trees are "true dwarf" fruit trees.

By adopting this terminology, the word "dwarf" may be retained in the original meaning of decidedly small, diminutive, and distinctive. Similarly, the word "dwarfed" reflects its broad meaning to include all plants, or fruit trees in general, that are smaller than the standard of comparison. It is in this sense that these terms are used in this book. And while much of this may seem like quibbling over small matters, it is nonetheless useful to grasp these distinctions.

Fig. 2. Seven-year-old Golden Delicious/EM IV on trellises at Okanagan Center, British Columbia, yielding over 1,000 boxes of fruit per acre. *Courtesy of Goodfruit Grower*

Classification of Fruit Trees According to Size

Since the rootstocks most commonly used for the dwarfing of fruit trees restrict them to a more or less definite and characteristic size, there has developed a definable understanding or classification of a dwarfed fruit tree in terms of size. In the case of the apple, for example, a truly dwarf tree is expected to develop scarcely taller than a man, say a limit of 6 to 8 feet in height. And usually this means working on the EM IX or perhaps EM 26 apple rootstocks, occasionally on EM VII with some varieties, or by double-working with special dwarfing interstocks, because these are the common rootstocks available that will produce so small a tree. In the case of the pear, a dwarf tree means one worked on the quince. For the cherry, peach, plum, and apricot, usage is less definite for the reason that there are as yet no thoroughly satisfactory and standardized dwarfing rootstocks to produce a truly dwarf tree.

To meet the need for some system of classification or terminology for different degrees of dwarfing, the East Malling Research Station, East Malling, Kent, England, has made four groupings for apple trees, as follows:

1. Very dwarf—characterized by EM IX
2. Dwarf—characterized by EM VII
3. Vigorous—characterized by EM II
4. Very vigorous—characterized by EM XVI

The terminology in use in American nurseries differs somewhat but agrees in general with the above four groups, based primarily on the height of the trees, as follows:

1. Dwarf—under 6 to 8 feet in height, characterized by EM IX
2. Semidwarf—under 12 feet in height, about the size of a peach tree in America, and characterized by EM II
3. Semistandard—slightly smaller than the standard or full-sized tree, about the size of a vigorous Montmorency sour cherry tree, characterized by EM XIII
4. Standard—full size, the standard orchard tree, characterized by French Crab seedlings.

The height given in this classification does not mean the highest single terminal growing point of the most vigorous tree; rather, it means the height of a fair proportion of the shoots and branches that make up the tree. It must be pointed out, too, that the term "standard" is not the same as that used in Great Britain, where the word refers to the method of training, that is, a high-headed tree. As used in North America it means the full-sized orchard tree grown on a nonrestricting rootstock—the "standard type tree of orchard culture."

P. Rémy, Station de Recherches d'Arboriculture Fruitière, Angers, France, has suggested an increase in the number of classes from four to six, designating different degrees of vigor, based on performance in France, as follows:

Class A   Weak   EM IX
Class B   Medium weak   EM 26
Class C   Medium   EM VII; MM 106
Class D   Medium vigorous   EM IV; MM 111
Class E   Vigorous   EM I, II, XIII; MM 104; A 2
Class F   Very vigorous   EM XII, XVI, XXV; Crab C; MM 109; M 779; French Crab

This system has the merit of providing additional classification to accommodate rootstocks that do not fall precisely into the original four groupings, including new rootstocks that are appearing and may appear in the future. It would be possible to adapt this system to American conditions and terminology by thinking in terms of three degrees of (a) the dwarf condition and three degrees of (b) the vigorous condition, as follows:

Class A   Very dwarf   EM IX
Class B   Dwarf   EM 26
Class C   Medium dwarf or semidwarf   EM VII
Class D   Medium vigorous   EM IV
Class E   Vigorous   EM II
Class F   Very vigorous   French Crab

This classification could be very useful and is recommended for consideration if a grouping of this kind is desired.

Refinement in Terminology

As will be shown more in detail later, scion varieties may differ in performance on the same rootstock. Thus, Northern Spy on the EM IX falls more nearly in Class C (medium dwarf or semidwarf) than it does in Class A (very dwarf); and Gallia on the EM VII might easily be placed in Class A (very dwarf). Because of the fact that a given stock-scion combination is in this way peculiarly itself in characters and performance, in contrast to the behavior of the scion variety alone or of the rootstock alone, the word "stion" has been introduced to designate a stock-scion combination. The word is derived from the first two letters of "stock" and the last three letters of "scion," coined by H. J. Webber, Citrus Experiment Station, Riverside, California. The term emphasizes the fact that the behavior of a scion variety is altered by the rootstock upon which it is worked, and vice versa, and that the combination is a unit that must be treated as such.

Thus, the performance and behavior of a given stion should be described and presented just as is now done for the scion variety alone, or for the rootstock alone.

If this were done, then instead of ordering a "dwarf," a "semidwarf," a "semistandard," or a "standard" tree, the purchaser would select the particular stion combination that would give him the performance he desired. He might prefer the McIntosh on the EM IX so as to have a very small, early-fruiting tree; on the EM VII, to have a tree slightly larger; or on the EM I, to have a more substantial semidwarf tree; and so on.

Undoubtedly, refinements of this nature will eventually become standard practice with dwarfed fruit trees, and they will be a great improvement when they do. As a matter of fact, the more progressive nurseries already list their trees in this way. The scion variety is stated first, followed by the name of the rootstock, separated from the scion variety by a diagonal bar. "McIntosh/EM IX" means the McIntosh scion, worked on the EM IX rootstock. Following this is a statement of the height to which the tree is likely to grow, its adaptability to soil and climate, and its general performance, just as is done with individual varieties of fruits.

The Compact or Size-Controlled Fruit Tree

Since there are characteristics other than size that are affected by growing a variety of fruit on a given rootstock, the terms "dwarf" and "dwarfed" may be properly criticized as inadequate. To meet this criticism several terms have been suggested, such as "compact," "size-controlled," "crop-controlled," and just "controlled." There is much merit in this thinking, and usage may eventually develop this concept just as it has adopted the "compact" automobile and "controlled atmosphere" storage, or "CA" storage, in place of "gas storage" or "modified atmosphere storage" where the carbon dioxide and the oxygen in the storage are controlled.

But here again there are limitations because one cannot speak of control in general without bringing in all the variations in performance and management that accompany the variable that is being controlled. There are matters of temperature, humidity, rainfall, pruning, orchard management, and pests. There are the intricate controls of flower initiation, flower development, fruit set, fruit development, abscission, and fruit maturity. There are countless controls in a tree that are not included in rootstock influences.

It should be pointed out, too, that this part of the discussion arises primarily with reference to commercial fruit production and the desire by the fruit industry for more complete control over both tree and fruit, such as control of tree size, age of fruiting, and regularity of cropping. Rootstock controls are especially applicable here. In other words, there may be need for a new term that will convey limited but useful meaning to the industry.

In order to meet the requirements of the present, but with an eye to the future, the term "controlled" is used freely throughout this book along with "dwarfed." Perhaps in time "dwarf" may come to be restricted to the "very dwarf" or "true dwarf" that is of special concern to the amateur and the small grower, and perhaps "controlled" or "compact," or some other designations, may be adopted by commercial fruit growers. It would be a reasonable and logical development.

2

THE HISTORICAL BACKGROUND
OF DWARFED FRUIT TREES

DWARFED plants have been known for centuries. They have been grown and cultivated for a very long time. In the third century before Christ, Theophrastus, one of the early Greek historians, recorded the growing of dwarfed fruit trees. He studied plants sent back to Greece from the conquests of Alexander the Great, and mentioned a small, low-growing type of apple as being among them. This was evidently received from Asia Minor, and was probably long known and grown there. The Roman agriculturists, too, were familiar with dwarfed trees, and used them in garden plantings. Subsequently dwarfed trees have been discussed in horticultural literature down through the years, chiefly from various parts of the Continent of Europe, from England, and from the Orient, and shifting in emphasis from one group of plants to another as the vogue or popular demand directed.

In China and Japan, interest has centered for the most part around ornamental forms, whether fruit-bearing or not. Dwarf pines and junipers were widely employed in China in the palaces of royalty during the Sixth Dynasty (first and second centuries after Christ) and were revered and respected as symbols of age. The development of dwarf plants reached its height during the Tung Dynasty (eleventh to thirteenth centuries after Christ), the forms most used being pine, juniper, apricot, pomegranate, and apple (Fig. 3). The growing of ornamental dwarf plants is a recognized feature of Chinese art and culture, with chief interest in pine, apple, lilac, crape myrtle, gardenia, red bud, apricot, azalea, and rhododendron.

Japanese culture of dwarf plants, or bonsai, is more recent than that of the Chinese. Early records tell of the growing of dwarf apricots about A.D. 1700 by the samurai. Between 1830 and 1844 dwarf plants gained

Fig. 3. **Thirty-year-old ornamental Crabapple tree of the type cultivated by the Chinese for centuries, now popular as Japanese bonsai.** *Courtesy of Dr. George Avery and the Brooklyn Botanic Garden*

great popularity, in which the apricot, cherry, orange, and pine were preferred, and are more fully described in Chapter 28. Bonsai culture has reached large proportions. Bonsai clubs and associations have been formed, and local and national exhibits are held with appropriate prizes and awards. Millions of ornamental plants are grown throughout Japan and touch every walk of life.

In Europe the interest in small-sized trees has been almost exclusively with the deciduous fruit trees—sometimes mostly for ornamental effect and at other times for the fruit. In either case, the culture of dwarfed trees has been dependent upon dwarfing rootstocks. The "Paradise" and "Creeper apple tree" are frequently mentioned in seventeenth and eighteenth century literature, the name Paradise probably coming from the Persian *pairidaeza*, meaning a park or garden, and first used with reference to apple rootstocks during the latter part of the fifteenth century. La Quintinye (1626–1688), gardener to Louis XIV, used dwarfing rootstocks for both pears and apples in the famous gardens he developed at Versailles. Duhamel du Monceau, in his *Traités des arbres fruitier* in 1768, discriminated between different degrees of "dwarfyness" as between *Le Pommier nain des Paradis*" and "Doucin." The term "Doucin" was first mentioned in 1519, probably referring to the characteristic sweetness of the fruit, from the French *douce*, meaning sweet. The quince was appreciated as a dwarfing stock for the pear, and the sweet cherry had been worked upon the sour cherry, but there was apparently little attention given to dwarfing classes of fruits other than the apple and pear.

THE DWARFED FRUIT TREE
IN EUROPE

Although dwarfed fruit trees are thus very old in human experience, major interest in terms of modern fruit production did not begin until the early part of the nineteenth century. This was in large measure, of course, a consequence of the Industrial Revolution and the resulting specialization that called for specialization in fruit growing as well as in other fields.

At all events, a review of horticultural literature in England during the 1830's and 1840's shows a great interest in fruit culture, in rootstocks, in dwarfing, and in stock and scion relations. Two principal types of dwarfing rootstocks were recognized at that time; namely, Paradise and Doucin, the former being very dwarfing and the latter being less dwarfing. Thomas Rivers in 1863 reported the origination of his Paradise stocks. His *Miniature Fruit Garden*, published in 1866, ran through many editions. Various dwarfing rootstocks, almost exclusively for the apple and the pear, appeared in France (Jaune de Metz in 1879), in Germany, in the Netherlands, and in England. As new forms were introduced into the trade, much confusion arose over their identity and trueness to name. They could no longer be classed simply as "Paradise" or "Doucin." English literature of the period mentions the problem frequently.

A number of attempts were made by various individuals to identify and standardize the rootstocks in use, including the Späth Nurseries in Germany, and Sprenger in Holland.

It remained, however, for R. G. Hatton, later Sir Ronald Hatton, of the East Malling Research Station in Kent, England, to accomplish the task. Continuing work that had been begun by Wellington in England in 1912, Hatton collected dwarfing rootstocks from various parts of the world—mostly from England and the Continent, and brought them to East Malling. Most of these had been used for considerable time, and several had been known for at least three centuries. There they were grown and found to contain mixtures. These were rogued, and the remaining pure clonal lines, sixteen in number, were named and numbered and introduced into the trade as standardized, true-to-name rootstocks.

The so-called "Malling," "East Malling," or "EM" rootstocks are widely used under name and number not only in England but also on the Continent and in most of the fruit-growing regions of the world, to some degree at least. Originally designated by the prefix "Type," as Type I, Type II, and so on, the accepted designation is "EM" followed by the number of the rootstock, in Roman numerals, as EM I, EM II, and so on. These standardized rootstocks account in large measure for the new interest in dwarfed fruit trees. They are used not only for trained ornamental forms but also in extensive commercial orchard operations throughout the world (Fig. 4). They are discussed more fully in subsequent chapters.

Trees Trained to Special Forms

In Germany, France, Holland, Belgium, and Switzerland, the dwarfed fruit tree trained to special forms, such as cordons and espaliers (Chap-

Fig. 4. Ronald Hatton, later Sir Ronald Hatton, Director of the East Malling Research Station, East Malling, Kent, England, and an open-center semidwarf tree of Bramley/EM VII which he then considered to be the ideal size and shape for Bramley in England. Photographed in 1935.

ter 26) has long been a characteristic feature of horticulture; in fact, the terms employed are the same in many regions in spite of differences in language. Not only can trained trees be seen in old engravings and woodcuts (Fig. 5), but they can still be found in gardens which have been perpetuated in original design and planting, such as the famous garden at Versailles, France, which is described more fully later in this chapter.

Trained trees have been developed in these countries in countless shapes and designs, as individual specimens or as group plantings, placed against the side of a building, trained along a walk, or grown against a wire or lattice frame. Some conception of the size of recent-day operations may be gathered from the large number of trees seen in

European nurseries. In a large nursery at Verrieres de Buisson, France, were growing 250,000 trained fruit trees in one nursery block a few years ago when the writer visited there. The variety of form and shape are detailed and intricate and exceedingly interesting.

Where summers are cool and where general climatic conditions do not favor the attainment of high quality, trees may be trained against a wall to advantage. With southern exposure, many otherwise unadapted varieties are grown to perfection. Peaches and nectarines, which typically demand summer heat for proper development, have proved especially suited to this culture. Excellent collections of varieties of deciduous fruits have been maintained along a walk in a garden when trained as single oblique or vertical cordons, thus saving much space and making possible what would otherwise be impossible.

The Garden of the King at Versailles

Adjacent to the great Palace of Versailles is the garden designed and built in the seventeenth century for Louis XIV of France by Jean de la Quintinye (1626–1688), who also served as the Director. It is often called the Jardin la Quintinye. Although planned as a kitchen garden for the production of fruits and vegetables for the king, fruit trees were a most important part. The dwarf fruit tree and trees trained to special forms were a feature.

The Ecole Nationale d'Horticulture is located here. The gardens are maintained as closely as possible to their original design. Some very old espaliers and contre-espaliers can be seen, as well as many interesting forms and methods of employing dwarf fruit trees in a garden.

The plan as prepared by Quintinye and published in *Instructions pour le jardins fruitiers et potages* in 1690 is reproduced in Fig. 6. In overall size the garden was about 1090 feet long and 865 feet wide, with a central area of 450 by 600 feet. Around it were terraces of gardens devoted to particular plants. Under some of these terraces were storage rooms and houses in which tender plants were kept during the winter.

Fig. 5. Portrayal of the erection of an espalier in the fifteenth century, appearing in *Jardinier français*, by Nicolas de Bonnefons, in 1651. The engraving is by F. Chauveau.

The free translation, which follows, is from the original publication by La Quintinye and conveys some idea of the elaborateness and completeness of the garden:

(1), (2), (3), (4) Gardens for strawberries.

(5), (6), (7), (8) Diagonal gardens located favorably for the sun's rays in four exposures.

(9), (10), (11), (12), (13), (14), (15), (16), (17), (18), and (19) Eleven small gardens which are enclosed by low walls and which are filled with different fruits and vegetables.

(20)   Forced asparagus.

(21)   Plum garden, with all sorts of plums in both bush and espalier form.

(22)   Common garden with forced asparagus.

(23)   Bush and espalier figs.

(24), (25)   Gardens for cucumbers and small salad beds.

(26)   Melon garden and various other beds.

(27)   Public gate where herbs and other vegetables are distributed.

(28)   Small sheds for herbs which are later distributed.

(29)   Fig garden with all fig trees in pots.

(30)   Greenhouse for potted figs.

(31)   House which the king has had the goodness to build for me.

(32)   Court of the above-mentioned house.

(33)   Gardener's house and house for his hands.

(34)   Lower courtyard of the above house.

(35)   Small garden for flowers.

(36)   Main entry for the king, bordered by high-headed pear trees.

(37)   Terrace which surrounds the central court and overlooks the main quarter.

(38)   The main quarter.

(39)   Cellar where roots and artichokes, cabbages, flowers, and so forth, are stored during the winter.

The espaliers of the main garden are filled with Admirable and Nivette peaches.

All the walls of the terraces of the main garden are covered with grapes. The north exposure of all the gardens is in pears.

The sixteen quarters (A, B, C) are in summer pears, (D, E, F) in autumn pears, and the others are in winter pears.

All the apples are around the eleven gardens on the Sataury side. In this plan of the garden, the only item missing is the garden of new varieties which is close to the garden numbered 23 which was previously the compost court.

Reading clockwise beginning with the King's Entrance (at six o'clock):

(1)   Early cherries in espalier.

(2)   Apricots in espalier.

(3)   Apricots in espalier.

(4)   Early peaches in espalier.

(5), (6), (7), (8)   All early peaches and White Madeleine peaches.

(9)   Mignonne peaches in espalier.

(10)   Mignonne peaches in espalier.

(11)   Red Madeleine and Bourbin peaches in espalier.

(12)   Persian and Chevreuse peaches in espalier.

(13)   Persian and early Violette peaches in espalier.

(14)   Early Violette peaches in espalier.

(15)   Purple peaches in espalier.

(16)   Admirable peaches in espalier.

(17)   Summer pears in espalier.

(18)   Azerole and autumn pears in espalier.

(19)   Winter and cooking pears in espalier.

(20)   Late purple, Violette, and Nivette peaches in espalier.

(21)   Plums in espalier.

(22)   Espaliers mixed with all sorts of good peaches.

(23)   Figs in espalier.

(24)   Nectarines in espalier.

(25)   Yellow peaches and other peaches in espalier.

(26)   Figs in espalier.

(29)   Figs in espalier.

(35)   Violette peaches and several other good peaches in espalier.

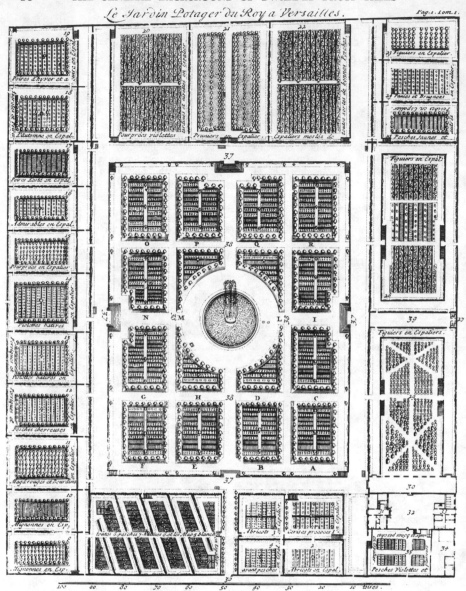

Fig. 6. **Original diagram of the famous gardens at Versailles erected for Louis XIV by La Quintinye (1626–1688), in which dwarfed fruit trees were a conspicuous feature.**

Trained Trees for Commercial Fruit Production

While trained dwarf fruit trees have been useful to the amateur and to garden planting, they have been used as well for commercial production of fruit in Europe. Large establishments have developed in various parts of Europe that produce fruit of fine quality and excellent color for special markets, all grown on trained trees.

One planting of pears on quince roots that the writer has visited near Paris comprised 140 acres. The plants were set 1 meter apart in rows 3 meters apart and were supported on 9 wires attached to iron stakes approximately 2½ meters tall, set in concrete. Well-managed plantations produce fruit of exceptional quality that commands good prices.

However, not all commercial plantings of trained trees are so elaborate. The writer has seen one in England, for example, consisting of apples on the EM II rootstock and grown as oblique cordons, set 4 feet apart in rows 8 feet apart, and supported on 2 wires about the height of a grape trellis. The trees were topped at about 6 feet and were crudely pruned. They were producing well and were considered the most consistently profitable planting among four hundred acres of fruit.

Commercial plantings of apple and pear in Europe are generally placed on dwarfing rootstocks (Fig. 7). In northern Italy, in the vicinity of Ferrara, not far from Venice, there are hundreds of acres of modern apple orchards grown on dwarfing rootstocks in a hedgerow system. The trees are planted approximately 6 feet apart in rows 15 feet apart and are pruned in such a manner that they become a wall about 2 to 3 feet thick and 9 to 10 feet high. This topic is more fully treated in Chapter 21.

Creeper Trees

In areas where winter cold is so severe that unprotected trees will not survive, various attempts have been made to grow trees low and close to the ground in order to take advantage of snow coverage and wind protection.

The Russians have developed what they call "creeping orchards" of apples for extreme northern conditions (Fig. 8). They are really severely dwarfed and low-headed trees grown closely in rows, with the tops bent and interlaced in the direction of the row, not over 3 feet in height. In winter they are nearly or completely covered with snow, which protects them from extreme low-temperature damage.

Even as far north as Omsk in Siberia, 10,000 to 12,000 acres of apples and sour cherries are said to have been grown after this fashion, with the

Fig. 7. Commercial orchards in Europe are commonly on dwarfing root-stocks. Thomas Neame, later Sir Thomas Neame, Faversham, Kent, England, and planting of six-year-old Coxe's Orange Pippin on the EM IX rootstock, which had already borne three crops of fruit when photographed in 1935.

branches "pegged down" close to the ground. The trees are planted with their trunks at about a 30-degree angle with the horizontal. Branches are held near the soil surface by hooked wires pressed into the ground.

Fig. 8. "Creeper" apple trees grow as far north as Omsk in Siberia, protected by coverings of winter snow. *Courtesy of Dr. J. R. Magness*

If any appreciable snow occurs, such trees are well protected. Even in the absence of snow such "trunkless" trees are said to withstand cold much better than trees grown with trunks.

THE DWARFED FRUIT TREE
IN AMERICA

Until the beginning of the nineteenth century, horticultural activities in America were very limited. Not until after the Revolutionary War was there much development of a commercial nature. Centering first in New England, the middle Atlantic States, and the Virginias, growing was largely patterned after European practices, especially those in England.

The early records of dwarfed fruit trees show a similar dependence upon European practices and interests, running almost parallel with them. First American accounts speak only of the Paradise stock, and then only very briefly. Our earliest book on pomology, published by William Coxe in 1817, merely mentioned the Paradise apple.

The First Cycle of Interest

Beginning about 1835, there arose a great interest in dwarfed fruit trees in America that continued until 1855 or 1860. This period coincided with the appearance of America's first horticultural periodicals, including the *Magazine of Horticulture* (1835), edited by C. M. Hovey, of Boston, and the *Horticulturist* (1846) edited by A. J. Downing, of Newburgh, New York, and subsequently by Patrick Barry, of Rochester, New York. The American Pomological Society, born in 1848, was dominated by Marshall P. Wilder, of Boston. Great activity and interest prevailed in pomology, in varieties, and in fact in anything along pomological lines. The dwarfed tree was no exception at that time.

The pages of these publications and the records of society discussions are filled with topics being treated at the same time in England. For example, the English interest in stock and scion relations of the 1840's had its counterpart in America.

C. M. Hovey, of Boston, and Patrick Barry, of Rochester, New York, were two of the principal advocates of dwarfed apple trees, both for commercial and for garden plantings. They wrote fluently, and because of their prominent editorial positions they had great influence.

It is interesting to note in passing that the more enthusiastic advocates of dwarfed apple trees were men who were themselves excellent plantsmen. Further, their orchards were located in some of the better horticultural areas of the country, factors that undoubtedly favored the performance of dwarfed trees.

About 1860 there came a decline of interest in dwarfed apples. Seemingly, part of this was due to relatively heavy fruit production and attendant low fruit prices. It was easy enough to grow apples in an extensive and half-neglected manner on standard trees without recourse to the greater attention required by dwarfed trees. Furthermore, new problems of summer pruning and training were introduced by the dwarfed tree that were not present with standard trees on seedling roots, and complicated the situation.

The decline of interest in dwarfed apples was coincident with avid interest in dwarf pears grafted on the quince. This was again, in part, a European influence—especially noticeable in the vicinity of Boston, Massachusetts, under the leadership of Robert Manning, Marshall P. Wilder, and C. M. Hovey. Demand for pear fruit was good, and the pear responded well to the special care that dwarfed trees demanded. This enthusiasm was not shared by other sections of the country, particularly the Philadelphia region and the western states, which reported unhappy experiences. Nevertheless, in 1858 Hovey was able to state that the most lively topic for discussion in horticultural circles was the dwarf pear. In 1859 the number of dwarf pear trees (dwarfed by working on the quince root) planted in Massachusetts during the preceding ten years was said to be from four to ten times the number planted on standard pear roots. Over half of the pears grown in the United States during this period were from trees on quince roots.

It must be noted that the happy experiences in Massachusetts with pears on the quince root during this period were vigorously and repeatedly explained by their New England champions as dependent upon six important points, chief of which was planting the tree with the union 3 inches or more below the soil surface so that the pear scion might become rooted. Otherwise, it was freely admitted, the dwarf pear on quince roots probably would not survive beyond fourteen years. The quince root was found tender to winter cold, and was frequently destroyed by temperatures that might not seriously injure the scion variety of pear worked upon it.

Accordingly, even this most flourishing period of dwarf pear trees in America can hardly be said to be strictly dwarf fruit culture. Rather it was a method of reducing overly vigorous tree growth, which is prone to the devastating attack of the fire-blight disease (*Erwinia amylovorus* Burr.-Smith). By this method the trees were not only brought into early bearing; they were also spared, to considerable degree, the ravages of this disease. Later the trees became scion-rooted and developed into high-yielding orchards on their own roots. It may not be out of place to remark that this practice still has merit for commercial pear production in the United States.

But to return to the dwarfed apple tree: there appears to have been no great interest during the period from 1860 to the early 1890's in adapting dwarf apple trees to commercial fruit production. In spite of much discussion and argument pro and con, no contribution to the solution of problems of the fruit industry of the period seems to have been made. One of the recognized difficulties was the unreliability and mixtures in dwarfing rootstocks. George Ellwanger, of Mount Hope Nurseries, Rochester, New York, spoke of the confusion in English nurseries regarding Doucin (or Doucain, as it was frequently spelled in American literature) and Paradise, saying that all were erroneously called Paradise. This situation extended to the American nursery industry.

Ellwanger, who, with Patrick Barry, had one of the leading nurseries in the eastern United States, was expert at distinguishing the various dwarfing stocks. This is not overly surprising, considering that he had the largest experimental planting of trees on dwarfing stocks in America during the nineteenth century, planted in 1845. Not only did he have the commonly used Paradise and Doucin; he also tested several of Thomas Rivers' selections, of which Nonsuch Paradise (now EM VI) was one, and which he rejected as unsatisfactory.

C. M. Hovey and Patrick Barry appear to have been the principal proponents of the growing of dwarfing stocks—Hovey by his active advocacy of their use and Barry by the detailed information he disseminated. In 1865 Hovey said: "We have, in our previous volumes, directed the attention of our readers to the importance of the introduction of Dwarf Apple Trees . . . and strongly urged their more extensive culture. . . . It is our intention to devote considerable space to the dissemination of information which will lead to their more extensive culture." Hovey's enthusiasm may be explained by his acquisition and success with two hundred to three hundred dwarf apple trees, each of a different variety, secured from Ellwanger and Barry.

Barry recommended both Paradise and Doucin for the commercial orchard. He considered the heavy investment in an orchard and the lack of return until about the twelfth year, and recommended as a remedy that dwarfed trees on both Paradise and Doucin stocks be used as filler trees.

Finally, Hovey acknowledged that the dwarfed apple trees were not "catching on," though he found the fact difficult to explain. He conjectured that it might be due to the fact that insufficient attention had been called to them. He added that fruit could be obtained very cheaply and that dwarfed trees "would not pay." And G. Jacques, evidently speaking of the true dwarf apple tree, wrote: "They are very pretty garden pets in the midst of a flower bed, or at the corners of alleys, or elsewhere where fancy may locate them. They seldom bear more than a dozen or twenty apples, and therefore the economical orchardist, looking to profit alone, ought not to consider them as worthy of his attention." And so, for the next thirty-five years—until the early 1890's—very little enthusiasm was engendered for dwarfed apple trees in America, although discussion continued. And in answer to a query in 1896 about his experiences with dwarfed apples, J. H. Hale, prominent Connecticut fruit grower, replied "Plenty of experience but no profit."

The Second Cycle of Interest

Then, quite suddenly in the 1890's, the menace of the San Jose scale appeared. There was no known control except by covering a tree with a canvas or tarpaulin and fumigating with hydrogen cyanide gas. Small trees were required for such an operation, and again interest in the dwarf tree and dwarfing rootstocks appeared.

State agricultural experiment stations, notably those in Massachusetts, New York, and Virginia, now became involved with dwarf fruit trees, and more critical studies were begun. In 1902 the New York State Agricultural Experiment Station at Geneva, New York, undertook cooperative tests with fruit growers throughout the state, under the leadership of S. A. Beach. The trees were planted in the autumns of 1903 and 1904.

The rootstock material was still loosely divided into two groups: (1) Paradise or French Paradise, and (2) Doucin or English Paradise. The unreliability of rootstock material and the need for proper identification and certification were reemphasized.

Fortunately for commercial fruit growing, but unfortunately for the furtherance of studies with dwarfed trees, lime-sulfur and oil sprays, which were introduced between 1907 and 1910, proved exceedingly effective in controlling the San Jose scale. Active interest in dwarfed apple trees ceased almost as quickly as it had begun.

By 1920 the dwarf apple tree plantings that had been made here and there around the country to meet the threat of the San Jose scale had been studied and appraised, and the dwarf tree was virtually discredited for commercial fruit production.

Nevertheless, some useful facts had been brought out. For example, problems of scion rooting, anchorage, pruning, fertilizer treatments, winter hardiness, and stock-scion relations were reviewed. Hedrick summarized the extensive tests conducted in New York State, saying that commercial crops had been borne little, if any, earlier than on seedling stocks, that many trees had been winter-killed, that surface-rooting and suckering had been serious, that many trees had blown over or failed to make good unions, and that there was great discrepancy in size of trees. It was believed that many of the so-called dwarf trees were not even on dwarfing rootstocks.

It must be remembered in this connection, and in defense of the dwarfed tree, that this was the period of large trees, tonnage production, low prices and low cost of production, and little concern for consumer demand and quality control. The dwarf tree was ill adapted to such gross competition; it required the special care that was not forthcoming during such an era.

The Third Cycle of Interest

A third wave of interest in rootstocks for fruit trees began in America in the late 1920's, brought about by steps the United States Government was then taking to prohibit the importation of seedling rootstocks upon which American fruit trees were almost exclusively propagated. The net result was the requirement for a supply of domestically propagated rootstock material, in anticipation of the embargo that was to go into effect in 1930 against rootstocks from outside the United States, which was aimed at foreign insects and diseases.

At just this time, beginning in the late 1920's, the favorable results reported from England with standardized dwarfing rootstocks at the East Malling Research Station was having its impact on America—especially in scientific circles. Accordingly, experimental lots of the Malling rootstocks were brought to this country. Early acquisitions were made by Anthony in Pennsylvania in 1920 and 1922, by Shaw in Massachusetts in 1922, by Tukey in New York in 1928, and by the Ontario Experiment Station in Canada some time prior to 1929.

With the enforcement of the federal embargo in 1930 against the importation of foreign nursery stock, the Malling rootstocks were in very short supply. Tukey and Brase at the New York State Agricultural Experiment Station at Geneva, New York, undertook to propagate and disseminate the Malling rootstocks to interested experiment stations, nurserymen, and orchardists throughout the country. Attempts were also made to induce nurserymen to propagate the Malling rootstocks and to introduce dwarfed fruit trees under Malling number rather than as "dwarf trees." In 1937 the nursery catalogue of the Maloney Brothers Nursery Company, Dansville, New York, listed for the first time in America the availability of trees on numbered Malling rootstocks. Kelly Brothers, also of Dansville, New York, and Jackson and Perkins, of Newark, New York, followed shortly.

Between 1938 and 1945 the Geneva Station distributed 142,669 dwarfing rootstocks and 15,820 dwarfed trees to 238 individuals and experiment stations in 36 states and Canada. These distributions had much to do with furthering the development of dwarfed fruit trees in the New World.

The new rootstock materials immediately showed promise. In New York State, for example, they propagated well, behaved well in the nursery, and performed well in early orchard tests. Some of the rootstocks were very dwarfing, some were semidwarfing, and some were scarcely dwarfing at all, thus providing a range of usefulness that had not before been experienced.

Further, a new set of circumstances had arisen in orchard circles that kindled interest in trees smaller than the standard. The adaptation, performance, and market demands of fruit varieties had reached a degree of standardization. The problems of spraying for insects and diseases had been solved, at least for commercial purposes. Fertilizer application and orchard-management practices had fallen into some semblance of recommended procedure. Handling and storage problems had received their share of attention.

The need now was for lowered costs of production, earlier fruiting, easier pruning, thinning, and harvesting, and easier spray penetration, all to meet changing market demands. The new standardized Malling dwarfing rootstocks seemed likely contributors to the solution of some of these problems. The combination of definite needs coupled with the promising means of satisfying these needs accounts for the revival of commercial interest in apple trees on dwarfing rootstocks in America beginning about 1935, and constituting the third cycle of interest.

Trees Trained to Special Forms

Early interest in America in trees trained to special forms, such as cordons and espaliers, reflected European tastes. It was patterned after varieties, customs, and habits brought from the Old World. Culture was naturally limited to estates and to wealthy landowners. Records of Mount Vernon and Williamsburg in Virginia, and of gardens in the vicinity of Philadelphia, New York, and Boston, show that trained trees were grown there during Revolutionary War days.

Yet trained trees did not succeed in becoming very popular in America, perhaps partly because of the large amount of care and intensive culture required in a country busy with the discovery, exploration, and development of its resources. Further, the climate of America is continental and severe—quite different from the equable maritime climate of favored horticultural sections of Europe. Thus, strong summer sunlight and severe winter cold coupled with bright winter sun produced serious scald on trunk and arms; cankers formed around spurs, crotches, and areas that did not mature well before the onset of winter; and heavy snow and ice caused severe breakage. Finally, trees trained to special forms, such as the popular Verrier palmettes and the Double-U, proved expensive and hard to come by. Their bulk, the difficulty of packing and crating, and their general awkwardness in transport have presented serious problems except where the nurseryman has been able to develop a local clientele.

A few trained trees may be found here and there about the country (Fig. 9), usually in old estates and gardens and more often than not under the tutelage of European gardeners, but they have never attained the popularity they enjoy in Europe.

Dwarfed and Trunkless Trees for Cold Regions

In northern Ontario, where temperatures of −40 degrees F. may be experienced, recourse has been to low-growing trees to avoid winter

Fig. 9. **Fan-shaped pear on Angers quince as grown in America.**

damage. Such trees may be protected by snow coverage where ample snowfall is common. In addition, it is possible to employ burlap, wooden frames, and other protective materials. Good success has been reported not only with apples but also with pears, cherries, plums, apricots, and peaches, as well.

In North Dakota a form of "trunkless apple tree" has been developed by heading the young trees at the ground line and shaping them into a bush. A planting of the hardy Hibernal variety showed no winter kill after fifteen years when grown in this manner, compared to 70 per cent loss for trees headed at 2 feet. At fifteen years of age the "trunkless" trees were 12.5 feet in height, with an estimated volume of 3,652 cubic feet. Production was 200.5 pounds of fruit per tree. While not recommended for a more favored climate, this method has merit for regions of severe winter cold and wind where fruit trees could not otherwise be grown.

Amateur Interest in Dwarfed Fruit Trees

Quite aside from the interest of commercial fruit growers in controlled trees, a vigorous new interest arose in dwarfed fruit trees among amateur horticulturists, gardeners, and small landowners about the time of World War II. This may have been brought about partly by the common attractions of anything "new," and the new Malling controlled fruit trees were in a sense new to that generation. But the greater share of interest came from the increase throughout the country in outdoor living and gardening in general. The inclusion of fruit trees in the national war-garden and victory-garden program in 1942 still further accelerated the trend. Demand for trees was high; a supply of dwarf trees was short; and some materials were sold as dwarf trees that were inferior and not true to name—with attendant dissatisfaction.

Although the dwarf apple has been most in vogue, especially trees on the EM IX rootstock, there has also been good interest in dwarf pears. Small trees of cherry, peach, and plum have been sought by customers, but no good dwarf types have been found to satisfy the need, and interest has waned.

The Future

There is much to be learned about how to grow different varieties of fruits on the different dwarfing rootstocks in America—soil preferences, planting distances, fertilizer needs, climatic requirements, pruning methods, and the general range of cultural operations. In time new rootstocks especially adapted to conditions in America will undoubtedly appear. But as with any development, success is associated with a thorough understanding of the problems involved and the pooling of a wide assortment of experiences. The organization of the Dwarf Fruit Tree Association in Michigan on March 4, 1958, and a similar Western Dwarf Fruit Tree Association in the State of Washington a year later, on February 20, 1959, may prove important and historic steps in this direction. It may in fact become so significant that it might be called the "fourth cycle of interest," in which grower interest for the first time dominated. Time alone will tell.

3

FEATURES, LIMITATIONS, AND POSSIBILITIES WITH DWARFED OR SIZE-CONTROLLED FRUIT TREES

DEMANDS OF MODERN FRUIT PRODUCTION

MODERN fruit production demands better control of fruit trees in growth and in fruiting, in contrast to some of the hit-and-miss, green-thumb operations of the past. It is no longer sufficient just to enjoy the growing of fruit; orcharding is a specialized business enterprise where competition for the consumer's dollar is keen, not only between neighboring orchardists but also between distant regions and with products other than fruit. Therefore the orchardist is interested in any suggestion that offers to reduce the gamble in fruit growing (Fig. 10). The hope has been expressed that orcharding could be placed on a basis something like engineering, in which with charts and handbooks and known facts the grower could operate with a fair degree of certainty. The terms "biological engineering" and "orchard engineering" have been coined, as though to imply order and consistency and predictability.

The more important of these desires may be enumerated briefly as follows:

1. Regular and annual production

Fig. 10. An example of the desire to prune fruit trees without the necessity for movable ladders is shown by the recourse to stilts, which place the worker three to five feet above the ground. Peach orchard in Colorado. *Courtesy of* American Fruit Grower *and Dr. Leif Verner*

2. Early return on the investment
   Early fruiting
   Greater number of trees per acre and improved stand of trees

3. Lowered cost per unit of production, and reduction in man-hours of labor

Easier and more economical orchard management, including pruning and thinning

More efficient and cheaper control of insects and diseases

Improved facility in harvesting and handling

4. Quicker and easier adjustment to changing conditions and market demands

Promotion of rotation orchards of "young" trees

Easier shift in varieties

Quicker orchard recovery from tree loss due to winter cold and other hazards

5. Higher percentage of high-quality fruit and reduction in culls

Improved color

Greater uniformity in size and color

Greater freedom from insect and disease blemishes

Reduced bruising

There are certain features and possibilities with dwarfed fruit trees that suggest they may be useful in satisfying some of these requirements. Yet it is apparent at once that there are other horticultural means of solving many of them. A number of years ago, for example, B. D. Van Buren, prominent fruit leader of the Hudson River Valley in New York State, said, in answer to the question whether or not low-headed standard apple trees would be as satisfactory as trees on dwarfing stocks: "The future development of the apple-growing industry in New York and other eastern states will in large measure depend upon the development of this type [low-headed] apple tree. It requires no changed method of nursery growing, but will be an adaptation of the now-grown standard apple tree to present needs."

And so one must be careful not to be carried away by wishful thinking about dwarfed fruit trees and what can be accomplished with them; nor on the other hand should one allow ignorance and inertia to close his mind to new opportunities. The pros and cons should be carefully balanced. There are places where dwarfed fruit trees are exceptional; and there are places where they do not belong. With these thoughts in mind, the various topics are discussed below in an attempt to present a fair appraisal.

REGULAR AND ANNUAL PRODUCTION

A variety of fruit tends to perform in a manner characteristic of that variety regardless of the rootstock upon which it is worked, although the general characteristics may be altered in degree. That is, a strongly biennial bearing variety, such as Baldwin, remains biennial on dwarfing rootstocks; but the biennial habit is accentuated. Similarly, a variety that is typically an annual bearer, such as McIntosh, tends to remain annual in production; but even the McIntosh variety can be induced to become somewhat biennial if grown on the EM IX rootstock and if no attention is given to avoiding this situation.

When a tree is brought early into fruiting, regardless of variety, there is a tendency for it to fruit unequally in successive years. However, by blossom thinning in a year of heavy bloom, it is possible to help a biennial variety to overcome this fault and to produce at least some fruit in the "off" year. And an annual variety can be helped to remain an annual producer. Furthermore, by withholding nitrogen fertilizer applications in the year of heavy fruiting, alternate bearing can be reduced.

Accordingly, dwarfing rootstocks in themselves do not overcome irregular and biennial bearing, and may even accentuate it. Yet, with proper attention, the problem is no more severe with dwarfed trees than with standard trees on seedling rootstocks. And the trend has been to replace biennial varieties, such as Baldwin and R. I. Greening, with annual producers such as McIntosh and Jonathan, so that the problem becomes less acute.

EARLY RETURN ON THE INVESTMENT

Early Fruiting

Early fruiting is a characteristic of trees on dwarfing rootstocks. How early some trees will fruit is sometimes not fully appreciated. Apple trees on the EM IX rootstock, for example, will usually blossom the first year they are planted in the orchard, depending considerably on the vitality and grade of the nursery stock and the nature of the variety (Fig. 11). Such naturally early-fruiting varieties as Wealthy, Rome,

Yellow Transparent, and Early McIntosh may be depended upon to perform this early, while such late-bearing varieties as Northern Spy may be a year later. Other rootstocks, such as EM VII, EM II, EM I, EM IV, MM 104, MM 109, and MM 111, do not induce such early fruiting; nevertheless they are earlier than the same varieties on standard seedling roots. Similarly, pears on the EM Quince C will carry a fruit or two the first or second year in the orchard. With peaches, plums, cher-

Fig. 11. **Dwarfed fruit trees come into bearing early.** Northern Spy/IX at four years of age. On standard rootstocks it is not uncommon for fruiting of this variety to be delayed until 12 to 18 years of age.

ries, and apricots, the dwarfing rootstock does not induce appreciably earlier fruiting, if at all.

But, to counterbalance this advantage of dwarfed trees, much can be done with standard fruit trees to bring them into early fruiting, such as careful selection of good nursery stock, low heading and little pruning of young trees, attention to orchard site, use of good cultural practices for young trees, attention to insect and disease and rodent control, and proper selection of varieties. In fact, there has already been a natural shift away from those varieties that are late and uncertain bearers, such as R. I. Greening, Northern Spy, Yellow Newtown, and Tompkins King, among apples. In their places have appeared more of the early-fruiting, small-treed varieties, such as Jonathan, Starkrimson, Cortland, Rome, and the spur-type trees. This trend may be expected to continue and to be an important characteristic in the appraisal of new varieties. Undoubtedly there are other practices that could be used to advantage to bring about early fruiting, but these suggestions will suffice for the moment.

Number of Trees per Acre and Stand of Trees

Dwarfed trees permit the planting of more trees per acre. Planting at 50 × 50 feet, as many of the older standard orchards were set, means 17 trees per acre. At 40 × 40 the number is 27 or 28; 30 × 30, 48; 20 × 30, 72; 20 × 20, 108; 16 × 16, 170; and 10 × 10, 435.

There is considerable evidence that this is, in the final analysis, the principal appeal of dwarfed fruit trees to the commercial fruit grower. That is, an acre of land fully occupied by bearing fruit trees will produce somewhat similar yield regardless of the rootstock employed. However, ten or twelve years may be required for standard trees to cover an acre completely; whereas dwarfed trees may cover an acre in three to six years.

The loss of a single tree is not so serious where there is a greater number of trees per acre. Mice, rabbits, tree borers, and the general run of accidents have a way of taking their toll in any planting, no matter how well managed it may be. On the other hand, it cannot be assumed that losses from these causes will be confined to individual trees here and there in a closer planting; they may as logically occur in greater number and on the percentage basis. Further, although close planting may be important in some other parts of the world where land is at a premium, there is some doubt that this situation has as yet been universally reached in North America. On the other hand, really good orchard sites are important, and a grower who has one will wish to make maximum use of it.

Obviously the cost is greater with a larger number of trees per acre, both for nursery stock and for the planting operation itself. But the extra costs can be paid for by even one good crop of fruit at an early date in the life of the orchard. For filler purposes, providing fillers are ever wise to use, the smaller trees on dwarfing rootstocks are seemingly ideal.

Dwarfed trees transplant easily and take hold well in the orchard. This is due not to the dwarfing nature of the rootstock, but to the fact that many of these rootstocks characteristically carry with them the ability to regenerate new roots freely. The stands of dwarfed trees in newly planted orchards are exceptionally high—attractive alike to nurserymen and orchardists (Fig. 12). And if a tree is lost, it may be replaced and got into production relatively soon.

COST PER UNIT OF PRODUCTION, AND REQUIREMENT IN MAN-HOURS OF LABOR

Efficient and Low-Cost Control of Insects and Diseases

Small trees are unquestionably more easily and more economically sprayed than standard trees. Where only two or three sprays were required during a season a generation ago, the number has risen to ten to fourteen in many regions. Further, the necessary spray equipment is expensive and elaborate, especially for large, thick trees into which the spray must be literally driven with powerful machinery.

To meet this situation, growers of standard trees have opened them by severe heading back and thinning out of branches. This permits easier spray penetration. Smaller equipment can then be used that does not require such high pressures or else carries the spray materials in a

Fig. 12. Dwarfed fruit trees are uniform and compact; a greater number may be planted on an acre. R. I. Greening/EM I, four years old.

blast of air rather than in water. Higher concentrations of materials can be used with greater facility—two to ten times the usual concentration —which means again both smaller equipment and more rapid coverage. Airplane and helicopter applications can also be used with greater likelihood of success.

All this is favored by small, open, closely planted trees. Where applications are made from the air, the trees may be spaced quite close together. And where spraying is done from the ground, and where space must be provided between rows for the movement of spray equipment, the hedgerow system of growing trees becomes of very great interest. The spray material is applied to what is virtually a wall of foliage, and the trees themselves are sufficiently narrow and open to permit easy penetration and drift.

What is not fully known, however, is what sort of local or microclimate may develop in orchards of these types. Past experience would suggest easier and more complete coverage of both fruit and foliage and much better control of pests. If carried on thoroughly and consistently, some major pests might be virtually exterminated, with corresponding reduction in spray costs. On the other hand, some pests might be favored by the new environment, and might present problems that would call for special consideration and a change in management.

Orchard Management, Including Pruning and Thinning

What has been said about efficiency and economy in spraying small trees for pest control applies equally to the application of fruit-thinning sprays and sprays for prevention of preharvest fruit drop.

With a larger number of trees and closer planting, the pollination problem is less acute. Rows of pollenizers can be planted closer to the trees they are to pollenize; or a few trees for pollenizing purposes can be easily planted in the orchard in place of the top-worked branch of a large standard tree.

In addition, if hand pollination ever becomes an accepted practice, the small tree has very much the advantage, in which solid blocks of a desired variety might be planted and the desired blossoms hand pollinated here and there as wished. Such a procedure would reduce the importance of vagaries in the weather at blossoming, reduce the spread of blossom diseases by bees, and free the tree from the overloading of fruit and from the consequent requirement for thinning.

As for pruning, the small tree can be reached more easily than can a larger tree, and the types of cuts are different. Instead of being a heavy operation, pruning becomes lighter and a different kind of labor. More detailed pruning and renewal pruning are required. Summer pruning, although largely discredited at the moment, may be found to have a place. How much, if any, economic saving there may be in the pruning operation has still to be determined.

Yet, if mechanical pruning or shearing becomes practical, the small tree, especially in the hedgerow, should be ideally suited to the practice. The procedure involves vertical arms of rotary saws or other cutting devices that shear vertically, and an overhead boom that shears horizontally at the desired height.

Protection against frost damage by overhead irrigation is more effective with small trees. In addition, orchard heaters are much more effective because of the compactness of the planting. On the other hand, the danger from stratified low temperature close to the ground can be, but need not be, an increased hazard.

The spreading of fertilizers, the application of foliar sprays, the distribution of mulch, and the control of weeds might not be much altered by the small tree in comparison with the large standard tree. However, with closely planted acreages, mulches have not been found so essential as with the wide spacing of standard trees, and there is the possibility that mulching can be either reduced or done away with.

All in all, whether the small tree will effect economies in overall orchard-management operations is something that has yet to be determined.

Harvesting and Handling Methods

One of the accepted features of the size-controlled tree is easier harvesting of the fruit. In fact, the expense and hazard of harvesting large trees from tall ladders has already demanded that such trees be brought down in height regardless of the crudeness and brutality of the method.

Much of the harvesting can be done from the ground by women and children, with less expense and less bruising (Fig. 13). The growing interest in "pick-your-own" operations is accelerated by the small tree. Customers seemingly enjoy bringing their children to the orchard and picking the fruit themselves.

Considerable emphasis has been placed on the possibility of "spot picking" from the small tree, in which trees are picked over more than once and only the properly matured fruit is harvested each time. If market demands favor this procedure, then the small tree has a definite advantage, but the increased cost of the operation places definite limitations upon it.

Attempts have been made to grow very small apple trees on the EM IX rootstock trained to wire trellises like grapes. At harvest, a low wagon or trailer is drawn between the rows, with pickers seated along the sides and harvesting the fruit as the outfit moves along. Such a procedure can also be adapted to slightly larger trees (EM VII and EM II) grown in the hedgerow system (Chapter 26).

While this is considered a new method of harvest, the writer recalls seeing a similar operation many years ago in Italy, where an ox-drawn wagon moved from tree to tree; and the women pickers, standing on the wagon frame, harvested the fruit as they very slowly and jerkily moved along. So does the primitive become modern.

Fig. 13. **Dwarfed fruit trees are efficiently harvested. McIntosh/EM I, five years old.**

Finally, there are indications that mechanical harvest of some kind may someday be realized. Walnuts are shaken from the tree and gathered by a vacuum pickup method. Prunes are also shaken. Cherries have been loosened from the tree with a chemical plant regulator and then shaken. Hydraulic lifts and platforms are in use. If this is to be the trend, the size-controlled tree has the advantage.

ADJUSTMENT TO CHANGING CONDITIONS AND MARKET DEMANDS

Rotation Orchards and Young Trees

Still another concern of fruit growers is the unprofitableness of old, large trees and the expense and difficulty of their removal. Records show that it is the young tree that produces the high-quality fruit, and at a low cost. Old, large trees are not only more costly and produce a poorer grade of fruit; they may even prove an economic liability. Twenty-five years has been given by many orchardists as the life beyond which apple trees are no longer profitable. An ideal has been suggested of full fruiting in six to ten years, fifteen years of production, and then removal. For this reason the trend in commercial fruit circles is toward "rotation orchards," in which new plantings are made at regular intervals to replace older plantings when they reach the unprofitable period.

Into this setting the smaller, earlier-fruiting, controlled tree fits well. Fruiting age is reached early; the orchard bears for a profitable period; and the trees are then quickly and easily removed and replaced with a new planting.

To offset the advantage of dwarfed trees in this respect, there is the possibility of maintaining rotation orchards of standard trees. With plenty of land available, and with the system once established, the standard tree can be handled in this manner.

Shift in Varieties

Closely tied with the idea of the rotation orchard is a shift in varieties to meet the changing conditions and market demands. The history of fruit growing shows that fruit sections arise largely with the performance of a superior and adapted variety and that they are then driven out by competition from other sections that, in turn, develop a superior and adapted variety. This history of the Northeast with the Baldwin and the R. I. Greening apples, the rise of the Pacific Northwest with Delicious, Jonathan, Rome, and Winesap, and the return of the Northeast to importance with the McIntosh, is a familiar story.

Added to this is the rapid turnover or change in varieties to meet the changing market demands that appear to be in keeping with the increased speed of modern living. The competition between fruits is great, and a variety that is in demand during one decade is unwanted the next.

The roadside stand with its call for a wide variety of fruit over an extended season and for "tree-ripened" fruit is another example of such a change, brought about largely by the automobile. Other changing factors, such as modern refrigeration, controlled-atmosphere storage, truck transportation, and the canning and freezing of fruits, have altered the usefulness of old varieties and have brought about the introduction of new kinds. The crabapples that are used for pickling have been partially replaced by a new process of pickling small-sized Jonathan fruits. The development of gift packages of fruit has placed a premium on large Comice pears and Golden Delicious apples. The Melba apple, handled delicately like a peach, finds an outlet where a high-quality early apple is desired. The controlled tree with its characteristics of small size and early fruiting is well adapted to this trend.

Quicker Recovery from Tree Losses by Winter Cold and Other Hazards

Major tree losses from early fall freezes, severe winter cold, and other climatic hazards are not uncommon even in the more favored horticultural areas of the world. The severe winter of 1933–1934 in western New York resulted in the death of thousands of acres of fruit trees and replanting with new trees. Similar catastrophes are not uncommon in the Midwestern sections of America; and the Pacific Coast regions have experienced severe injury from sudden fall freezes before trees had properly matured.

Generally speaking, injury from cold is more damaging than at first appears, and trees that might seem able to recover often merely linger listlessly and succumb a few years later. Much time is thus lost in returning the orchard to profitable production. Experience has shown that removal of the injured tree and prompt replanting would often have been the better procedure. Under such conditions, the dwarfed tree offers real possibilities. In fact, some practical fruit growers have suggested that this feature of dwarfed trees is in itself sufficient reason for their use.

YIELD AND QUALITY OF FRUIT

Yields of Fruit

There is no question but that for the first few years of the orchard the yield from a planting of dwarfed fruit trees is higher per acre than that from a planting of standard trees.

To cite a few examples, the total cumulative yield of 50-pound boxes of fruit in Oregon from a planting of Golden Delicious/IX (363 trees to the acre) for the first ten years of the orchard was 5,082, compared with 1,488 for the same variety on standard seedling roots (48 trees to the acre).

In another planting of Golden Delicious/IX (363 trees per acre) 1,000 boxes of fruit were produced per acre the sixth year, 1,600 boxes the eighth, 1,700 boxes the tenth, and 2,400 boxes the twelfth.

This difference in early yields in favor of dwarfing rootstocks is very important, since this is the critical period in terms of economics in the establishment of an orchard. This becomes perhaps the most appealing aspect of dwarfed fruit trees, and is more fully discussed in Chapter 25.

It is, however, a matter of common knowledge that standard trees will produce 300, 500, 700, and even as high as 1,700 or 2,200 bushels of apples per acre in exceptional years. There are few large-scale records for trees on dwarfing rootstocks in North America that approach these figures, although there are some as high, or higher, from Europe. In Massachusetts fifteen-year-old McIntosh/EM VII trees planted 20 × 30 feet, or 72 trees per acre, have produced 864 boxes of fruit per acre. The trees could have been 15 × 20 feet, which by calculation reaches 1,600 or 1,700 boxes per acre. However, any such theoretical calculation is questionable because it disregards such factors as competing roots, and water and nutritional requirements. Cortland and McIntosh in Michigan on EM II and EM VII planted 20 × 30 feet have yielded 600 to 800 bushels per acre at fourteen years of age. Individual trees at twelve to fifteen years of age have borne 12 to 21 boxes of fruit. Limited acreages in the Pacific Northwest have produced 3,000 bushels per acre on trellises.

At the same time, it is not yield alone that determines profit, as has already been mentioned. It is the cost of production per unit plus the quality of the fruit, and early reports in America indicate high early yields per acre of high quality fruit. Only time will give the answer to this question in North America, but figures from abroad favor the dwarfed fruit tree. All this will be more fully treated in subsequent chapters.

Color of Fruit

Because of the openness of the trees, which permits more sunshine to reach the fruit, dwarfed trees produce fruit of high color for the variety. In many instances the color is exceptional. Not only this, but a very high percentage of the crop is of good color, and cull fruit because of poor color is very much reduced. This is one of the outstanding features of the dwarfed tree. On the other hand, proper pruning, thinning, and fertilizer management of standard trees can do much to accomplish the same ends. And, depending on the site and the layout of the orchard, the dangers of frost rings and russeting from low temperatures may be increased on dwarfed trees, as is discussed in Chapter 16.

Uniformity in Size and Color of Fruit

It has often been remarked that dwarfed trees produce fruit of exceptional uniformity in size and color. This seems to be the case (Fig. 14). The fruit on very young trees is often oversized and lacking in uniformity, but this is a characteristic of young fruit trees in general. As the tree reaches full bearing age, the uniformity in size and color becomes apparent. The fruit may be only medium in size and medium

Fig. 14. **Dwarfed trees produce high-quality fruit. Delicious EM IX.**

in color on a given tree, but all of it is remarkably similar in size and color. This may be brought about by easier thinning and better spacing of the fruit throughout the tree; but regardless of how it is accomplished, the uniformity in size and color is noticeable.

Insect and Disease Blemishes

Because of the ease and efficiency with which dwarfed trees are sprayed for pest control, both as single trees and as hedgerows, the crops are exceptionally free from blemishes of this nature. There are no high tops of trees to reach with spray, and there is not much fruit inside the tree that may be missed.

Bruising

Dwarfed trees are low and relatively closely planted so that they tend to protect each other from the wind, and bruising from wind whipping and rubbing is reduced. Bruising from picking bags and ladders and other harvest operations is also lessened.

FACTORS OF SOIL, CLIMATE, AND MANAGEMENT

Adaptation to Soil and Climate

Quite aside from the size and fruiting characteristics of dwarfed fruit trees, there is the promise that some of the dwarfing rootstocks may be useful in adapting fruit trees to special soil and climatic needs. For example, it appears that the EM XIII rootstock for apples will succeed in relatively heavy and wet soils, and the quince will stand wet land even better than the pear. The Coxe's Orange apple, which is susceptible to crown rot (*Phytophthera cactorum*), can be grown successfully on resistant rootstocks, such as EM IX, VII, II, and XXV, and MM 104, 106, and 111.

Some varieties of pears and apples frequently do not mature their fruit on standard trees in the length of season that prevails in a given region. They may be brought to perfection when grown on very dwarfing rootstocks. The Grimes Golden apple, for example, will develop good quality in western New York on the EM IX rootstock, whereas on standard roots it is often inferior in both flavor and color. Passé Colmar, Glou

Morceau, Beurré Superfin, and many other choice pears develop excellent size and quality on the quince root in regions where the fruit is almost worthless on standard pear roots.

Spring Frost Hazard

It is frequently observed that the blossoms on the lower part of a tree may be destroyed by a spring frost, while blossoms in the tops of the tree may be unhurt. This is due to the settling and stratification of injurious cold air. Accordingly, it would not be surprising if dwarfed trees in low areas, with poor air drainage, should lose a crop in such a situation. This means that special attention should be given to placing dwarfed trees on sites relatively free from spring frosts and where air drainage is good.

Trouble can also be experienced if small, closely planted trees or hedgerows of trees are planted across a slope so that air drainage is impeded. One of the worst locations is a slight slope with an embankment or row of tall trees across the lower end so that downward-flowing cold air is literally bottled up in the orchard.

On the other hand, in the proper situation dwarfed trees have borne good crops following severe spring cold where standard trees have not. The explanation is that the warming radiation from the earth is blanketed by the small, closely planted trees; whereas with the large, more widely spaced standard trees the warming radiation is dissipated upward. Mulching tends to increase frost danger because of reduced radiation from the soil. All this is more fully discussed in Chapter 16.

Hardiness to Winter Cold

Apple rootstocks used for dwarfing are in themselves hardier to winter cold than, or at least as hardy as, the tops of many commonly cultivated varieties of fruits. Among apples, for example, they are hardier than the Baldwin, Gravenstein, and R. I. Greening. Further, the root, being in the ground, is less likely to suffer damage. There are great areas in North America where these varieties would not survive winter cold even as standard trees. The quince rootstock is more tender to winter cold than the varieties of pear worked upon it. Winter injury may be expected where frost penetrates deeply into the ground, unless snow cover or mulch provides protection.

Leaning and Blowing Over

Trees on the more dwarfing rootstocks are not firmly rooted and are prone to lean and even to blow over (Fig. 15). Staking is good practice

Fig. 15. Each type of clonal dwarfing rootstock must be critically appraised; some types possess undesirable characteristics, as shown in this six-year-old McIntosh planting on the EM VI rootstock, which tends to lean badly.

for the first several years, and sometimes need only be enough to steady the tree against the prevailing wind.

Each rootstock presents individual problems peculiar to itself. Thus, the very dwarfing EM IX apple root is brittle and easily snapped. Breakage seldom, if ever, occurs at the union, but on the rootstock below the union. The semidwarfing EM IV has a high top/root ratio so that it is easily blown over by high winds, especially on light soils where anchorage is not good. On heavy soils, with better anchorage, less blowing over occurs.

There are practices that reduce leaning, blowing over, and breakage. If a tree is headed low, at 12 to 18 inches, the leverage on the top by wind is greatly reduced. Even trees on the Malling IX apple rootstock can be grown without staking if they are grown as low bushes.

The hazards are also lessened by planting in protected spots. Shelter hedges of willow or other tall-growing trees give good protection to relatively large plantings. With the initial protection on the windward side by the shelter hedge, one row of fruit trees tends to protect the next.

The more vigorous of the dwarfing rootstocks anchor the tree firmly and require no special staking. Included are the EM XIII and EM XVI apple rootstocks. Trees on EM XII, which is not at all dwarfing, are as sturdy as trees on standard seedling rootstocks.

Suckering and Scion-Rooting

Some of the dwarfing rootstocks tend to send up suckers from the roots, as EM III and EM VI, and they should be avoided where suckering is severe. Others, such as EM I and EM IX, appear susceptible to crown rot, in some locations, usually associated with winter injury, and they should be avoided in adverse locations. There are other rootstocks that will provide the degree of dwarfing desired and that are not subject to these difficulties.

Attention must also be given to depth of planting. If the trees are planted deep, so that the scion is below the ground level, the scion may take root and the dwarfing effect of the rootstock be lost. This matter of scion-rooting has, however, often been overemphasized. If the union is placed at the ground level or slightly above, little difficulty will be experienced.

Susceptibility to Drought

Because of the restricted root system, dwarf trees are more subject to injury from drought than are standard trees. The practice of mulching with straw, hay, peat moss, or some other protective material reduces this handicap.

Garden Care and the Need for Information

In spite of all the instructions that may be given, a natural "feel" for a plant is necessary to succeed with it. How many fruit growers and gardeners there are in the United States who have developed or who will develop this touch for dwarfed fruit plants remains to be seen. Hopefully it is large and growing; but without it, the dwarfed tree does not thrive abundantly well. The dwarf tree, especially in the very dwarf form, as an EM IX for the apple and quince EM C for the pear, is essentially a garden plant and responds best to the gardener's watchful eye and expert care. It will not tolerate neglect. It cannot stand competition with other plants. It seems especially attractive to mice, rabbits, and borers. It will not respond favorably in an unfavorable horticultural environment. Since it prefers intensive culture, it is not likely to be found suited to extensive operations where trees may sometimes be left to shift for themselves.

As with any new development, success is associated with a thorough understanding of the problem involved. A new variety of fruit succeeds only when an understanding has been reached as to how it should be grown. The McIntosh apple, for example, succeeded only when growers pooled their knowledge and showed how the culture of this variety differed from that of some of the old favorites. The organization of the Dwarf Fruit Tree Association in Michigan (1958) and a similar one in the State of Washington (1959) should prove exceedingly helpful in this regard.

Only when sufficient knowledge has accumulated from a wide assortment of experiences will real success be assured with the dwarfed or size-controlled fruit tree.

ADDITIONAL POSSIBILITIES WITH DWARFED TREES

Although most of this chapter has dealt with dwarfed fruit trees for commercial fruit growing, there are other places where they are useful, especially the very dwarf forms. Much of what has been said about dwarfed trees in commercial orchards applies equally here.

They may be used as miniature ornamental house plants, as greenhouse plants in tubs and pots, as ornamentals in landscape design, and as wall coverings (Fig. 16), points of interest, espaliers, cordons, and other special forms. They may be used in small garden areas, in subsistence gardens, in formal fruit gardens, and in mixed vegetable and fruit gardens, as well as in commercial fruit plantations. The particular use to which they are put will depend upon a number of factors, including climate, soil, the area at disposal, the type of rootstock available, the varieties that will be grown, the system of management, and the personality and likes and dislikes of the planter.

Fig. 16. **Trained fruit trees are very effective as wall covers.**

For Restricted Areas

The very dwarf fruit tree is valuable around the home grounds or where space is at a premium. These small trees may find themselves along a border, in group plantings, and in the kitchen garden where their small size and restricted root spread will not shade other crops appreciably or compete for soil moisture and nutrients. And if a home-owner or tenant moves, it is possible for him to transplant very dwarf trees and take them with him to the new location.

Subsistence or Suburban Fruit Growing

Dwarfed fruit trees, particularly EM IX and EM VII, are ideal for the subsistence or suburban fruit grower. Five acres of orchard can be tended weekends, evenings, and holidays. New insecticides, fungicides, herbicides, and plant regulators, coupled with modern small power equipment, make all this possible. Harvest may be handled on the increasingly popular "pick-your-own" basis, in which the customer harvests the fruit himself, and often furnishes his own container.

Part Two

TREE STRUCTURE AND PHYSIOLOGY AND HOW A TREE IS DWARFED

4

STRUCTURE AND PHYSIOLOGY
OF A FRUIT TREE AND HOW
IT GROWS AND FRUITS

SOMEONE has defined a tree as a plant "that stands of itself and can be climbed in." In simplest terms a tree is an axis (root and stem) and its appendages (leaves, flowers, and fruits), as shown in Figure 17. Like other plants, it is a living organism, and subject to the same forces and processes in growth and development. It is composed of individual cells that by division and differentiation constitute the growth of the plant and bring about the development of specialized tissues and organs. It is concerned with both growth and reproduction, in which man exerts a managing hand and shapes and adapts to his needs and his desires.

Yet there are differences as well as similarities between a fruit tree and many other plants. First of all, a fruit tree, unlike an annual plant that lives only one season, lives year after year and stores food materials within itself so that the performance in one season is dependent considerably upon what went before. This is why two or three seasons may be required before a treatment expresses itself in a fruit tree, whereas with a tomato plant it may show in a few weeks. It is why the effects of drought or of winter injury sometimes do not appear until a year or two later. It is also why the fruit crop of one season is often as much dependent on the preceding season as on the season in which the crop is borne, or even more so.

Second, the fruit tree with which we are dealing belongs to a group of plants which has a cambium layer, as shown in Figure 17. By cell division this layer gives increase in diameter of root, trunk, branch, and stem, and produces the conductive system of the tree. Such plants are termed dicotyledons, for the reason that the seed develops two cotyledons, or true seed leaves, familiar to all in a germinating bean seed. This is in contrast to another group, monocotyledons, like corn, which does not have a continuous cambium and has but one conspicuous cotyledon, or seed leaf.

But it is not the cotyledons that concern us; it is the difference in the structure of the stem and the root of the two groups. The monocotyledon does not have a continuous cambium. The conductive bundles of the monocotyledon are separate, and scattered in the stem, as in sugar cane, corn, and palm. On the other hand, the fruit tree, which is a dicotyledon, does have a continuous cambium, and the vascular system is in concentric rings. This is why girdling the stem of a palm, as discussed later in this chapter, does not kill the plant, whereas a deep girdle in a fruit tree destroys the cambium and leads to the death of the plant unless some method is devised to bridge the girdle. It is also why a fruit tree can be grafted and budded, whereas a monocotyledon cannot, as will also be discussed later.

Third, a typical cultivated fruit tree is not a single individual, like a tomato plant or a forest tree, but is composed of two individuals growing together as one. The forest tree and the tomato plant grow from seed, so that the root and the top are of the same origin; and the top is literally growing on its own roots. Not so with the cultivated fruit tree. Onto some special rootstock material, which provides the root system, a desired scion variety is budded or grafted, which supplies the top and the fruit. This is a most important difference, and one that is usually so much taken for granted that it is almost entirely lost sight of.

Fig. 17. (Opposite) Diagram A, showing the principal organs and tissues of the body of a seed plant; B, cross section of stem; C, cross section of root. (After Robbins and Weier. Botany: An Introduction to Plant Science, John Wiley and Sons, New York, 1950)

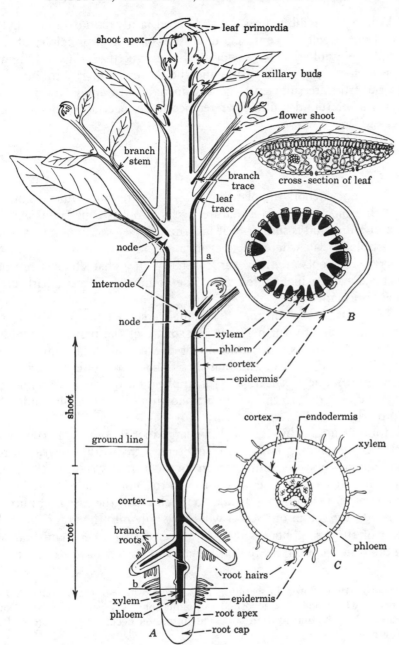

While these differences must be constantly borne in mind, they should not needlessly confuse a discussion of general principles of structure and physiology. The diagrams of the structure (Fig. 17) showing the principal organs and tissues and the flower (Fig. 18) that develops into the fruit are sufficiently general in nature to apply to a fruit tree. They will be found useful for reference in connection with subsequent discussion.

THE ROOT

The root anchors the plant and absorbs water and mineral nutrients from the soil. Because in a mature tree the root is continuous with the stem or trunk and is similar to it in structure (see section on stem), it is impossible to say where one begins and the other leaves off. Typical underground roots are, however, covered with a characteristic brownish suberized bark, which differs from the commonly fissured bark of the trunk and branches.

The root system mirrors the branch system and extends outward horizontally slightly further than the spread of the branches. The bulk of the feeding roots are within 2 feet of the soil surface and outward toward the periphery of the branches. The point is illustrated by the injunction frequently heard for the application of fertilizer; namely, to apply the materials to the soil in a circle around the tree out under "the drip of the branches."

Another interesting fact is that a large root is commonly continuous with a corresponding large limb. In effect, a tree with a few main scaffold branches finds a counterpart in the main root system in the soil, so that a mature tree has often been likened to a collection of rooted shoots held together in a common trunk. So much is this the case that nutrients applied to the soil in one sector will move only into the branches of that sector of the tree. There is little cross-transfer to other main branches. On the other hand, trees that are small and do not have strong scaffold

Fig. 18. Apple flower in section: (A) anthers, which produce pollen; (B) stigma, where pollen is deposited in pollination; (C) petal; (D) sepal or calyx lobs; (E) fertilization and seed formation occur here. (After A. E. Murneek)

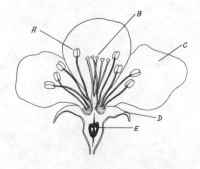

branch development have root systems that are less extensive and that anastomose and tend to fill the soil mass beneath the tree. The EM IX rootstock is an example of the latter; it develops many fine roots rather than a few main roots as does the strong-growing EM XII rootstock.

Lateral roots arise from the pericycle of both the root and the stem (Fig. 17). This tissue lies just outside the phloem layer. Some root materials, such as the EM VII rootstock, give rise easily to new roots; whereas other materials, such as the EM II, are less free in this regard. Why this difference exists in rooting response is not well understood.

It is an interesting fact that roots grow downward. While still a matter of conjecture, it is postulated that a hormone (indoleacetic acid) which is manufactured in the growing tip of the shoot may be the controlling agent. It flows downward from the tip of the shoot, suppressing shoot development or suckering from the root. When a root is placed in a horizontal position, the hormone tends to accumulate in the underside of the root. The cells on the upper side of the root elongate more than those on the underside, and the root grows downward. This is exactly the opposite of stem response.

Further, although the root is commonly thought of only in terms of absorption of water and nutrients from the soil, it appears that roots also synthesize materials that may be returned to the aboveground parts of the plant. The tobacco plant is a good example; the nicotine is manufactured in the roots. If a tobacco plant is grafted onto the root of a tomato plant, the combination will grow but the tobacco scion contains no nicotine. Conversely, if a tomato plant is grafted onto a tobacco plant, the tomato scion may contain nicotine.

All these processes have a bearing on how a tree is dwarfed, the rooting of dwarfing rootstocks, suckering, root extension, and stock-scion relations, as will be shown later in more detail.

THE STEM

The stem supports the plant and conducts water, mineral nutrients, and various organic materials both upward to the leaves and fruit and downward to the roots. In cross section, in a young tree, the outermost layer is a single layer of cutinized protective cells, termed epidermis.

Immediately inward is the cortex or cortical region, including a layer of endodermal cells, which is often a storage region and which also differentiates supporting fibers. Next inward is a layer of cells known as pericycle, which also differentiates fibers and in addition is very important in propagation for the reason that roots frequently originate here.

Inward from the pericycle is the vascular cylinder of the plant, composed of phloem and xylem, with a cambium region of active cells separating the two. This is the transport system of the plant.

The phloem is primarily concerned with the downward movement of elaborated plant food and other organic compounds manufactured in the leaves. When the downward movement to the roots is interrupted, as by mechanical girdling, the performance of the tree is markedly altered, as we shall see.

The xylem is the principal region of upward movement of water and mineral nutrients, together with some organic materials.

The cambium is a region of active growth, and is responsible for the increase in diameter of the stem by dividing and differentiating xylem toward the inside and phloem toward the outside. Because of its high regenerative capacity, it is a most vital part of the stem in the closing of wounds and in promoting union of scion and stock during budding and grafting operations.

In the center of the stem is the pith, which serves mostly as a storage region.

As the stem increases in diameter, some of the outer layers are sloughed off and replaced by a cork layer that develops from a second cambium-like regenerative tissue. In mature trees it is not uncommon for epidermis and cortex to be entirely sloughed off, leaving only the phloem, cambium, xylem, and pith surrounded by the protective cork layer.

Shoots arise from leaf buds that are formed during the preceding season and remain dormant all winter. In early spring they begin rapid growth, often at a phenomenal rate, and reach maximum length for the season within a few weeks after full bloom.

The shoot grows upward, in contrast to the downward growth of roots. It is now generally accepted that this upward growth is controlled by hormones or growth substances (auxin or indoleacetic acid and gibberellins) which are manufactured in the tip of the shoot and move downward, causing elongation of cells below. When the hormone is distributed equally on all sides, elongation is symmetrical and the shoot grows straight upward, much as when an automobile is raised by jacking up all four wheels at once. But when the distribution of auxin is unequal, the cells on one side elongate and those on another do not, so that the shoot "bends" much as an automobile is tipped when only one side is jacked up. When a shoot is placed in a horizontal position, the auxin tends to accumulate on the lower side and causes the cells to elongate and so bring the shoot back into an upright position.

Similarly, light affects the direction in which a shoot grows. Light causes auxin to accumulate on the dark side of the shoot and promote cell elongation on that side so that the shoot "bends" toward the light. In high altitudes where sunlight is strong, especially the ultraviolet waves, auxin is rapidly destroyed and cell elongation is reduced, so that the plants become stunted or dwarfish. Further, dwarf forms of corn have been shown to have lower content of elongation hormones than tall corn. The crown gall organism, which causes galls to form on plants, manufactures auxin and stimulates the growth of adjacent cells so as to produce the galls.

In the artificial control of growth and fruiting of fruit trees, it is mechanisms such as these that are brought into operation.

THE LEAVES

The leaves, in the presence of sunlight, manufacture food materials from the carbon dioxide of the air, from water, and from other materials supplied to them. The process is called photosynthesis. The principal manufactured products are the carbohydrates that supply both the energy for growth and also much of the actual constituents of the tree and its fruit.

If leaves are shaded, if they are injured by pests, and if the movement of materials is interfered with, the growth and fruiting of the plant are affected. Thus, the leaves of trees that are open to sunlight are more efficient per area of leaf. About 100 square inches of apple leaf are required for the development of an apple fruit of good size and quality, or about 30 to 40 leaves per fruit, depending upon the size and exposure of the leaves. Similarly, fruit buds are promoted on spurlike growths with vigorous leaves. If leaves are removed, the processes are reversed and the plant is checked in growth and fruiting.

The leaves also give off (transpire) substantial quantities of water—as much as 15 to 20 tons of water per year from a large fruit tree when it is making good growth and producing a good crop. The loss of water from an acre of trees is greater than from an acre of pond or lake. For each pound of dry matter produced by the tree, approximately 500 pounds of water are transpired. Fruits contain 85 to 90 per cent water. Here again is the explanation of why the water supply and the transporting system within the plant are so important and why any interference with them has such pronounced effects.

While the leaf has commonly been thought of solely as an organ concerned with photosynthesis and with transpiration, it is also able to take up both water and nutrients as well as to lose materials through leaching of rain and dew. This is the basis of so-called "foliar" or "non-root" feeding, which is resorted to as an emergency measure or supplemental measure at critical times in growth and development. Thus, when winter injury has injured or damaged parts of the conduction system, so that materials from the roots cannot be transported to parts where they are needed, it has been possible to aid in recovery by applying sprays of urea and other nutrient materials to the leaves. Materials may even enter through the bark during the dormant season, especially as the plant begins new growth in the spring. These techniques are of value in controlling tree performance.

THE FLOWERS AND FRUITS

The flower is a group of specialized leaves concerned with sexual reproduction. A typical flower (Fig. 18) consists of an outer whorl of sepals that are usually green in color and are known collectively as the calyx; a whorl of petals that together are known as the corolla; the stamens that produce the pollen from which the male germ cells (microgametes) develop; and the pistil, which is composed of one or more sections or carpels bearing the female germ cells (megagametes) and enclosing the ovules that later develop into the seed. The fruit develops from the carpels; and it may consist variously of a single carpel, as in the peach, cherry, and plum; of several carpels, as in the grape; or of several carpels and accessory tissue, as in the apple and pear.

Flower Bud Initiation and Development

Flowers of a fruit tree develop from flower buds that are initiated and differentiated during the preceding growing season. They are distinct from leaf buds, which give rise to shoot growth; yet flower buds may consist solely of flower parts, as in the peach and cherry, or of mixed leaf and flower parts, as in the apple and pear. At all events the distinction lies between buds that contain flower parts and those that do not.

The first visible symptoms of flower-bud differentiation, as seen under the microscope (Fig. 19), is a flattening at the terminal growing point of the shoot. This occurs about three to four weeks after full bloom in the Wealthy variety of apple. In other varieties, such as Rome Beauty, differentiation may be delayed until as late as mid-July or early August. Differentiation in the pear starts about the middle of July. For the peach the time is usually given as the latter part of July; for the cherry, the first week in July; for the plum, the first and second weeks in July; and for the quince, late summer or early fall.

However, four to five weeks prior to visible differentiation, certain physiological conditions have been set up in the plant that induce the process of differentiation. The point to emphasize is that initiation of blossom buds for the next year's fruit crop begins much earlier than is frequently believed—usually some time around full bloom for many varieties of apples and pears.

Induction of flower buds depends a great deal on a large leaf surface of new leaves near the stem tip or bud during the period from two to four weeks after full bloom. Thus, good early growth is of most importance; and if the leaves are small or the tree is too heavily loaded with fruit, blossom buds may fail to form.

Defoliation or injury to leaves during the season, or poor growing

Fig. 19. Development of the flower bud of apricot at the following dates: (1) August 1; (2) August 10; (3) August 20; (4) September 15; (5) October 1; (6) October 15; (7) November 1; (8) December 1; (9) January 1. (After C. B. Wiggans)

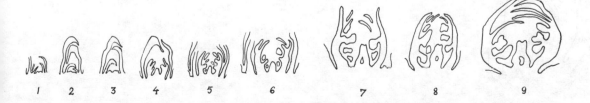

conditions, may affect the development of the flower bud after it has begun to form. These factors do not, however, affect the onset of initiation inasmuch as the time for initiation has passed.

Accordingly, any practices intended to affect blossom bud formation, such as ringing or blossom thinning, must be done prior to this time of initiation.

After the flattening at the apex of the shoot that indicates that the shoot is beginning to differentiate as a blossom bud, the first real flower parts to be seen are the sepals or calyx lobes. They arise as slight protuberances or "humps" near the outer edge of the flattened area (Fig. 19). A few weeks later, another ring of protuberances appear just inside the sepals, and are the petals. Later in the season, the stamens arise; and still later the carpels appear, which are to be the fruit. The sequence is shown in Figure 19.

In the fall of the year, all the flower parts are not yet fully differentiated and developed; yet if you take a razor blade and slice lengthwise through an apple, you can determine whether or not it is a leaf bud or a blossom bud. The leaf bud in longitudinal section appears crescent-shaped, whereas a blossom bud appears conelike or V-shaped (Fig. 20).

During the winter the blossom parts still further differentiate, and by the time of full bloom all parts are present and fully developed. From this it can be seen why the statement is made that the preceding season has so much to do with a fruit crop. If growing conditions are favorable, and if the tree has been maintained in a healthy and vigorous condition, the fruit buds will have good strength and vigor and will be more likely to set fruit, all other things being equal.

Pollination and Fruit Set

The next step is the pollination of the flower parts, the fertilization of the egg cell (megagamete), and the setting of the fruit. In the pollina-

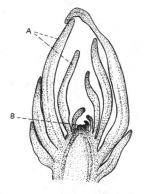
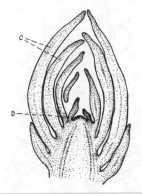

Fig. 20. **Apple buds in early August, the season prior to fruiting, cut length-wise to show differences between fruit buds and leaf buds as they develop: (Above), fruit bud; (A) protecting scales, (B) growing tip where flowers develop; (Below), leaf bud; (C) protecting scales, (D) growing tip where leaves and shoot growth will develop. (After E. J. Rasmussen)**

tion process, pollen grains carrying the male cells (microgametes) are transferred to the tip (stigma) of the pistil by bees, insects, or other agencies. The pollen grains germinate, and the resulting pollen tubes grow down the style of the pistil, into the ovarian cavity, and into the ovule. There the male cell fertilizes the female cell to give rise to a new plant (embryo) enclosed within the seed coats. Fertilization of the flower in this way causes the fruit to set and develop. This is brought about by naturally occurring hormones from the developing embryo and associated endosperm tissue, which prevent the development of an abscission zone in the stem of the flower or fruit and prevents them from dropping.

The factors associated with good set are spring temperatures favorable to bee flight and pollination, good nutritional condition in the plant, and freedom from damaging frost, cloudy weather, or drought. An application of nitrogen fertilizer made either the preceding fall or early in the spring, just as the frost comes out of the ground, helps to "stick" the blossoms as fruit.

Fruiting Habits

Much is made of the ability to distinguish leaf buds from fruit buds by their general appearance; but usually this ability is dependent upon a knowledge of the fruiting habit of the tree plus keen observation. Thus, leaf buds of a growing peach tree are found at the node of the shoot (one to a node), having formed in the axil of a leaf during the previous season. But when the peach tree forms blossom buds, three buds are formed at a node—the center one of the three being a leaf bud, and the ones on either side being fruit buds (Fig. 21).

On the other hand, apple- and pear-blossom buds are typically borne on short stubby growths called spurs. Such spurs usually make a short growth one season and terminate as blossom buds. In the next year this blossom bud develops into a fruit. In the third year the spur again makes a short growth and terminates as a blossom bud. This accounts for the zigzag growth of spurs (Fig. 22). Yet there is no certainty about this;

Fig. 21. **Peach flowers are borne singly on each side of a leaf bud on one-year-old wood. (After E. J. Rasmussen)**

a spur may blossom year after year and set no fruit, owing to frost, lack of fertilization, or unfavorable growing conditions. Conversely, a spur of some varieties, as the McIntosh apple, may fruit for several years in succession. In addition, some varieties may form blossom buds terminally on vigorous shoots. The R. I. Greening and Northern Spy varieties of apple may do this. Finally, some varieties, such as the Wealthy apple, may form blossom buds in the axils of the leaves on vigorous one-year shoots, much as the peach does. Yet these are the exceptions and not the rule; most apples and pears are borne on spurs. Cutting a bud, as has been discussed, is positive identification and is simply done.

Cherry trees produce flower buds both laterally on one-year wood and on short spurs. Each blossom bud contains more than one flower but no leaves. When the tree is of fruiting age and is growing vigorously, the blossom buds are found on spurs, so that most of the crop is found on spurs; and when the growth is short and weak (under eight inches) many of the blossom buds are formed laterally on one-year wood. The plums and the apricot similarly form blossom buds both laterally on one-year wood and on spurs. The European plum and the apricot bear relatively more of their crops on spurs than do the Japanese plums and the peaches.

Factors Associated with Blossom-Bud Initiation

To return for a moment to blossom-bud initiation, their are several explanations as to what set of factors is responsible. One explanation is that an accumulation of storage carbohydrates (sugars and starches) brings it about. Certainly there is a close relationship between blossom-bud formation and carbohydrate accumulation. A large, healthy leaf surface, as we have seen, is associated both with the accumulation of storage carbohydrates and the formation of flower buds. In addition, the

Fig. 22. Apples and pears are most commonly borne on spurs. Practices are employed that will induce spur formation, which favors high production: (1) Young apple shoot (two years old) pruned to induce spurs shown in (2) older apple spurs (five years old) that have been induced to form from young apple shoot (1). (After "A glossary of terms used in pruning fruit trees," *Scientific Horticulture* 11: 67–74, 1952–1954)

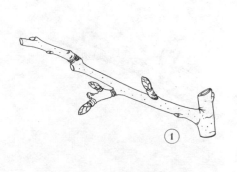
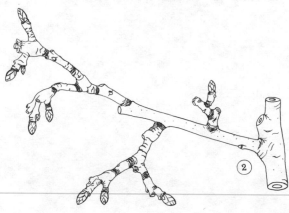

presence of a large number of fruits that compete for the carbohydrate supply is associated with a reduction in the number of blossom buds that form for the next year's crop. But whether it is the storage carbohydrates that bring about blossom-bud formation or whether it is the formation of blossom buds that brings about an accumulation of carbohydrates is not yet known. Nevertheless, an understanding of this association is helpful in controlling the growth and fruiting of a fruit tree.

Another explanation is offered by the so-called carbohydrate-nitrogen or C–N relationship that has been observed in plants. Simply stated, nitrogen is secured by the plant principally from the soil, and is influenced in amount and availability by temperature, soil moisture, root penetration, and other soil environmental factors. On the other hand, the carbohydrates are manufactured in the green portion of the plant from the carbon dioxide of the atmosphere in the presence of light. The general idea is that the ability of a plant to fruit is indicated by the relationship between carbohydrates and nitrogen in the plant. While this may represent an over-simplification of a series of complicated processes, yet it has its value in helping to explain fruiting and gross factors affecting it.

On this basis, fruit trees may be placed in one of four classes as regards C–N relationship and fruiting:

Class I is the plant that has an abundance of nitrogen but a deficiency of accumulated carbohydrates. It is the thin-wooded, spindly, weak tree that grows in the shade. It is like an undernourished child. It does not develop and does not fruit.

Class II is the adolescent, which, supplied with an abundance of both nitrogen and carbohydrates, keeps on growing exuberantly, with large dark green foliage, but does not get down to the business of fruiting.

Class III is the highly productive individual, in which sufficient reserve carbohydrates have accumulated to provide for fruit-bud formation and fruiting. This is the age of maximum production, and the joy of the successful fruit grower.

Class IV is the age of senescence and decline, where carbohydrates have accumulated to such a degree that the foliage is yellowish in color, the tree biennial in bearing, and the characteristics of age appear—not unlike the man who has accumulated sufficient avoirdupois to be termed "a bit heavy" or even "corpulent," and who is content to sit by the fire in his slippers, except for sporadic outbursts of enthusiasm from time to time that slowly diminish.

With these four classes clearly in mind, it has been shown that much can be done in the fruit plantation to control the behavior of a tree (Fig. 23). The weak tree of *Class I* needs to be brought into the sunshine, where it may manufacture more carbohydrates, pass into the adolescent stage of *Class II*, and there accumulate a sufficiency of carbohydrates to settle down into the productive period of middle life (*Class III*). Conversely, the aging tree of *Class IV* can be rejuvenated and brought back into the productivity of *Class III* both by severe pruning, which removes excess carbohydrates, and by application of nitrogenous fertilizers, which increases the nitrogen balance.

Further, ringing and girdling are helpful in bringing a tree from *Class II* of adolescence into the productive *Class III* by virtue of the fact that carbohydrates are induced to accumulate above the ring or girdle (Chapter 5). Root pruning is similarly helpful, but for the reason that it reduces the intake of nitrogen (Chapter 5). High summer temperatures may delay fruiting in young trees because the tree respires carbohydrates too rapidly to accumulate a sufficient amount for fruitfulness. Pruning a young tree excessively removes the carbohydrates, prevents accumulation, and tends to keep the tree in the weak, nonfruiting *Class II*.

Insect and disease attack and caustic sprays that injure the foliage all likewise tend to reduce the carbohydrate supply and delay fruiting.

Many men have recognized the similarity of all of this to the behavior of animals, and have dreamed wistfully, but forlornly, upon some method or source of rejuvenation such as Ponce de León sought in the Fountain of Youth several centuries ago.

Again like the quest for the Fountain of Youth, there has long been a hope that someday some flower-inducing hormone might be found in plants that would provide the proper explanation of fruit-bud development as well as provide a method of control. There is evidence that such a hormone does exist and that it is manufactured in the leaves and transported to the buds. When there is a heavy set of fruit, the explanation is that the flower hormone is utilized by the young fruit, and therefore few if any blossoms are initiated for the next year's crop. The next year, since there is no competing fruit, the hormone induces abundant flower-

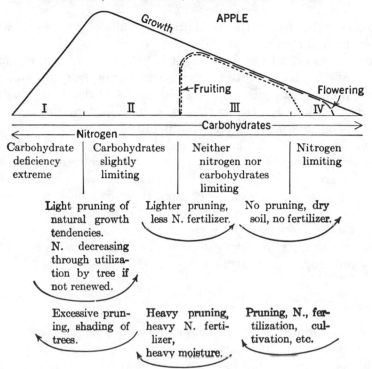

Fig. 23. Relationship between the nitrogen and carbohydrate composition in the apple and its response to various cultural treatments with respect to flowering and fruiting. (From *Modern Fruit Production* by J. H. Gourley and F. S. Howlett, The Macmillan Company, New York, 1941)

bud formation. In this way, it is explained, alternate bearing is brought about.

Be this as it may, there is no doubt that a heavy set of blossoms definitely reduces the initiation of flower buds for next season's fruit crop. At the same time, if blossoms are removed flower buds will again form. Over one hundred years ago, Robert Manning, of Salem, Massachusetts, removed all the blossoms from every other tree in a row of Baldwin apples. The de-blossomed trees bore no fruit that year, but did form blossom buds that developed into a crop of fruit the next year. On the other hand, the trees from which the blossoms were not removed cropped that year but formed no flower buds, and were barren the next year. By this method Mr. Manning set up a system of biennial bearing in which every other tree in the row carried fruit one year but not the next. Thereby he had fruit each year.

These facts tend to explain why it is that thinning of the fruit frequently fails to change a strongly alternate-bearing tree into an annual bearer. Thinning may affect the size and quality of the developing fruit, but it may be done too late in the season to induce blossom-bud initiation for the next year's crop. On the other hand, chemical or mechanical thinning of blossoms is done sufficiently early to have an effect on blossom-bud formation. The result is not only to aid in the development of fruit of good size and quality by reducing an overload of fruit but also to promote annual bearing. Blossom thinning is now recognized as one of the most effective treatments to promote annual production of fruit and an orderly market supply.

These principles are of great value in attempting to control the growth and cropping of a fruit tree.

Development of the Fruit

The development of fruit of good size, color, and quality is dependent upon sufficient foliage per fruit, as we have seen, coupled with adequate moisture and nutrient supply and favorable sunlight and growing temperatures. High color is related to an accumulation of carbohydrates in the fruit plus sunlight, especially the ultraviolet light in sunlight. It is not just imagination that leads to the impression that the highest-quality apples are also the best colored for the variety.

Open trees that expose the fruit to the sun aid materially in coloring. An interesting "stunt" in this connection is to stick a piece of black paper or adhesive tape onto the side of an apple that is exposed to the sun. The paper or tape may be cut out in the shape of letters, words, or some design. The covered portion of the fruit will be lacking in red color, and will stand out conspicuously.

Time of Fruit Harvest

The proper picking date for fruit is not exact. A number of factors are taken together in making the determination. The seed coats typically become brown; the ground color of the fruit changes from a grass green to a slight touch of yellowish green; and the fruit separates cleanly from the tree when it is twisted upward. Pressure testers are available that correlate the degree of maturity with the pounds of pressure necessary for indentation of the fruit by plungers of certain diameters. In a "normal" season, the number of days from full bloom to fruit maturity is a surprisingly constant factor; and even with the variations that exist from year to year, it is a helpful guide to the proper time of harvest. For example, the average time lapse in days for several major varieties of fruits in the latitudes of western New York and central Michigan is:

Apples		*Cherries*	
Baldwin	138	Bing	71
Cortland	128	Black Tartarian	57
Delicious	132	Carnation	65
Golden Delicious	138	Coe	50
Gravenstein	114	Early Purple	42
Grimes	134	Early Richmond	50
Jonathan	134	Empress Eugenie	55
Lodi	76	English Morello	70
McIntosh	127	Florence	52
Northern Spy	127	Governor Wood	47
R. I. Greening	135	Kirtland	52
Rome	140	Knight	51
Stayman	145	Lambert	69
Wagener	137	Montmorency	62
Wealthy	109	Napoleon	68
York	138	Reine Hortense	53
		Royal Duke	62
		Schmidt	72
		Windsor	75
		Yellow Spanish	67

Peaches		*Pears*	
Belle of Georgia	122		
Carman	113	Bartlett	121
Elberta	128	Clapp Favorite	116
Greensboro	91	Elizabeth	98
Halehaven	114	Flemish Beauty	133
Hiley	126	Kieffer	147
J. H. Hale	125	Lawrence	147
Mikado	88	Seckel	137
Redhaven	98	Sheldon	145
Richhaven	111	Souvenir du Congress	130
Rio Oso Gem	134	Tyson	113
Rochester	104	Wilder Early	93

These figures will vary for different regions and locations within the region. By keeping a record of blooming and maturity dates for his orchard, a grower may develop a chart of his own that will be found useful in planning his orchard operations.

5

HOW DWARFED FRUIT TREES DIFFER FROM STANDARD TREES; CONDITIONS AND PRACTICES THAT INDUCE DWARFING

DWARFED fruit trees have essentially the same structure and parts as do the standard trees discussed in the pages immediately preceding. Likewise, they carry on the same physiological processes that have been enumerated, and they respond similarly to various stimuli. They are governed by the same rules and regulations of nature; the mechanisms of response are the same.

The difference between a dwarfed and a standard tree, then, lies in the manner in which the various structures and physiological processes are acted upon or manipulated. They may be controlled by genetic factors or by natural environment. They may also be altered by changes in environment and by artificial and mechanical means. With an appreciation of the typical structure and physiological processes in plants, much of the old-time mysticism is eliminated, and plant reactions become rational and predictable.

Not only may a dwarfed tree be smaller in stature than a standard tree, but, as has already been pointed out, it may differ in many other characteristics. The shape and spread of the tree may be changed; also, the age of fruiting, time of blossoming, and time of fruit harvest. The diameter of new shoot growth may be affected, as well as the longevity of the tree, its bearing habit, its leaf/fruit ratio, and its nutritional requirements. In other words, although one of the most significant features of a dwarfed tree is its controlled stature, there are many other characteristics that are changed and that may be very important in terms of commercial fruit production. Furthermore, the mechanism of dwarfing is exceedingly varied, as the following paragraphs will show.

GENETIC CONTROL OF DWARFING

With most classes or groups of plants there are forms that are inherently or "by nature" dwarfish. Thus there are dwarf varieties of bean, pea, and corn (Fig. 24); dwarf forms of petunia (*Petunia*), lobelia (*Lobelia*), aster (*Aster*), canna (*Canna*), dandelion (*Taraxacum*),

Fig. 24. Seedling of Swatow Dwarf peach, four years old, about four feet tall, representing a plant which is genetically or inherently dwarfish. Note the rosetted appearance of the plant, owing to the short internodes. Compare with Fig. 25.

alder (*Alnus*), billberry (*Vaccinium cespitosum*), palmetto (*Saba*), poinciana (*Poinciana*), pomegranate (*Punica*), and willow (*Salix*). There are dwarf forms of walnut (*Juglans fertilis*), of sour cherry (*Prunus cerasus*), of Mahaleb cherry (*Prunus mahaleb dumosa*), of peach (*Prunus persica*), of pear (*Pyrus communis arbe courbe*), and of raspberry (*Rubus* sp.). All these come fairly true from seed (Fig. 24).

There are at least six hundred known kinds of dwarf conifers. Some of these are very dwarf and grow only a few inches or even as little as one-quarter of an inch a year. Among these are the Spreading hemlock (*Tsuga canadensis* Minuta), Cole's Prostrate hemlock (*Tsuga canadensis* Cole's Prostrate), and the Pygmy Deodar cedar (*Cedrus deodara* Pygmaea). Others are more vigorous but still dwarf in the broad sense of the term, such as Weeping Japanese Red pine (*Pinus densiflora* Pendula) and Dwarf Hinoki cypress (*Chamaecypris obtusa* Nana Gracilis). Some dwarf conifers have arisen as spontaneous mutants or bud sports on isolated branches of standard forms. They can be propagated by cuttings or by grafting. Others, such as the Pygmy Deodar cedar, have arisen as seedlings.

Breeding for Dwarfness

It is common for most forms of plants to produce a certain number of dwarf variants in a seedling population. Among a large number of apple seedlings a few dwarf forms are always to be found, ranging characteristically from 2 to 3 per cent in some progeny to less than 1 per cent in others. The genetic makeup of the plant in these cases determines the habit of its growth and its dwarf characters. They may be as healthy and as full of vitality as standard or full-size specimens in spite of small stature. But the genes control the mechanism within the plant so that it is by nature dwarfed.

Many more dwarfed forms of plants are available than is commonly recognized. Man's desires have been mostly in the direction of "bigger and better," and dwarfish forms have been discarded. The author recalls distinctly several very attractive dwarf forms of sour cherry (*Prunus cerasus*) that appeared in a breeding program for improved varieties of cherries. The plants were rogued out as being undesirable by the standards of selection then in operation. Undoubtedly much could be done to breed dwarfed trees and dwarfed woody ornamentals by focusing attention on smaller forms and by breeding in the direction of "smaller and better."

In fact, the way has already been shown for several plant materials. Thus, bean breeders have produced dwarf or bushy bean plants (*Phaseolus vulgare*) suited to mechanical harvesting. Tomato breeders have developed small, compact upright tomato plants (*Lycopersicon esculentum*) that also are adapted to a mechanical harvester. The inheritance of dwarfness in corn (*Zea mays*) has been worked out, and strains can be bred at will of desired plant characteristics. And in Europe excellent results have been obtained in breeding new and improved dwarfing rootstocks for fruit trees.

It is also well known that some varieties of fruit trees are by nature smaller than others. The Delcon apple, on its own roots, is a semidwarf form. Varieties with blood of the Siberian Crab apple (*Malus baccata*) tend to smaller size, such as the Haralson apple. And there are other species of apple, as Toringo Crab (*Malus sieboldii*), that are small-growing. Size differences appear also in various species of pear, peach, cherry, plum, and apricot. Theoretically at least, although perhaps not practicable because of the many generations and the length of time involved, it would be possible to breed varieties of tree fruits dwarfed to the desired degree, just as is done for herbaceous crops. At all events, it would seem reasonable to expect that small-tree varieties may be introduced in greater number, inasmuch as interest has arisen in such forms. It must be remembered that emphasis in the past has been toward polyploid and gigas forms, and against small and dwarfish forms.

Inherent Dwarfing Mechanisms

It is interesting in this connection to observe the various mechanisms within the plant that induce dwarfing. In some instances, natural dwarfing is due to inherent physiological or structural makeup. Thus a naturally weak, shallow, or restricted root system may result in a natural dwarfing. A procumbent or spreading habit of growth, as with the EM IX apple, results in some degree of dwarfing. Willowy, thin-branch types of growth produce smaller trees than vigorous upright growers. This is in part due to the fact that weak branches tend to bend downward from

their own weight and from a weight of fruit. Varieties that fruit terminally on young wood also bend and spread and remain small trees for
the same reason. The Cortland, Rome, Gallia, and Ben Davis apples are
examples. And early fruiting, as with the Yellow Transparent, Fenton,
and Oldenburg apples, results in smaller than average trees, since fruiting is in itself a dwarfing process.

The relative abundance of various native hormones (augins and gibberellins) in a plant may also affect its size and form. For example, it
has been shown that certain genetically dwarf strains of corn (*Zea
mays*) contain a much smaller amount of auxin than do tall-growing
strains. Interestingly enough, the low auxin level is due to a natural high
rate of auxin destruction rather than a low rate of production. From
what has been said in the preceding pages about auxin being responsible
for cell elongation, this is what might be expected.

Gibberellins also promote cell division and cell elongation, especially
just below the apical meristem. This increase causes stem elongation.
Both the internodes and the leaves may become more elongate. Gibberellins allow dwarf mutants to develop into plants of a normal size,
indicating that they occur naturally in plants, dwarfism often being the
expression of a lost power of gibberellin synthesis. When cabbage plants
have been treated with gibberellin, they have developed dramatically
into plants 16 feet tall (Fig. 25).

The number of sets of chromosomes a plant possesses may affect its
size and performance. When the number is doubled, the new plant is
known as a tetraploid. When the number is tripled it is called a hexa-
ploid, and so on. These polyploid forms are frequently characterized by
larger leaves, larger stomata, larger fruits, and shoots that are relatively
large in diameter.

Again, differences have been found between varieties of apples in
the proportion of phloem tissue to xylem tissue in their stems. Associated with a high phloem/xylem ratio is a large number of thin-walled
parenchyma cells in the xylem. Very dwarfing apple rootstocks, as the

Fig. 25. Cabbage plants lose their typical compact character (left) when
treated with gibberellin, which induces stem elongation (right). Note that
treated plants are tall primarily because internodes are lengthened. Compare with Fig. 24. *Courtesy of S. H. Wittwer*

EM IX, contain a higher proportion of living tissue than is found in vigorous rootstocks. It has been suggested that because of this situation the efficiency of transport of water and metabolites is reduced in both a horizontal and a vertical direction in the dwarfing rootstock. Further, a greater share of the total metabolites is required by the dwarfing rootstock than by the scion top, so that there is competition between the two. And finally, the oxygen supply to the proportionately larger number of living cells in the dwarfing rootstocks is deficient and limits respiration. As a consequence, the energy requirements for water and nutrient uptake are reduced.

The number of growing points that appear on a plant affects its habit of growth. When there is one dominant leader, the growth is typically vigorous, and lateral branching is suppressed. It may be noted in passing that such trees develop strong taproots and relatively few lateral roots. On the other hand, those plants that by nature develop a number of growing points are inclined to be more spreading and of smaller stature (Fig. 24); the root system, too, is more shallow and branched.

One of the characteristics of very dwarfing apple rootstocks is a tendency to branching, as well as to assumption of prostrate form when permitted to grow naturally. It is not unlikely that this situation is associated with auxin level and transport in the plant. Thus, dwarf corn, with low auxin content, has a much more branching habit than does vigorous corn, with higher auxin content. Furthermore, dwarf corn tends to become prostrate. Similarly, a low auxin level has been found in varieties of aster that branch freely, whereas those with high auxin level develop a relatively unbranched habit. While these are only associations and speculations, they are highly suggestive. Undoubtedly, much of the explanation of genetic dwarfing will eventually be explained in terms of biochemistry.

Plants differ in the behavior of their roots toward the uptake of certain nutrients. Thus one form of apple has been shown to be characteristically selective against molybdenum. Again, some apple varieties have a higher requirement for certain nutrients, such as potassium, magnesium, and nitrogen, than do other varieties. As long as the environment provides enough of these nutrients, the plant will be vigorous, but if there is a shortage the effect will be seen by reduced growth. Though the environment becomes the determining factor, it is the genetic makeup of the plant that fundamentally controls it.

ENVIRONMENTAL FACTORS THAT AFFECT DWARFING

Alpine Conditions

A plant may be dwarfed by its natural environment. For example, Alpine plants are characteristically low-growing. This is now explained as due to strong light, especially the ultraviolet waves of light, which are relatively intense at high altitudes. As has been previously explained, the hormone produced in the tips of the shoots is responsible for cell elongation. Ultraviolet light destroys auxin. Accordingly cell elongation is reduced, and the plant is dwarfed.

Associated with Alpine conditions are low availability of nutrients and frequently of water, both of which tend to check a plant. The Chinese and Japanese take advantage of these effects of the environment in producing the ornamental dwarf plants for which they are famous. They seek out diminutive trees, often several hundred years of age, growing among the rocks or on some mountain crag, and transplant them to domestication.

Water, Nutrient Supply, and Climatic Adaptation

A shortage of water available to the plant in the soil also checks growth. This is largely because the nutrients the roots absorb are dissolved in water and are carried in solution in the plant, and any reduction in water is observable in the nutrient status of the tree. Under conditions of extreme water shortage, of course, the tree may be seriously injured or killed. But by keeping a plant "on the dry side," the amount of nutrients absorbed by the roots is reduced. When trees are grown in pots or small containers, this is what happens, and the trees are maintained in a desired dwarfish condition.

Similarly, if the orchard floor is in grass or some other herbaceous crop, the water may be reduced by the cover, and the trees dwarfed. Usually such trees have a yellowish cast to the foliage, characteristic of nitrogen deficiency.

Deficiencies of nutrients other than nitrogen may also slow growth. Mild deficiencies of potassium, magnesium, manganese, calcium, boron, zinc, and iron may result in smaller leaves and reduced shoot growth. Severe deficiencies of any of these elements may cause characteristic chlorotic and necrotic areas of the leaves, and reduced shoot growth. In other instances, as in severe zinc deficiency, the length of internodes may be so markedly reduced that a shoot terminates in a rosette of leaves.

Further, as will be shown in later paragraphs in this chapter, fruiting is in itself a dwarfing process. Any natural environmental condition that promotes the accumulation of carbohydrates in the plant tends to induce blossom-bud formation and fruiting. A light soil, which may be low in fertility and moisture, will bring about earlier fruiting and a smaller tree than will a heavy, highly fertile soil supplied with abundant moisture.

Severe winter cold may also check growth. If the injury to roots or to the conductive system of the tree is not severe but is by chance just the proper amount, the trees may be held small and be brought into early fruiting in a region where severe winter cold prevails. The author has seen McIntosh apple trees in northern Maine that suggested this situation.

The length of the growing season is also a responsible factor in controling tree size. The same variety of tree at the northern limits of its adaptability is smaller than in the more optimum and longer-growing conditions a little to the south. The Grimes and the Stayman varieties of apple, for example, are trees of good size in the latitude of Virginia and Missouri, but become much smaller as they are grown farther north, as in Michigan and northern New York. The length of the growing season may not be the only factor responsible, but at all events the climate in which a variety is grown may tend to cause a more dwarfish tree.

Conversely, a variety adapted to a northern climate may become dwarfed in a warmer, southern climate. The Baldwin and the McIntosh apples are examples. Both of these varieties have a high rate of respiration. They make large trees in a temperate northern climate. But in a warm climate they respire their supply of carbohydrates so rapidly that the growth is severely checked and the performance of the tree is poor.

Diseases and Insects

Disease and insect attack may also dwarf a plant. Aphids in particular, and leaf hoppers to some degree, are likely to produce a dwarfish effect. The tarnished plant bug, by destroying the active terminal bud of a shoot, checks the growth; and the Oriental peach moth, by stinging and destroying the terminal portion of vigorously growing shoots, also reduces the stature of a tree. The X-disease of peaches, which is due to a virus, produces a shortening of the internodes of the choke cherry, resulting in a decidedly dwarfish habit of growth.

Plants dwarfed in these ways are not properly dwarf plants—certainly not in the horticultural sense. And the presence of an insect or a disease may perhaps not properly be considered part of a natural environment. Yet it is hard to draw a line of distinction. In fact, although there is at present no scientific evidence to back the statement, it is entirely possible that some of the so-called natural dwarfs may owe their characteristics to some as yet unrecognized diseased condition. Examples are the broken-color types of tulips and the self-blanching types of celery, each of which has been found to carry a virus disease responsible for the varietal characteristics displayed.

HORTICULTURAL PRACTICES THAT INDUCE DWARFING

Fruit growers employ a number of horticultural practices to promote early fruiting and at the same time control to some degree the size of the tree. Because the two are so closely related, it is important to understand this relationship.

Fruiting as a Dwarfing Process

Fruiting is in itself a dwarfing process. Before a tree has reached the physiological age or condition at which flower buds differentiate, the nutrients available to it are utilized in growth of the vegetative parts—the leaves, the shoots, the branches, and the roots. As long as the supply of nutrients is used in this way, the tree continues to increase in size.

Typically, as the tree becomes older, it accumulates carbohydrate reserves that are associated with the initiation of flower buds. In turn, the tree slows down in vegetative growth as it blossoms and fruits. The question is naturally raised as to which comes first. Is it the slowing down of vegetative growth that brings about flowering and fruiting, or is it the flowering and fruiting that brings about the slowing down of growth?

The two processes are, of course, closely related. As soon as flower buds are initiated, the balance in the tree is noticeably altered. It seems almost as though a balance arm had been tipped. Whether this actually occurs, and some hormonal balance in the plant is shifted, is yet to be proved. At all events, the vegetative growth is suppressed as blossoming and fruiting begins, and nutrients are now diverted in part to the fruit. In fact, the reduced tree growth in terms of dry weight has been shown to be only slightly less than the dry weight of the fruit that has been produced. It is this situation that has given rise to the conclusion that fruiting is a dwarfing process and that "wood," meaning vegetative growth, is antagonistic to fruit, and vice versa.

Once the balance has been attained that promotes flowering and fruiting, the tree is not easily upset. Many practices that formerly were of major consequence are no longer of concern. A Northern Spy apple tree, for example, which is typically late in coming into fruit production —often twelve to twenty years—cannot be heavily pruned before it has reached the stage of blossoming and fruiting without suffering further delay in blossoming and fruiting. But let the tree begin to produce flowers and fruits, and it can be pruned almost with impunity without upsetting the fruiting habit that has now become established.

Blossoming and fruiting tend in this way to control the size of a tree. Those trees that come into bearing early are as a rule smaller than those that come into bearing later. The EM IX apple rootstock, for example, induces very early fruiting and a tree of small stature as well (Fig. 11). Progressively, there is a correlation between larger tree size and delayed fruiting for each size-controlled rootstock in succession. The EM VII apple makes a larger tree than EM IX and is later in coming into bearing, and the EM XVI makes a still larger and still later-bearing tree.

Not only do rootstocks show this relationship; so also do different scion varieties. Thus the Rome apple, which is not a large-growing tree, comes into bearing earlier than the McIntosh, which is larger. And the Northern Spy, which grows still larger than the McIntosh, is notoriously late in fruiting. There are, of course, many exceptions to these generalizations as one becomes more refined and critical, but there is merit in the general concept.

Most of the practices that are discussed in the balance of this chapter are aimed at early fruiting, but they also have an effect upon the size of the tree.

Low-Heading of Tree

Perhaps the most significant and simplest practice in promoting early fruiting is low-heading of the young tree, that is, cutting off the leader to a height of about 24 to 30 inches from the ground. The scaffold branches then develop in an area below this point, with the lowest branch at about 12 or 18 inches. Unfortunately, it is the practice in nurseries to rub buds off the main stem of yearling trees to a height of about 24 inches or even higher, with the idea of developing a tall tree that appeals to the uninitiated buyer. But when this is done, the diameter, or stockiness, of the tree trunk is reduced. If the buds and small shoot growths that typically develop on a young tree are allowed to grow, they help to thicken the stem and aid materially in the training of a small, compact low-headed tree. Such trees can be brought into bearing in five to seven years, even on standard seedling rootstocks (Fig. 25).

This is in contrast to the high-headed trees of a generation or two ago when sheep and other livestock were grazed in the orchard or when orchards were regularly plowed spring and fall, kept clean cultivated, or interplanted with other crops (Fig. 26). The trunk of necessity was free of branches to a height of 6 or 8 feet, and the trees were often delayed in fruiting until an age of fifteen or twenty years.

Little Pruning

Little pruning of young trees also promotes early fruiting and dwarfing. It must be remembered that the tree manufactures its own foodstuffs and that the leaves are the factories. Every shoot that is removed reduces the food-manufacturing ability of the tree. The earliest-fruiting trees are those that are virtually neglected so far as pruning is concerned. Obviously, some training and shaping of the tree are essential in successful orcharding; but the less the cutting, the quicker the tree reaches fruiting age.

The author once saw a most striking effect of excessive pruning of

Fig. 26. **High-headed, grassed cherry orchard in England with sheep grazing beneath the trees. Such high heading may be useful for special purposes but it markedly delays fruiting.**

apple trees of the Wealthy variety. Wealthy is typically an early-bearing variety, and without much pruning it will fruit at four or five years of age in the orchard. These particular trees, however, were fourteen years of age and had not yet borne fruit. They were growing on the estate of a wealthy man who had secured foreign gardeners to train his plantings. The gardeners had tried to keep the trees small by heading each year's growth back severely. In this they had succeeded, and the trees were no taller than a man. But by having their tops reduced, the trees were never able to accumulate sufficient carbohydrates to induce blossom-bud formation and fruiting. Had the nutrient supply to the roots been reduced, or had the tree been worked onto dwarfing rootstocks as is common in Europe, the trees would have responded to treatment. But as it was, an abundant nutrient supply to the roots plus the severe reduction of the tops kept the trees in a vegetative and unfruitful condition.

Suppression by Pruning

Trees can be suppressed, however, by judicious pruning. A vigorous shoot can be checked by tipping it back during the dormant season. Lateral buds are forced into growth, and a more compact habit of growth is produced. If overdone while the tree is young and before it has begun to fruit, the results will be similar to the experience related in the preceding paragraph. But once the tree has begun to fruit, strong branches can be cut back more freely without upsetting the fruiting habit of the tree, while at the same time the tree is kept within bounds. Growers who make it a practice to keep trees small in this way also thin out entire branches so that the tree does not become too thick and so that light may enter the tree more easily. A variation is to cut out branches in four segments of the tree something like a four-leaf clover. Not only does such training permit better entry of light and spray materials; it also provides spaces where ladders may be placed for easier and more efficient fruit harvesting.

From results the author has seen, it is apparent that not every grower can handle trees in this way to keep them small and yet maintain high production; but some growers are highly successful.

Maintaining Healthy Foliage

Trees may be stunted and dwarfed by attacks of insects and diseases that injure and reduce the foliage; but fruiting is then delayed. Since early fruiting is so closely related to the accumulation of carbohydrates, it is important that the leaves be kept in a healthy condition for the manufacture of carbohydrates. Young trees are too frequently neglected in the spray program merely because they are bearing no fruit.

Use of Plant Regulators

Certain chemicals are effective in controlling growth processes in the tree. Maleic hydrazide has been found to have a pronounced but temporary inhibitory effect. Respiration is reduced and the growth of the tree is checked. The material is used at 1,000 parts per million, applied after the leaves have appeared, just enough to wet the leaves. With some woody plants, such as privet, new growth has been suppressed for at least a month, and growing has been retarded for an additional several months. No injury has been reported at this concentration. At higher concentrations the foliage may be injured.

When maleic hydrazide has been applied to apple trees before full bloom, the most pronounced effect has been to destroy fruit set. Similar results have been recorded with peaches. While theoretically it would seem that this material would not only retard growth but also promote blossom-bud formation and fruiting, and so produce the ideal dwarfed fruit tree, such has not been the case.

Scientific interest has steadily increased in chemical treatments with which to develop dwarfish or compact plants, especially ornamental and fruit plants. Major plant-growth retardants have been found among compounds built around quarternary ammoniums, such as sulfonates, carbamates, and cholines; also among the phosphoniums, which contain phosphorus, instead of nitrogen, as their key ingredient. Certain choline derivatives have proved very effective as depressants. Common names for several of these materials are Amo-1618, B-9, Carvadan, Phosfon, and CCC.

The outstanding change brought about by these materials is a shorter length of internodes. In addition, the stem often thickens, leaves develop deeper green color, flowers may be more vivid, and time of flowering and maturity of the plant may be delayed. While these materials have immediate application in floriculture and in the production of herbaceous pot plants, what value they may have on the dwarfing of woody plants and fruit trees is not yet known.

Attempts have been made to control the time of blossoming in the spring so as to escape late spring frosts. Naphthaleneacetic acid and related compounds have been applied to apple, peach, cherry, and plum trees during midsummer. In some instances fruit buds have been delayed in opening the following spring by as much as fourteen days, and vegetative buds have been delayed nineteen days. But other results have been contradictory. The trees generally have been injured and the leaves and fruits deformed.

On the other hand, chemicals are used successfully in the thinning of blossoms and fruits to prevent an overload of fruit and to promote annual bearing. Naphthaleneacetamide and naphthaleneacetic acid have been found very effective. At 10 to 40 parts per million, depending upon the variety, the temperature, and the vigor of the tree, good commercial thinning has been obtained with applications made ten days after full bloom. The removal of blossoms completely can be accomplished at concentrations of 60 to 80 parts per million. While not in themselves effective in controlling the size of a tree, these chemicals bear an indirect relation through their effect on blossoming and fruit set, as we have seen.

Reduction in Nutrients

Trees that are strongly vegetative can be checked in growth and induced to early fruiting by withholding nutrients, especially nitrogen. The same effect is produced by seeding down the orchard to a grass cover. This general practice need not be elaborated on further here, but it has been used very widely and very successfully for a great many years.

Root Pruning

As we have seen, the root is the organ especially concerned with the uptake of water and mineral nutrients, besides anchoring the tree in the soil. Reduction or restriction of the root system interferes with these processes and tends to reduce vegetative growth, check the size of the tree, and promote flower-bud initiation and fruiting.

Deep plowing and heavy cultivation of an orchard may so reduce the feeder roots as to check growth and induce flower-bud formation. The breakup of a heavy mulch, in which many feeder roots are found, may act similarly.

A laborious practice not much used in this country, but formerly widely employed in Europe with individual trees, is to prune the roots by spading up the ground around a tree and deliberately severing some of the main roots. The procedure is especially suited to the stone fruits (cherry, peach, plum, and apricot), which cannot be ringed so successfully as can the apple and the pear.

The operation is most useful with young, vigorously growing trees up to five or six years of age (Fig. 27). It is best done in fall, begun by digging a trench 12 to 18 inches deep in a circle around the tree, at a distance appropriate to tree size. A useful guide is a radius of 9 to 12 inches for each inch of trunk diameter. For example, for a 2-inch tree the trench should be 18 to 24 inches from the tree. Strong lateral roots should be severed. If no strong roots are found at this distance, work should progress in toward the trunk until they are found. Strongly descending roots immediately under the tree should be cut as well. The trench is immediately refilled. In actual practice, with trees three to four years of age, this procedure amounts virtually to digging and resetting. Though it may sound like a severe operation, it can be done most efficiently and with excellent results.

In fact, the nursery practices of undercutting or of lifting and transplanting trees and shrubs in the nursery every other year are root-pruning operations. Not only is the plant promoted in flowering and fruiting at an early age; the main feeding root system is concentrated in the soil mass that accompanies the tree in moving, thus reducing the shock of transplanting.

In 1840 Thomas Rivers, a leading English pomologist and nurseryman of his day, described root pruning to his contemporaries as a useful practice. While obviously of value for a few trees only, as around the home and where space is limited, the description is interesting and suggestive. In his method he first of all pointed out that trees with a vigorous root system are the ones to be preferred for treatment. His procedure was to train a tree in the nursery for two or three years before planting it in the garden in fall. It was then left undisturbed for three years until it was well established. The third year, a circumferential trench was dug 10 inches from the stem and 18 inches deep. Every root was cut, including all roots directly under the tree, to a depth of 15 inches. A sharp spade was recommended, since to use a knife was too tedious. In the fourth year a trench was dug in similar fashion 14 inches from the stem so as not to injure the mass of fibrous roots that resulted from the previous trenching. In the fifth year the trench was made 18

Fig. 27. **Root pruning induces early fruiting and also keeps a tree small.** *Courtesy of F. L. S. O'Rourke*

inches from the stem, and so on. It was suggested further that a slight depression be left around the tree in the trench, where fertilizer could be placed. In this way terminal growth was limited to 4 to 8 inches, and a tree was kept firmly in hand.

In Europe it is not uncommon to grow fruit trees to four or five years of age in the nursery, root pruning them every other year. The trees are then lifted and sold as transplanted trees, as is done with many ornamentals, either with bare roots or "balled and burlapped" so as to carry some of the soil with the roots. The practice could be followed to much greater degree in this country, and with equally good results. An individual who wishes quick returns from a dwarf tree or two could well afford to pay a higher charge for trees of this kind, and an enterprising nurseryman could develop a good business in his immediate neighborhood.

With larger trees—eight to ten years of age—it is impractical to lift and replant. Instead a trench may be dug in a circle around the tree just inside the drip of the branches where heaviest root feeding occurs. Make the trench as wide as the spade and 18 to 24 inches deep. Any strong roots that are uncovered should be cut through, and an effort should be made to cut under the tree and sever any strongly descending roots. The soil may be merely turned and replaced in the trench, or it may be placed to one side for a day or two and then refilled into the trench.

The best time for root pruning, as has been said, is in late fall when the trees are mature. It may also be done in early spring just as the frost is coming out of the ground but before new growth has started. This gives the tree a chance to develop new feeding roots and become reestablished either before the ground freezes up or before spring growth begins. In regions where severe winter cold is not a hazard, late fall is the best time; but where cold winters prevail the tree may be so weakened by fall root pruning as to result in serious cold injury. On the other hand, if done too late in the spring the new shoot growth may suffer from water shortage.

In any event, good judgment and common sense are essential in any of these operations, remembering that the tree is living and that the root system must be re-established so as to supply the tree with the necessary water and nutrients from the soil if it is to survive. If well done, results are surprisingly good. If done awkwardly and at the wrong time, they can be disastrous.

Girdling, Ringing, Scoring, and Notching

Girdling is an all-inclusive term that refers to a constriction, obstruction, or mechanical cutting that completely encircles or "girdles" a plant axis—shoot, twig, branch, or trunk. It includes cuts made into the wood as well as cuts made only through the bark. It obstructs or interferes with movement of materials in the phloem. It may be done accidentally by mice or by an unremoved nursery label, or deliberately with a string or wire or by severe notching into the sap wood with an ax so as to destroy the tree. It may be done by drawing a knife or a saw around the plant part so as to cut just through the bark and down to the wood but not into it. This treatment, though a form of girdling, is more frequently called "scoring." Or it may be done by removing a strip or ring of bark of varying widths, when it is more commonly called "ringing" (Fig. 28). In any event, ringing and scoring are horticultural terms referring to cuts made through the bark that then heal over as the plant grows during the season.

The practice of ringing the bark of fruit trees to induce earlier fruiting is a very old art, having been described by Virgil and Columella in the first century B.C. and the first century of the Christian Era, respectively. Thomas Andrew Knight, in 1822, wrote that the nutrient sap elaborated in the leaves passes down the bark, and when checked by girdling the bark "the repulsion of the descending fluid therefore accounts . . . for the increased produce of blossoms. . . ."

When properly performed, ringing tends to dwarf the tree, induce blossom-bud formation, and promote fruiting. This is commonly observed with trees that have been girdled by mice, injured at the collar by the collar-rot organism, or barked by a lawn mower, by a tractor,

Fig. 28. **Bark ringing promotes early fruiting and smaller trees: (Left) complete ring; (Right) two half-rings, two to three inches apart, less drastic than complete ring. (After W. F. McKenzie in** *Fruit Culture for the Amateur*, **Garden Publications, London, 1947)**

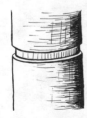 

or by some other means. One commonly hears the explanation that this is nature's attempt to survive by reproducing through seed just before the plant dies. While this may be the result, it is not the explanation.

Ringing also tends to increase the size and sugar content of the fruit and to cause the fruit to mature a few days to a week earlier. The practice has been standard for grapes in many parts of the world, in order to increase the sugar content and size of the fruit and to improve the set of varieties that are inclined to be shy bearers. The response in the grape is reflected immediately during the same season in the set and fruit and in the crop. But with fruit trees the effect is first of all to induce the formation of blossom buds for the next year's crop; this is a fundamental difference. It is true that the fruit tree responds during the same season by the initiation of blossom buds, but no visible expression is evident until the next. There are many things that may occur during the two seasons to upset markedly the expected favorable results of ringing, such as a severe winter, spring frost, and drought.

Ringing of tree fruits has never been considered very favorably by commercial orchardists except as an emergency measure, and largely because there are more reliable methods of accomplishing the same results, as by the use of dwarfing rootstocks. Severe damage has sometimes resulted with young trees, weak trees, and where too wide a ring of bark was removed so that the wound did not heal. But for filler trees that are tardy in fruiting and that are not expected to live long, for overly vigorous trees, and for trees that are exasperatingly late in coming into fruiting, as with Northern Spy trees that may not have borne fruit at fifteen to twenty years of age, ringing and scoring have a place. This is especially so for individual trees or for a few trees in the garden. The practice is seldom, if ever, used for peaches, cherries, plums, or pears— almost exclusively for apples.

As has been indicated, the best time for ringing and scoring of fruit trees is in late spring or early summer before the fruit buds have formed and while the tree is still growing vigorously. In terms of plant development this means from about the time of full bloom to a week later. In the latitude of New York State and southern Michigan, this usually means sometime in late May or early June.

In "ringing," a strip of bark about ¼ inch or less in width may be peeled from around the branch or trunk. A double-bladed ringing knife is available that can be set the desired spacing. The bark should be peeled carefully so as not to scrape or scratch the cambium layer beneath. Injury to the cambium will retard or prevent healing of the wound and may result in serious damage and even death of the tree. Wounds should be covered with some protective material that will aid in healing the wound and closing the gap. Tire tape, surgical tape, and cotton strips soaked in grafting wax and rolled into a ball for convenience in handling have all be found effective. Scoring wounds need not be protected.

In "scoring," a knife blade or saw is drawn around a limb or trunk so as to sever all the bark tissues as far in as the wood. No peeling is done, and the wound is slight, is not objectionable, and need not be protected.

A good practice in ringing or scoring is to treat alternate limbs rather than the main trunk. This takes more time but is safer. With alternate-bearing varieties, such as Baldwin, the result is further to induce some branches to bear one year while leaving others to bear the next and thus developing a tendency to annual fruiting.

An instance of successful ringing and scoring has been reported from Ohio where 10,000 apple trees of several varieties were ringed, and 15,000 others were scored. Both treatments were effective, and the scoring was substantially easier and cheaper than ringing. Treated trees averaged 2.75 bushels of fruit per tree the year following treatment, compared with practically no fruit on nontreated trees. The results were good with all varieties treated, but were better with Baldwin, Delicious, Grimes, Jonathan, Rome, and Stayman than with Liveland, Sutton, and Spy.

Notching is a practice confined to young shoots for the purpose of either inducing fruit-bud initiation in a given location or encouraging the development of a vegetative shoot. It is useful in training dwarf trees to special forms and controling individual branches.

The procedure is to cut a slight notch through the bark either just below a bud or just above it, as the case may be. If the cut is made above a bud, the result is to force the bud into vegetative growth. This is explained by the reduction in growth hormone from the tip of the shoot reaching the bud and suppressing it through terminal dominance. On the other hand, a notch just below the bud causes carbohydrates and other materials to accumulate above the notch and promote fruit-bud initiation.

Bark Inversion

A new ringing technique, known as "bark inversion," has been tried by Sax and his co-workers at the Arnold Arboretum for the dwarfing of young trees. The method is of considerable interest and value not alone because it dwarfs a tree but more especially because it sheds light on the physiology of dwarfing.

The technique consists of the removal of a ring of bark from the trunk of a young tree and replacing the ring in an inverted position. The inverted ring is bound tightly with a rubber band until it is united with the wood. Because of its inverted polarity the ring of bark inhibits the flow of materials downward through the phloem to the roots of the tree just as in ringing.

The effect is not permanent, however, because new bark is regenerated at the vertical seam of the ring, and the new phloem so formed is reoriented and permits normal phloem transport downward.

A double bark inversion, with the vertical seams on the opposite sides of the trunk, increases the duration of the dwarfing effect. But nutrient sap may diffuse laterally or vertically if normal phloem transport is checked. Therefore lateral diffusion of nutrient sap soon results in a lateral orientation of the new xylem and phloem to bridge the normally polarized tissues regenerated at the seams of the two inverted rings of bark, and the dwarfing effect is lost.

Using the modern radioactive tracer technique, radioactive phosphorus has been introduced into young apple trees through the leaves and followed in its movement. Radioactivity, indicating the presence of phosphorus, was found to pass up the xylem, become incorporated in organic nutrients, and then move downward through the phloem. Where a section of inverted bark had been placed, the activity accumulated above the ring; and, indeed, some moved across into the xylem and back up the stem as has already been suggested in the discussion of ringing.

Although the dwarfing effect of a ring of bark thus appears to be related to the checking of nutrient transport by the phloem, it has been suggested that other factors may be involved, such as the movement of hormones. It is pointed out that auxin transport normally also occurs in the phloem and that the movement is downward from the tip of the tree. Auxin suppresses the development of lateral buds. A ring of inverted bark would prevent such downward movement. That such actually must occur is indicated by the fact that dormant buds below the inversion are no longer suppressed but are stimulated into active growth.

Further, cambial activity in an inverted bark ring is suppressed, so that the ring is overgrown by the portion of the stem above the ring. However, if a ring of bark is taken off and replaced in the upright, rather than in the inverted position, the cambium of the upright ring grows freely. This would seem to strengthen the view that an inverted ring prevents the downward flow of auxin, even to itself.

Twisting, Bending, and Spreading

A tree may be checked in growth and induced to early fruiting by twisting, bending, and spreading of the branches. For example, it is a common observation that when a strongly upright growing apple or pear tree carries its first crop of fruit, the weight spreads the branches and bends them downward. Thereafter, production is improved.

In fact, as long ago as 1729, Langley observed, "The nearer branches are laid to a horizontal position the velocity of the sap is the more retarded; and the nearer to a perpendicular position, the more freer; therefore, branches that are inclinable to luxuriency, may be checked by being nail'd horizontally; and those that are weak helped by being nail'd perpendicularly." And Thomas Andrew Knight, in 1803, discussed the fruiting of horizontal branches, and explained that it was "by no means improbable, that the formation of blossoms may, in many instances, arise from the diminished action of the returning system in the horizontal or pendant branches."

Following these ideas, growers have tried various artificial means of bending upright branches downward. Such branches should not be over $1\frac{1}{2}$ to 2 inches in diameter. One system is to run a stout string from a peg or stake in the ground to the tip of a branch and pull it downward toward the horizontal, or to run a light wire from a metal "eye" screwed low on the trunk to one placed out near the end of the limb that is to be bent. Another method is to suspend a weight, such as a common brick, from a branch. At one time, horseshoes were used for this purpose because of their appropriate weight and convenience in fastening.

In England a planter with overly vigorous young trees resorted to bending rather than cutting out vigorous shoots. He selected pairs of vigorous shoots from opposite sides of the tree and brought them down and under the tree to form a circlelike festoon (Fig. 29). He then fastened these branches together at their tips with adhesive tape. A year after bending, these shoots were filled with fruit buds and bore a heavy crop. The branches not so treated did not fruit immediately. The fruiting process checked the growth of the entire tree, and it soon came into bearing. Eventually, these festooned branches were shaded out and removed, but the operation was considered highly effective.

In the Orient and in parts of Europe where labor is available, upright branches are held spread apart with wooden sticks. Light bamboo and lathing strips are convenient. The sticks are notched at the ends so that they do not slip, and are forced gently between two strong growing shoots.

While these procedures are not usually considered adaptable to commercial orcharding, there are nevertheless some growers who from time to time have employed them to advantage. For example, young Bartlett pear trees frequently send up one to four strong branches. It is not difficult to attach a cord to the tips of these few main branches and bend them off-center. The result is to induce blossom-bud formation and fruiting; once the tree has started to fruit, it is checked in growth and induced to general fruiting. Mr. Arthur Karr, of Yakima, Washington, at one time employed this method most effectively.

In the Caldwell system of training pear trees practiced in California in the 1920's, upright-growing one-year-old shoots were bent downward to an angle somewhat below the horizontal (Fig. 30). This caused new shoot growth to arise just below and behind the highest point of the bend, while the growth beyond this point became reproductive in character and rapidly developed a good system of fruit spurs (Fig. 31). At each dormant season all the new shoots were again tied down in such a way as to shade lower branches as little as possible. The tying was done to any convenient point on the trunk or on any branch. At ten

Fig. 29. Bending promotes fruiting and tends to dwarf the tree. Shoot bent down and fastened (festooned). A year later, bent shoots will carry flower buds and may be cut back as indicated. (After Raymond Bush in *Tree Fruit Growing*, Penguin Books, Middlesex, England, 1943)

years of age, trees were large and highly productive. They had never been touched by shears except for the first winter.

Nine-year-old nonfruiting trees of Northern Spy apple and Kieffer pear have been brought into fruiting on a rather large scale by this treatment. Whether or not the practice is used extensively in commercial orcharding, it unquestionably has a place where a few trees are involved.

The same principles apply to procumbent forms of fruit trees and varieties that are characteristically slender and willowy in shoot growth. The weight of the branch itself tends to pull it out of the vertical position and induces early fruiting, which still further accentuates the trend. With naturally weak-growing varieties worked onto a very dwarfing rootstock, the trees become sprawling and bushlike and difficult to handle. To meet the problem, fruit growers attempt to develop one branch as a central leader and keep it growing erect by removing any blossoms or fruits that may form.

The training of trees to special forms, in which trees and branches are variously inclined, bent, and spread, had its origin in the improvement and control of fruiting (Chapter 26). The trained tree is often thought of as an artistic, odd, or even freakish method of growing fruit trees. It is misunderstood in this respect. As an example, when a yearling apple or pear tree is planted at an angle of 45 degrees, its behavior is quite different than when it is planted erect. Almost every lateral bud on the slanted yearling tree will start into growth, whereas in the erect tree it is usually only the buds on the upper half of the stem that will do so. Further, in the erect tree the buds near the top will develop into

Fig. 30. **Young pear tree with strong upright branches that have been bent and tied at slightly below the horizontal to induce early fruiting, according to the "Caldwell system," named for W. A. Caldwell of El Dorado County, California (pictured), who developed and practiced it.** *Courtesy of* American Fruit Grower *and W. P. Tufts*

strong-growing shoots, whereas in the slanted or oblique tree the buds develop into shorter, spurlike growths that come into bearing early.

The same situation applies to horizontal cordons and to forms in which horizontal or oblique arms are developed. The bending at right angles still further controls shoot elongation and spur development.

Apparently what happens is that the materials manufactured in the tips of the branches, which normally flow downward in the phloem, tend to move down less easily in a bent branch. They accumulate in the horizontal or bent portion of the branch and induce blossom-bud formation.

Twisting and bending young shoots at sharp angles have the same effect. They are useful procedures with trees trained to special forms, as with ornamental fruit trees and espaliers. Japanese gardeners have long employed them, including the rather spectacular procedure of tying a young tree or branch into a loose knot, or weaving the branches one under the other in festoon fashion.

Dwarfing or Size-Controlling Rootstocks

By far the most useful and reliable single method of dwarfing a fruit tree and controlling it in size and in fruiting is by the use of dwarfing or size-controlling rootstocks. Rootstock materials are numerous and specific. Since the rootstock is such an important feature of how trees are dwarfed, the subject requires special treatment in the chapters that follow.

Fig. 31. **Six-year-old Bartlett pear tree that has been trained by tying down new shoots each year according to the Caldwell system. Note new branches that arise at bend. These in turn will be tied down to a position a little below the horizontal.** *Courtesy of* American Fruit Grower *and W. P. Tufts*

6

STOCK AND SCION RELATIONS;
HOW ROOTSTOCKS MAY DWARF

T HE method by which a rootstock brings about dwarfing is not well understood. A variety of explanations has been made, with very little agreement. More than likely this failure to agree indicates the complexity of the problem. No one single factor seems alone responsible. The factors are many and interacting. One may operate in one instance, another in another, and a combination of several in still another. The discussion which follows may appear elementary and overly simplified; certainly it is incomplete. But it has seemed worthwhile to present something of the present understanding of stock and scion relations.

In any analysis of the problem it must never be forgotten that fruit trees, unlike cereals or vegetable crops, are composed of two individuals growing together as one. Cereals and most vegetable crops are propagated from seed, so that the root and the top are from the same original source and are, in fact, one and the same plant. With fruit trees, on the other hand, this is not true. Instead, a rootstock is propagated from seed or by some vegetative means, as from layers or cuttings, and onto this rootstock is budded or grafted the scion desired for the top of the plant. The result is a plant composed of two different individuals growing together, the one dependent upon the other, each for particular functions necessary for survival, and each affecting the performance of the other.

And while this statement may seem repetitious, one cannot work long with dwarfed fruit trees without being impressed by the importance of the stock-scion relationship rather than merely the effect of a rootstock alone. H. J. Webber, Director of the California Citrus Experiment Station, Riverside, California, was evidently impressed with this idea when he coined the word "stion" to represent a plant made up of a stock and a scion growing as one. He derived the word from a combination of st(ock) and (sc)ion.

Stion is a very useful word. It implies something new and different; indeed, a given stock-scion combination is different than any other stock-scion combination and again different than either the scion or the stock separately. It is doubtful if there ever was a stock-scion combination made in which one part did not affect the other. We have fallen into the habit of speaking of the stock as one unit and the scion as another, giving the impression that we are dealing with interchangeable parts, as with an automobile. But McIntosh/EM IX does not bear as much relation to Northern Spy/EM IX as one would like to imagine. Each combination is a unit by itself. Each one is, indeed, a STION.

We know, for example, that the stock may affect and alter the performance of the scion, as in the case of a dwarfing rootstock dwarfing the scion. We know also that the scion may alter the rootstock, as in the case of the evergreen sweet orange (*Citrus sinensis*) worked upon the deciduous trifoliate orange (*Poncirus trifoliata*), in which case the evergreen top causes the deciduous rootstock to grow more rapidly than it does with trifoliate orange top. Or, in less dramatic fashion, the Oldenburg scion transmits a red coloration to the rootstock upon which it is worked. We know also that an intermediate stempiece may affect both scion and rootstock, as in the production of a dwarfed tree by interposing a dwarf interstem between a vigorous rootstock and a vigorous scion. One part affects the other. It is not the individual parts but all parts taken together that define the performance of the stion.

LIMITS OF GRAFTING

Various degrees of dwarfing are often associated with mild forms of uncongeniality. Not all plants can be combined by budding and grafting, as an apple tree on a pine tree. They are then said to be incompatible.* In fact, the relationship between consorting parts must be fairly close botanically. As a general rule, success is most frequent with plants belonging to species that will hybridize. Thus, the common cultivated varieties (clones) of apple (*Malus domestica*) are generally compatible with other varieties (clones) of apple, and pears (*Pyrus communis*) are compatible with other pears. Closely related species may usually be worked one upon the other with varying degrees of success, but with less certainty, as the sour cherry (*Prunus cerasus*) on the mahaleb cherry (*P. mahaleb*). And even some genera can be combined, as the apple (*Malus*) on the pear (*Pyrus*) and the pear (*Pyrus*) on the quince (*Cydonia*). Combinations between different families are rare but have been reported for *Garrya eliptica* (*Garryaceae*) on *Aucuba japonica* (*Cornaceae*).

Yet here again one cannot generalize too freely because there are numerous accounts of closely related plants that fail to unite. R. J. Garner, of the East Malling Research Station in England, tells of an interesting experience with four plum seedlings. They were all raised from the same cross; then each seedling line was propagated as a clone and several of each budded onto peach. Two of the four plums were congenial with peach; one lived until the fifth year, and one died the first year. Karl Sax, of the Arnold Arboretum in Massachusetts, tells of crosses he made between *Syringa lacinata* and *S. vulgaris*, from which more than half of the seedlings died the second year. Yet when some were budded on *S. amurensis japonica* they grew vigorously. Still further, Tukey and Brase have reported that buds taken from one tree of McIntosh united with apple rootstock USDA 227, while McIntosh buds taken from another tree failed to unite. There is an explanation for this last experience, probably involving a latent virus present in one source of McIntosh budwood and not in the other, but it illustrates the point.

The tristeza disease in oranges and the pear-decline disorder in pears

---

* Barbara Mosse of the Rothamsted Experimental Station, England, has defined incompatibility as "The characteristic interruption in cambial and vascular continuity which leads to the spectacular smooth breaks at the point of union." She has further classified incompatabilities into two major groups, namely, (1) translocated incompatabilities and (2) localized incompatabilities. In the first group are incompatabilities which may be transported along the stem of another plant (intermediate stempiece) without the two incompatable parts being in contact. In the second group are incompatabilities which are dependent upon immediate contact of the two incompatible parts.

are additional instances of apparent incompatibilities that are not inherent but are induced by viruses. They are discussed more fully later in this chapter.

There have been many attempts to develop a reliable method of predicting incompatibility, including seralogical techniques, but none has been successful. Dependence must still be placed on the empirical method of trial-and-error. Whether a combination will express itself as unsuccessful, successful under some conditions, sickly, severely dwarfing, moderately dwarfing, slightly dwarfing, or even invigorating must still be determined by attempting to make the combination and observing what happens.

THE NATURE OF THE ROOTSTOCK

A rootstock may markedly affect the stion by virtue of its own inherent characters. Thomas Hitt, in A *Treatise on Fruit Trees* published in 1757, admonished that "stocks are in some measure a sort of soil to the kinds of trees raised on them." At one time it was commonly held that dwarfing rootstocks achieved their control by starving the scion of the mineral nutrients they secured from the soil. But this is no longer tenable, at least as far as mineral elements are concerned. The mechanism is much more complex.

Thus, a rootstock may limit growth because of a shallow root system, as the EM VII apple rootstock. Or its roots may be subject to breakage, as the EM IX. It may be limiting by being poorly adapted to light soils or heavy soils or high water table or droughty conditions. It may limit the range of adaptability, as in the case of the quince, which is more tender to winter cold than many pears worked upon it; the quince root may be killed by severe cold, and so cause death to the plant. It may be subject to disease organisms such as root rots or crown gall. It may be susceptible to nematodes or an insect pest, as is the grape. The European grape, for instance (*Vitis vinifera*), is susceptible to the grape-root aphis, whereas the American fox grape (*Vitis labrusca*) is resistant. If American varieties of fox grapes are grafted onto susceptible rootstocks of the European grape, the plant fails because of the aphid attack. The reciprocal combination succeeds—namely, European scion upon American rootstock—for the reason that the American rootstock is resistant.

It has also been shown that a given rootstock may exert a degree of selective absorption, and so affect the performance of the stion. The boron content of the scion has been found to be controlled by the rootstock in some citrus combinations in California.

Very dwarfing apple rootstocks have been shown to have a high proportion of phloem to xylem and a large number of thin-walled parenchyma cells. As a result, as explained in Chapter 5, uptake and transport of nutrients by such rootstocks are reduced. On the other hand, dwarfing cherry rootstocks fail to behave as do apple rootstocks in these respects.

The auxin content of some forms of dwarf plants has been shown to be low (Chapter 5). Further, such dwarf plants with low auxin level are inclined to branch more freely than vigorous plants with high auxin level, and they tend to a prostrate development. This is suggestive of very dwarfing apple rootstocks.

In fact, it is frequently said that the scion takes on the stature and vigor of the rootstock, and to prove the point numerous instances are cited of typically strong-growing scions that are dwarfed by being grown on weak-growing rootstocks. Both the EM VIII and EM IX apple rootstocks, for example, are relatively weak-growing rootstocks, and they produce a very dwarfing effect upon the scions worked upon them.

But this is not the complete answer to dwarfing, as shown by the fact that the EM XII apple rootstock is less vigorous in the propagation bed than are the EM I and EM XIII, yet EM XII makes the larger tree. Furthermore, the pear and the apple are both vigorous plants, yet when the apple is worked upon the pear, or vice versa, a dwarfish tree is produced. While it is doubtless true that a weak-growing rootstock may induce weak growth and dwarfing in the stion of which it is a part, there are additional factors that influence dwarfing.

Fritz Kobel of Switzerland has explained the dwarfing effect of some weak-growing rootstocks by noting that weak-growing stocks require small reserves of carbohydrates and extract less of minerals from the soil. The vigorous scion, on the other hand, tends to accumulate carbohydrates and is deficient in nitrogen. The result is a high carbohydrate-nitrogen relation in the scion, hence dwarfing, and early fruiting.

A given rootstock may be able to synthesize substances the scion may not, as in the case of tobacco used as a rootstock for the tomato. Tomato foliage does not normally contain nicotine. On the other hand, it is in the root that nicotine is synthesized in the tobacco plant. Accordingly, the tobacco rootstock synthesizes nicotine, which is then translocated to the tomato scion-foliage, where it is found.

These, then, are some of the many ways in which the rootstock of itself contributes to the performance of the stion.

THE NATURE OF THE SCION

As has been shown, a weak-growing scion may dominate a stock-scion combination, as does the EM IX apple when used in this way. Or, a dwarf form of bush bean (*Phaseolus vulgaris*) tends to dwarf the stion when it is grafted onto a vigorous climbing form, whereas the reciprocal is less dwarfed. This is mostly a quantitative effect, in which a scion with characteristic large leaves and long growing season may synthesize more energy materials than a scion with small leaves and short growing season, and thereby favor a more vigorous plant. Again the example of the trifoliate orange (*Poncirus trifoliata*) is called to mind. When used as a scion grafted onto the sweet orange (*Citrus sinensis*), which is an evergreen, the resulting stion is more dwarf than the reciprocal combination. The effect is largely qualitative, in which the longer persistence of the evergreen leaves results in the synthesis of a greater amount of energy substance.

Or, as will be shown in greater detail later, the foliage of one plant elaborates materials that are often different from those the stock would elaborate with its own leaves, and without which it cannot survive. Such an example is the muskmelon (*Cucumis melo*) grafted onto the Malabar gourd (*Cucurbita ficifolia*). When the Malabar gourd is used as the scion, the graft succeeds; but when the muskmelon is used as the scion the graft fails. Another example is the lemon (*Citrus lemon*) grafted onto the sour orange (*Citrus aurantium*) in Italy. In this case, the combination is subject to the gummosis disease, but is made resistant and succeeds when a few leaf-bearing scions of the sour orange stock are grafted onto the lemon top. The mineral composition of the plant may also be altered by the scion. Thus the magnesium content of the scion has been shown to affect the magnesium content of the stock with citrus in California.

Finally, the grafting of a scion of one apple variety onto the top of another variety may result in the death of the entire tree. Such an occurrence is perhaps due to the transmission of a virus disease from the one plant to the other in the grafting process—exactly as in an inoculation of a healthy but susceptible plant with a virus from a diseased but resistant plant. Some of the incompatibilities and difficulties with the Virginia Crab as a rootstock and a body stock have been traced to the introduction of latent viruses into the Virginia Crab when grafted with a virus-infested scion. In some instances a characteristic pitting of the Virginia Crab wood develops, which is not unlike the tristeza disease of citrus, which is also due to a virus. While in some respects not exactly a "scion effect," these experiences do at least indicate some of the complex factors that are present in stock-scion relations and why many observations defy explanation from anything we now know.

THE NATURE OF THE GRAFT UNION

A great deal of attention has been focused upon the union between the rootstock and the scion as the controlling factor in dwarfing (Fig. 32). This has taken the form largely of inquiry into the mechanics of the union, with special reference to the factors that promote a good union and to possible interruptions in the movement of materials from rootstock to scion, or the reverse, something akin to the effect of ringing, scoring, or constricting.

It has been found that when the stock and scion "unite," there is no actual fusing of cells in the sense that two cells become one. The individual cells of stock and of scion remain distinct, although a collection of cells taken at the point of union may include cells of both stock and scion. In other words, a so-called "union" is actually a dovetailing of elements of both stock and scion. The wood or xylem elements may be arranged in continuous rows across the line of union, yet the cells of stock and of scion are separate, and the union is merely an anastomosis of tissues.

Types of Unions

It is important to recognize that there are varying degrees of continuity of tissue. A strong union is one in which the vascular elements of both stock and scion are dovetailed strongly together, as illustrated by

Fig. 32. The stock-scion combination is emphasized in a case such as this in which the rootstock has outgrown the scion. Bartlett on French pear rootstock, seemingly unaffected at 25 years of age.

the French prune on Muir peach. On the other hand, at least four types of structural defects in stock and scion relations have been noted by E. L. Proebsting of California and some would seem to have a relation to dwarfing.

First, there is the defect in which the vascular connections between the stock and scion are to a considerable extent interrupted by the deposition of a relatively soft cushion of wood or "wood parenchyma" at the line of union, as in the case of the apricot on the Myrobalan plum. The Mahaleb cherry rootstock forms a somewhat similar union with the sour cherry. The result is a frequent breaking-off of the scion at the point of union. This may occur when trees have reached fifteen to twenty years of age or older. The interruption in the movement of materials past the layer of wood parenchyma may produce some degree of dwarfing.

A *second* type of defect is that involving distortion of the vascular tissues at the point of union. While it may not cause either mechanical weakness or disturbance in function in mild cases, it is an important factor in severe cases. Distortion in some instances involves the formation of whorls and loops of vascular elements. Vessels may be seen at right angles to one another within the space of a millimeter. This break in continuity of the vascular system definitely impedes the movement of water and nutrients upward from the roots, as has been shown by applying suction to the stem of apple trees and pulling water through such unions. The very dwarfing EM IX apple rootstock characteristically forms this gnarly and contorted type of union with many varieties. It is strongly suggested by these facts that the interrupted continuity of vascular tissue is responsible for many instances of severe dwarfing where a rootstock is involved.

A *third* type of defect is that in which gummy masses are formed by the degeneration of the xylem between the medullary rays at the line of union. This is restricted largely to stone fruits. The blocking of movement may be so severe as to result in severe stunting or death of the plant.

The *fourth* and final type of defect is that in which a cork layer is formed between the phloem of stock and scion, thus resulting in an interference in the downward movement of materials from the top of the tree to the roots. Unions between the apple and the pear are of this nature, and to some degree between the mahaleb cherry and the sour cherry. Apparently what happens is that as secondary thickening progresses, and the stems of both stock and scion increase in diameter, the continuity of bark and phloem is interrupted and islands or layers of cork form here and there at the union. These cork formations increase in number and extent with each succeeding year. Such a union is, of course, sooner or later severely dwarfing. Eventually the bark and phloem of the stock and of the scion may be completely and effectively insulated from one another around the entire circumference of the tree, resulting in a starving of the roots and death of the plant. A delayed uncongeniality has been reported from California for the walnut, in which disruption and uncongeniality may not appear until plants are forty years of age.

H. Jiménez, Venezuela, has reported a very interesting case of "delayed incompatability," described by Barbara Mosse of England. A graft combination of *Carica goudotiana/C. cauliflora* (papaw) grows normally until the male inflorescences are produced, when both stock and scion decline. Female trees of *C. goudotiana* are quite compatable on *C. cauliflora*, and male trees continue to grow vigorously if the inflorescences are removed.

Arthur Eames and L. G. Cox of Cornell University have described a remarkable case of graft-union failure involving a white fir tree (*Abies concolor*) in which there was no swelling and every outward indication of congeniality. The tree had reached a height of 25 feet, was 12 inches in diameter, and was approximately forty years of age when it suddenly fell in a mysterious manner. There was a clean line of separation between stump and top, as though sawed or chewed cleanly off. Examination showed that the tree had been grafted but that the cambiums had failed to unite because of faulty positioning. Yet the tree grew. Most remarkable, however, there was no interlocking of tissues or continuity of vascular elements. Further, there was no callous formation or parenchyma layer between the cells of the stock and scion. Rather, the vascular elements of each had developed more or less at right angles to the normally ascending and descending elements of the tree. The tree stood in a position well protected from wind and had lived in this unusual manner for forty years.

Swelling at the Union

Many of the unions between stock and scion show swellings. Some may be overgrowths of the scion; others may be overgrowth of the stock. In some instances they are striking in appearance. The popular notion has quite naturally arisen that uncongeniality and dwarfing are directly related to the degree of swelling or overgrowth at the union.

This may or may not be the case. Actually, some of the more severe cases of overgrowths and swellings, as in citrus and pears, apparently represent complete congeniality and no dwarfing.

Weber has pictured seven stock-scion relations involving overgrowths, which he has observed in citrus (Fig. 33). He has felt that with the more extreme overgrowths, such as C+3 and C—3, there must eventually in time be mechanical failure. He considers C, C+1, and C—1 as especially successful, long-lived combinations. Each union must be examined as to its nature, bearing in mind the four types of defects in unions and the interruption in movement of food materials and nutrients past the union.

Girdling Effect of a Union

For practical purposes, then, the union in itself may often operate as a natural girdling operation, interfering with the downward movement of carbohydrates and thereby tending both to reduce the growth of the root and to favor accumulation of carbohydrates above the union. Apparently, many desirable and successful dwarfing and early-fruiting combinations of scion and rootstock are to be associated with this natural girdling phenomenon, in which each year there is just sufficient interruption in movement of materials between stock and scion past the union to bring about the observed and desired results. The dwarfing of some pear by some quince roots may be this type, in which there is virtually a perpetual self-semigirdling of partial congeniality.

Differences in growth rates of stock and scion, differences in time of starting in spring, and differences in time of maturation of tissues could all produce effects similar to mechanical girdling. In fact, it seems not only possible but highly probable that fruit-bud formation, early fruiting, and the dwarfing that accompany fruit production may in some cases be due not only to an interruption in total quantity of materials between stock and scion but also to a shift in the time at which interruption occurs. That is, if the interruption occurred at about the time of full bloom or shortly thereafter it might naturally favor fruit-bud formation at just that critical time, yet might not otherwise appreciably alter the vigor or growth status of the tree.

VIRUS–INDUCED DISORDERS

A disorder known as pear decline has occurred in the Pacific Coast region of the United States with Bartlett pear trees. It is suggestive of

Fig. 33. **Rootstock reactions in citrus arranged in "minus" and "plus" series according to stock size where "C" represents "normal," good congeniality of stock and scion. C, C + 1, and C + 2 are considered successful, long-lived combinations. (After H. J. Webber)**

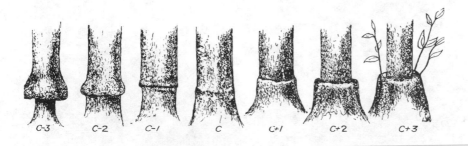

C-3   C-2   C-1   C   C+1   C+2   C+3

certain incompatibilities in tree fruits, and it may well be indicative of what occurs in some so-called incompatibilites.

The trouble is associated with stock-scion relations, appearing as a bud-union disorder with certain pear rootstocks. Pear trees on the Oriental rootstocks *Pyrus serotina* and *P. ussuriensis* are highly sensitive to decline. Trees on imported French pear seedlings (*P. communis*) are intermediate, and trees on Bartlett seedlings (*P. communis*) are highly resistant. Symptoms, according to Batjer and Schneider, result from a series of anatomical changes initiated by death of phloem cells (sieve tubes) immediately below the bud union.

A relatively low level of starch is found in the roots of affected trees, and starch tends to accumulate above the union and close to it. This suggests that phloem degeneration at the union prevents translocation of carbohydrates to the roots, resulting in a weakening or death of the root system and eventually in loss of the entire tree. Evidence indicates that the disorder is not an inherent incompatibility of stock and scion, but an induced one, perhaps by a virus.

Evidence is lent to this view by the similarity of this disorder in pears to the tristeza disease in oranges caused by a virus that is aphid-transmitted. It occurs with sweet orange (*Citrus siensis*) on sour orange rootstocks (*C. aurantium*). Phloem cells (sieve tubes) of the sour orange rootstock immediately below the bud union become necrotic. Translocation of carbohydrates to the roots is interfered with. Schneider has reported that reserve starch of the roots is used up, the roots rot, and the trees wilt and die.

PHYSIOLOGICAL FACTORS

Relation of Anatomical and Physiological Factors

While the explanation of various stock-scion relations has in the past been largely centered upon the mechanics of the graft union itself, this does not satisfy all the requirements for what occurs when stock and scion are grafted. In fact, there is considerable evidence for the belief that various biochemical and physiological factors operative between the stock and the scion may control and bring about the various types of union, congeniality, and dwarfing that are observed. The nature of the resulting union may be dependent upon these biological processes; and the mechanical action of the union may be secondary. Webber concludes that lack of congeniality may be considered the important factor in poor unions, even though the physiological reactions are not completely elucidated.

Congestion of Nitrogenous Materials at the Graft Union

Silberschmidt has stated, following extensive experiments with herbaceous grafts, that stock-scion differences are largely anatomical and quantitative, but brought about by physiological processes. The processes involved in the formation of a union are similar in many cases, but the number and size of the groups of parenchyma tissue initially connecting the two plants become fewer and smaller, the more remote the plant relationship. With plants of poor affinity there occurs a congestion of nitrogenous materials at the union that alters the metabolism of the plant. In cases of close affinity, there is no such congestion of nitrogenous substances.

Lignification at Stock-Scion Junctions

Buchloch has studied compatible and incompatible unions between the pear and the quince. He has concluded that incompatibility is associated with lignification of the adjoining cell walls in the line of union. In incompatible unions the processes of lignification are interrupted. Typically, when a graft is made, a callus develops that is formed of meristematic tissues from both stock and scion. Cell walls consist essentially of cellulose fibers, and there is a mutual middle lamella between them consisting of pectic materials. The next step in the development of the union is the formation of secondary cell walls of cellulose and hemicellulose in two or three distinct layers. Up to this stage, the mechanical condition of the union is weak. The pectic substances of the middle lamella now disappear and are replaced by lignin, which gives a strong union between compatible stock and scion. On the other hand, in the case of incompatible stock and scion, the lignin does not develop, and the union remains weak.

He has hypothesized that incompatibility is due to interaction between the enzyme systems of incompatible graft components.

Transport of Materials Across a Graft Union

It has already been noted that nicotine will pass across the union from tobacco roots into a tomato scion grafted thereon. Atropine will pass from a rootstock of *Atropa belladonna* across the graft into a potato scion. On the other hand, anthocyanin, which is a soluble glucoside, does not cross from a red bean (*Phaseolus lunatus*) to tissue of the Navy bean (*P. vulgaris*). When red beets are grafted to sugar beets, the tissues are clearly defined and the pigment does not move from the red beet into the sugar beet.

The apple tree characteristically contains the phenolic substance phloridzin, and the pear tree contains arbutin. Yet when the apple is worked upon the pear, neither of these substances passes the graft union into tissues of the other.

The tracer technique, employing radioactive isotopes, has corroborated these observations. Radioactive phosphorus, for example, in some instances involving a dwarfing rootstock, may move downward toward the union and accumulate there (Fig. 34). On the other hand, just the reverse is true in some other cases, as reported for peach dwarfed on *Prunus tomentosa*. No overgrowth appeared at the union, and radiophosphorus moved from the scion across the graft into the rootstock.

It has also been noted that auxin (a native hormone in plants responsible for cell elongation and growth), which normally moves downward in the phloem, does not move across a strip of inverted phloem. It is postulated that dwarfing may be brought about in some instances by the inability of auxin to move across certain unions so as to reach the root where it is needed in growth. But this observation is not constant with all materials.

Polarity in Relation to Transport Across a Graft Union

Another illustration of the relation of the graft union to the transport of nutrients from stock to scion and vice versa is shown by grafting experiments with the tomato (Fig. 35). A lateral shoot from a potted tomato plant was grafted at its tip into the tip of a similar lateral shoot from a second potted tomato plant. Thus, the tissues of the two shoots at the union were in reverse position to each other in polarity. The tissues united, but the transport of nutrients across the graft union differed with the material being transported in the plant. For example, phosphorus was supplied to the roots of one plant, tagged with radioactive phosphorus (P^{32}). This material moved freely up to and across the graft union and into the plant to which it was grafted. However, with the calcium (Ca^{45}) no such crossing of the graft union occurred. The calcium moved upward from the roots of the potted plant to which it was applied, but it was blocked by the union. Calcium did not cross the union. This again shows the complexity of stock/scion relations.

ELABORATION AND TRANSPORT OF ANTAGONISTIC AND ESSENTIAL MATERIALS

There are many instances of the elaboration and transport of either antagonistic or essential materials, both by and between stock and scion. They are further evidence supporting the biochemical and physiological factors in stock-scion relations. For example, Karl Sax conducted experiments with the Baldwin variety of apple grafted onto a rootstock known as No. 33340, which was from a cross between *Malus sargenti* and *M. astracanica*. The combination failed. On the other hand, Baldwin succeeded when grafted onto rootstocks of *M. sikkimensis*.

Now, if Baldwin was grafted onto a stempiece (not a root) of No. 33340 and if this in turn was grafted onto a rootstock of *M. sikkimensis*, the combination succeeded and developed vigorously. In other words, it would seem that some material essential to growth of the Baldwin/No. 33340 stion could not be provided by the roots of No. 33340, but was supplied by roots of *M. sikkimensis* (Fig. 36).

Again, as reported by Barbara Mosse, both the Hale's Early peach

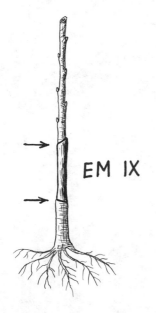

EM IX

Fig. 34. (Left) The concentration of radioactive phosphorus (arrows) in double-worked apple trees, three days after application of the isotope to the top of the tree. Isotope accumulates (arrows) in the intermediate stempiece of EM IX, especially near the points of union, indicating the restriction of downward transport of organic nutrients in the phloem. (After A. G. Dickson and E. W. Samuels in the *Journal of the Arnold Arboretum*, July, 1956)

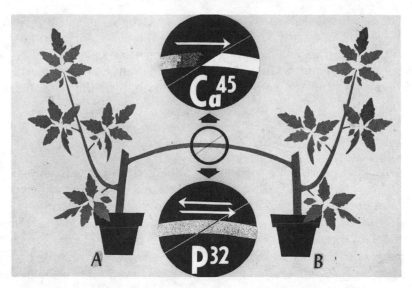

Fig. 35. The nature of the graft union may affect the transport of nutrients from stock to scion, and vice versa, as shown by this graft in tomato plants. Lateral shoots from two tomato plants were grafted end to end (center circle). When radioactive phosphorus (P^{32}) was supplied to plant in pot A, the material moved freely across the union into plant B (lower circle). However, when radioactive calcium (Ca^{45}) was supplied to pot A, the material moved upward to the graft union, but failed to cross the union into plant B against the reversed polarity (upper circle). *Courtesy of John Bukovac*

and the Myrobalan B plum are compatible with the Brompton plum; but Hale's Early is incompatible with Myrobalan B. Accordingly, when Hale's Early peach is grafted upon Brompton plum, the resulting union is satisfactory; but if a ring of bark of Myrobalan B is now placed on the Brompton rootstock of the successful Hale's Early/Brompton combination, symptoms of incompatibility appear on the Hale's Early scion.

Fig. 36. (Right) An illustration of the complexities of incompatibility. The Baldwin apple makes very little growth on rootstock 33340 (left), but it makes normal growth on 33340 as an intermediate stempiece which is worked in turn upon a rootstock of *Malus sikkimensis* (right). (After Karl Sax)

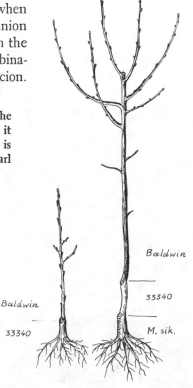

Baldwin

33340

Baldwin

33340

M. sik.

Even though the ring of Myrobalan B is placed as much as 4 inches below the Hale's Early/Brompton union, the symptoms appear. Apparently some principle causing these symptoms of incompatibility can be transported across a 4-inch stempiece of mutually compatible material (Brompton) from the bark ring of Myrobalan B to the scion (Hale's Early).

Suggestions from Stock-Scion Relations in Cucurbits

One final experience, with interspecific grafts of cucurbits, may be related that tends still further to accentuate and help explain the chemical and physiological factors involved in stock-scion relations.

In the 1940's W. G. Van der Kroft, at the Research Station, Wageningen, Holland, attempted to control the fusarium wilt disease in cucumbers (*Cucumis sativus*) by grafting the cucumber onto the Malabar gourd (*Cucurbita ficifolia*), the latter being resistant to the disease. The combination was successful, the plants grew well, and the fusarium wilt was controlled.

Subsequently, S. J. Wellensick and later H. C. M. de Stigter tried a similar method for controlling the fusarium wilt in melons (*Cucumis melo*), by grafting the melon similarly onto the Malabar gourd (*Cucurbita ficifolia*). The melon is closely related to the cucumber. Although the plants started off hopefully, they suddenly wilted and died, the stock seeming to fail first (Fig. 37).

Most important, however, if some of the leaves of the gourd stock were permitted to remain, the combination of melon/gourd was compatible and continued to grow. When stock foliage was present, a good union resulted, involving xylem as well as phloem connections. With no stock foliage, the sieve tubes of the phloem collapsed, the stock failed, and the stion died. In fact, a successful melon/gourd combination would

Fig. 37. An instance of physiological incompatibility, in which *Cucurbita ficifolia* (F) is successfully grafted (left) onto muskmelon stock (M); but the reciprocal combination M/F is incompatible (center) and dies unless leaves of F (F + 1) are left on the stock F (right). This would seem to indicate that the leaves of F supply some material (perhaps hormonal in nature) without which the combination between M and F fails. (After de Stigter in Landbouwhogeschool te Wageningen, 1956)

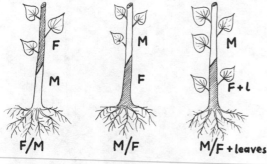

F/M
Successful.

M/F
Fails.

M/F + leaves
Successful.

die in four or five days if the gourd stock was defoliated. And here again decline began with rapid and specific collapse of the sieve tubes of the gourd stock.

Still further, the amount and the vigor of the scion growth was found related to the number of leaves on the gourd stock. If only one stock leaf was left, the scion growth was very slow, and starch tended to accumulate in the scion, above the graft. With an increasing number of leaves, growth and vitality of the scion increased, and starch tended to disappear from the scion.

De Stigter concludes that "the presence of stock leaves (gourd) is required to provide the stock with some 'specific substance' that enables the phloem of the stock to perform its functions in the proper way. In case of shortage or even complete absence, some enzymatic process in the sieve tube-companion cell complex is assumed to be affected, which upsets metabolism as shown by local starch accumulation.

"The specific substance is presumed to be of enzymatic or hormonal nature. Its function, however, may be of quantitative rather than of absolute character, this depending on the plant's growth activity.

"The stock leaves (gourd) in their turn depend on the vitality of the root system, which again is determined by the interaction with the scion. The growth-regulating activity of the stock leaves (gourd) thus appears to be connected with the very nature of the interactions between the melon top and the cucurbit (gourd) rootstock."

Subsequent to this work, De Stigter has reported that when the incompatible melon/gourd is grown in water culture where microorganisms do not attack the weakened roots of the rootstock, the combination gradually recovers as new roots arise from the rootstock. He has concluded that there is an adaptive mechanism in the rootstock by virtue of which it is able to synthesize the "translocation factor" from the melon's supply of food materials. The rootstock needs time to build up an enzymatic apparatus of sufficient capacity to meet the demands of the melon/gourd combination. The process is slow. In soil, the microorganisms destroy the rootstock before this adaptive mechanism can be developed and the combination fails.

This type of phenomenon has been reported many times in the literature down through the years, in which the presence or absence of foliage on one of the consorting parts markedly affected compatibility and growth. An example is the insertion of sour orange shoots into lemon tops that have been worked onto sour orange stock. While heretofore largely discounted, such observations apparently have some basis in fact.

Further, here is an instance where two consorting parts are completely incompatible; but when a deficiency is supplied, as through the addition of stock leaves (gourd), the incompatibility disappears.

Next, the combinations with varying numbers of stock leaves (gourd) may properly be called dwarfed plants. With more stock leaves they are less dwarfed than with fewer dwarfed leaves. In short, here is a situation that runs the entire gamut from complete incompatibility, to compatibility with severe dwarfing, to compatibility with less dwarfing, to compatibility and vigorous growth.

Theory of Mutual Supply and Utilization of Materials

Little by little the theory proposed by Daniel in 1894 is being more favorably considered; namely, that a good union comes about when both the stock and the scion are able to use the materials produced or stored by the other and when each can supply the materials needed by the other. Kosloff apparently supports this idea when he refers to the chemical specificity of the plants rather than the "union" as being responsible for the accumulation of materials above or below the union. If the biochemical nature of the two plants is sufficiently different, they are incompatible. Webber says that compatibility depends upon similarity of elaborated materials and of by-products in the two consorting parts. And Thiel suggests that the degree of compatibility depends on the ability of stock and scion to reabsorb the primary isolating layer found between them at the union.

This hypothesis takes on new significance in the light of the increasing appreciation of the root as an organ of synthesis. Formerly thought of mostly as an organ of uptake and of storage, it now appears that various complex organic compounds may be synthesized in the root and translocated to other parts of the plant, where they are utilized and stored. Also, this hypothesis places emphasis upon the concept of the plant as a unit. One part bears a relation to another, and the word "stion" takes on added meaning.

INCOMPATIBILITY, UNCONGENIALITY, AND DWARFING

It becomes clear from all of this why it is so difficult to determine what is compatible and what is not, and where incompatibility leaves off and where dwarfing begins.

To say didactically that a given combination is "incompatible" is to say that the two components are intolerant of each other, are incapable of living together in harmony, and cannot be united or conjoined. "Compatible," on the other hand, implies mutual tolerance and acceptability. There are what seem to be clear cases of incompatibility, as in plants far removed from each other botanically. Yet when by the modification of a single factor, as the presence of stock leaves, the "incompatible" combination becomes "compatible," then the sharp distinction between "compatible" and "incompatible" fades. One wonders how many other "incompatible" combinations might become "compatible" if more of the controlling factors were understood and could be controlled by new techniques.

Perhaps the word "congenial" would be a good word to use in bridging the gap. "Congenial" means suited or adapted in nature or character, tolerant, kindred. It is possible to have degrees of congeniality, whereas it is not possible to have degrees of incompatibility and compatibility. Either a combination is compatible or it is incompatible. A stock-scion relation may be more congenial or less congenial under a given set of circumstances. It could even be rated on a scale of, say, ten, as to the degree of vigor or happiness. Thought of in this sense, there are instances where degrees of dwarfing and degrees of congeniality may be synonymous, and where various degrees of congeniality may even bring about various degrees of dwarfing.

The stock-scion relationship is certainly one of symbiosis, or partnership of dissimilar organisms, as with nitrogen-gathering bacteria growing on the roots of legumes and the algal-fungal relation in lichens. Symbiosis may be antagonistic or antipathetic. It may be intimate, helpful, and mutualistic. There may be degrees of mutualistic symbiosis.

Much the same relationship can be seen in stock-scion relations. The rootstock itself, because of certain inherent characters, is a contributing factor. The scion and an intermediate stempiece also play parts. Then there are various types of unions and degrees of congeniality that are both mechanical and physiological in nature and that markedly control performance. Added to this are virus and other diseases with differing resistances and tolerances, insect pests, length of growing season, sunlight, temperatures, rainfall, soil conditions, parasitic soil organisms, nutrient levels, time of bud start, time of onset of growth, growth rates, length of growing season, hormones, enzyme systems, protein reactions, biosynthesis in stock and scion, and undoubtedly many more contributing factors not yet apparent.

About all that can be said is that a multitude of factors come into play when stock and scion are brought together and that only by the old technique of trial and error can the relationship and performance be determined from anything that is now known.

Part Three

DWARFING ROOTSTOCKS

7

DWARFING ROOTSTOCKS
FOR THE APPLE

FOR many centuries apple trees have been propagated by budding or grafting a desired scion variety onto some root system other than its own. The reason for this practice is essentially that apples do not come true from seed. They are extremely heterozygous.

To illustrate, some botanists have placed the apple in the genus *Pyrus* (*Pyrus malus*), which includes the pear and the quince. By others it has been placed in a genus of its own, *Malus*. Within this genus are many species, subspecies, and horticultural forms of the apple. The cultivated forms have variously been designated *M. pumila*, *M. sylvestris*, and *M. domestica*. Related species are *M. baccata* (Siberian Crab apple), *M. ioensis* (Prairie Crab apple), *M. floribunda* (Showy Crab apple), and many others. Horticultural forms include upward of five thousand varieties (cultivars).

Because of this complexity and natural hybridization between species and forms in nature, it was long ago found essential to propagate a desired form by some vegetative means, as by cuttings or layers or grafting.

One may immediately question why propagation by cuttings, for example, was not used for the apple as is done for the grape. The answer is very simple: Because very few varieties of apple root readily from cuttings as does the grape, it became necessary to obtain some easily available rootstock material. This has been provided in America by seedlings

123

raised from seed of French Crab or domestic apple varieties, as Delicious, Rome Beauty and McIntosh. Most of the available rootstock materials, when grafted, produce the very vigorous standard trees commonly met with in commerce.

However, there are dwarf forms of apple just as there are dwarf forms of most other plants. Some of these have been known for centuries. The production of small apple trees by propagation of a desired variety (cultivar) upon a suitable rootstock is a very old practice. The discovery and development of these dwarf forms is an interesting chapter in horticultural history.

ORIGIN AND HISTORY OF THE PARADISE APPLE

Alexander the Great, who lived during the fourth century B.C., sent plant material home to Greece from his conquests in Asia Minor. Theophrastus, the great Greek historian of that time, records "Spring Apple" among them, which was a dwarf, self-rooting form of apple.

Roman agriculturists seem to have been thoroughly familiar with this dwarf apple. Descriptions suggest the dwarf, low-growing, self-rooting forms of apple that have been subsequently described repeatedly down through the years by a great number of writers and that have come to be called "Paradise" apples. E. A. Bunyard, in his history of the Paradise rootstocks (1919), felt that this dwarf apple was the prototype of the well-known French Paradise dwarfing rootstock.

Just how the name "Paradise" became attached to dwarfing apple stocks is not entirely clear, but there is some reasonable conjecture. Thus, the word comes from the Persian *pairidaeza* (Sanskrit *paradeca*), meaning a park or garden. The French *parvis*, meaning "small," derived from Old French *paradise*, is used for an enclosed courtyard, especially one in front of a church or cathedral. The word is found in English and is used in the same sense today.

The rather common desire to associate plants with the Bible and with biblical teachings, as Bunyard suggests, may have led to placing the origin of this dwarf apple in "Paradise" or the "Garden of Eden."

Certainly the apple was well known in the eastern Mediterranean countries, and certainly Alexander the Great sent plant material of this type back from Persia, with which the original Garden of Eden is associated.

At all events, "Paradise" is the name that has been used for a particular type or types of apple, much used for dwarfing various varieties (cultivars) of apple worked upon them. They are found growing in large numbers in eastern Asia and the Caucasian Mountains under such conditions as to leave no doubt that they are indigenous. So distinctive was this form considered several centuries ago that it was even made a separate species of apple and variously called *Pyrus acerba* D.C., *Malus acerba* Merat, and *Malus pumila* var. *Paradisiaca* Schneider.

First mention of a Paradise apple in horticultural literature is by Champier in 1472, speaking of apples in Normandy. Dalechamps described in 1507 what is probably the French Paradise. Charles Estienne in 1540 mentioned the Paradise in his *Seminarium* (1540), and considered it identical with the "Petisienne" of Pliny the Younger (A.D. 62–113).

Olivier de Serres (1600) said that "every tree whose branch will root will make a dwarf." This observation, as will be shown later, is not necessarily true, but it is a point worth remembering in this discussion. There are many forms of apple known today that have this ability. They most frequently exhibit small root knots or excrescenses on the stem above the ground. These are growths of adventitious roots, often called "burr knots." And while all apples that root readily in this way are not necessarily dwarfing rootstocks, nevertheless the most common dwarfing rootstocks do possess this character.

From this one is led to believe that the original conception of the Paradise apple was that it would root readily. That it was also useful as a dwarfing rootstock was a secondary consideration. Indeed, it must be again pointed out that easy rooting was a real difference and something of a distinction, inasmuch as most apples do not propagate easily by vegetative means. Yet because the early forms of Paradise were apparently both self-rooting and dwarfing, the association was erroneously made that all self-rooting or "Paradise" types were dwarfing apple rootstocks.

To complete the story, the "French Paradise" is said to have reached England about 1696 and was widely grown by 1747. It was known in America early in the nineteenth century.

Origin and History of the Doucin

Although the word "Paradise" thus became almost synonymous with "dwarf," other rootstock material was found from time to time that would dwarf the apple. One of these types was the so-called "Doucin" (sometimes spelled "Doucain"), which was for a long time, and is even sometimes today, included as a "Paradise" stock and called a subspecies or variety of Paradise.

That there is a recognizable difference between the Paradise and the Doucin is shown by the description given in *Encyclopédie horticole* by E. A. Carrière, in which he distinguishes between plants of *Les Paradis* and plants of *Les Doucins* as follows:

"*Paradise stocks* [*Paradis*]—propagated other than from seed; less vigorous than Doucin, more bushy, more scraggly; branches more slender; bark darker and shiny; roots finely divided, numerous, slender, shallow-rooted; fruits longer than wide, ribbed, whitish, shiny, blushed; flesh soft, insipid; ripening in July.

"*Doucins*—vigorous, more upright; branches larger; bark grayish, downy or tomentose; roots going deeper into the soil, less branchy, not hairy; fruits wider than long, not ribbed, green; flesh greenish, agreeably aromatic; ripening later."

The origin of the Doucin is not well documented. It appears that various types or varieties of dwarfing rootstocks arose down through the years here and there and from time to time. Among these types were some to which were given the name "Doucin," derived from the French *douceur*, meaning "sweetness." The term referred to the fruit borne by the plant itself, as contrasted to the unpleasant fruit of the Paradise. But this cannot be the type of Doucin known today because the fruit of the latter is not sweet.

The first mention of Doucin appeared in European literature in 1519. It was known in America early in the nineteenth century. The distinction between Paradise and Doucin is made consistently by horticultural writers down through the years, and following the same general pattern. The Doucin was recognized as being more vigorous than the Paradise and as producing a tree only semidwarf in stature for the variety (cultivar) worked upon it. It was devoid of "burr knots," deeper rooted, and less easily propagated.

Yet because Doucin types were included by some writers of that time (nineteenth century) under the common name Paradise, the well-known Doucin used in England to produce a semidwarf fruit tree came to be known in the trade as "English Paradise." To distinguish this from the shallow-rooted Paradise stock used on the Continent to produce a very dwarf fruit tree, the latter came to be known as "French Paradise."

Confusion of this type is found throughout horticultural literature, continuing until recent times. In fact, in the classic publication by R. G. Hatton of the East Malling Research Station, made in 1917, to clarify and standardize the various forms of dwarfing rootstocks, the Common Doucin ("English Paradise") is included among the nine "Paradise Apple Stocks" described, although the author pointed out the difference.

Present usage does not make such general distinctions between Paradise and Doucin. The designations are gradually being replaced by different and more exact terminology, as will be shown later. Nevertheless, "Paradise" and "Doucin" have been used for centuries and have some historical significance.

THE STANDARDIZATION OF CLONAL DWARFING APPLE ROOTSTOCKS

As horticulture developed through the centuries throughout the world, more dwarfing rootstocks were found and developed. New forms arose spontaneously and independently in Germany, in France, in the Netherlands, and in England. They were variously named according to the place or individual associated with their origin, or by some descriptive term, as "Holstein Doucin," "Rivers' Paradise," "Holly Leaf," and "Red Paradise."An excellent example of what has happened occurred near Metz, France, in 1879. A chance seedling arose there that was introduced as Jaune de Metz (Yellow Metz), which has since become a very important dwarfing rootstock.

As these rootstocks moved into commerce, they became mixed and misnamed. The fact that they were valued first of all for their ability to root easily was a major factor in increasing the confusion, because a plant that was inadvertently dropped to the ground somewhere in the nursery could easily take root and perpetuate itself. In addition, these types suckered readily from the root, so that any replanting or changing of beds undoubtedly added more mixtures.

At all events, by the middle of the nineteenth century in England it was recognized that there was confusion in the rootstock materials found in the nursery trade. Articles appeared in horticultural journals reciting experiences with different rootstocks. In 1869 Barron enumerated eight Paradise stocks used in England, reporting trials and descriptions. In 1870 Thomas Rivers, one of the leading horticulturists and nurserymen of England, mentioned fourteen kinds of Paradise in his widely used book *The Miniature Fruit Garden.*

It was not until about 1910, however, that the situation with rootstocks became so serious, and demand for correction became so great, that standardization really got under way. This occurred not only in England but also in Germany, the Netherlands, and France.

In Germany, the L. Späth Nurseries, near Berlin, began tests with rootstocks about 1910. It was at East Malling, in Kent, England, however, at the Wye College Fruit Experiment Station, later to be named the East Malling Research Station, that the most significant work in standardization of rootstocks was undertaken. In the fall of 1912, the director of the East Malling Station, R. Wellington, undertook a study of the Paradise forms of apple rootstocks. World War I broke out in 1914, and Wellington joined the armed forces and turned the work over to his associate, R. G. Hatton. Seventy-one collections of "Paradise" stocks were secured from thirty-five sources—29 British, 3 French, 1 Dutch, and 1 German. The last named furnished twenty collections.

Examination of the materials bore out the confusion that was already suspected. For example, Hatton, reporting in 1917, said "Many groups of nine stocks purporting to be similar contained two or more distinct types." And speaking of the "Broad-Leaved English Paradise," he continued, "Whilst we nominally received it in twenty-one collections called 'Broad-leaved English,' we actually received it only three times as true samples." It was indeed time for clarification and standardization.

The EM or Malling Apple Rootstocks

The various collections that have been mentioned and that were received at the East Malling Research Station were planted out in the fall of 1913 for propagation. They were allowed to stand for one year to become established. They were then cut back to the ground the second spring, encouraged to send up shoots, and propagated by mound layering. The rooted shoots from each plant served for study and comparison, and became the progenitors of standardized, named types. It must be emphasized that these rootstocks were not propagated from seed, but from rooted shoots (layers) of the mother plant. Accordingly, each plant propagated by this means from a given initial plant was identical to another. Such plants form a clone, and such rootstocks are called clonal rootstocks.

Since the nomenclature was so confused, it was decided to abandon common names and use Roman numerals for the specific, selected plant types.

Accordingly, in the first publication, made in 1917, Hatton named and described nine types of vegetatively propagated dwarfing rootstocks. These nine types, numbered, I, II, III, IV, V, VI, VII, VIII, and IX, include the types that are still the most important dwarfing rootstocks. Subsequently, the number was increased to sixteen types and later to twenty-six, including several that are not at all dwarfing.

The prefix "Type" was later dropped and replaced by "Malling," indicating the place where the types were standardized. Still later, at the Twelfth International Horticultural Congress held at Berlin in 1938, it was suggested that the proper designation be "EM," this being the abbreviation for East Malling. And so "Type I," "No. I," and later "Malling I," is now "EM I." Further, for complete accuracy, it was suggested that the old term "Paradise" and "Doucin" be dropped and that the botanical name *Malus* be used for the apple, *Pyrus* for the pear, *Cydonia* for the quince, and *Prunus* for the apricot, cherry, peach, and plum.

In practice, then, the proper procedure in naming a rootstock is to give the generic name, followed by one or two capital letters to identify the source of the material, and this in turn to be followed by the type number. For example, the Type I, which Hatton selected at the East Malling Research Station as the true Broad-leaved English Paradise, is properly called "Malus EM I." This system has been well accepted and is being adopted generally throughout the world. Obviously there is no purpose in using the generic name "Malus" in a discussion dealing with apple rootstocks only, and so the designation is commonly shortened to merely "EM I" when speaking of apples.

This, then, is the story of how the well-known "Malling" or "EM" rootstocks came to be. They are standardized lines of well-known vegetatively propagated rootstocks, some of them several centuries old. They represent a most important step in the progress of horticulture, and one with which the names of R. G. Hatton, later Sir Ronald Hatton, and his associates at the East Malling Research Station in England are permanently identified.

DESCRIPTION OF THE MALLING
APPLE ROOTSTOCKS

Detailed descriptions and photographs of the Malling apple rootstocks have been published in a number of places. Of these, the book by Maurer gives the most complete and most detailed information on rootstocks EM I to XVIII, including photographs of early shoot growth, mature shoot growth, dormant shoots, leaf outlines, and fruit in color. The bulletin by Shaw describes and pictures rootstocks EM I to XVI under American conditions. Tydeman presents both detailed descriptions and photographs of the entire series, namely, EM I to XXV and EM Crab C. The publication by the Ministry of Agriculture, Fisheries and Food of Great Britain pictures the most popular of the Malling rootstocks, together with clear and helpful aids to identification.

Following is a list of the Malling apple rootstocks, as well as information the author has found significant and useful, including briefly the name, origin, plant characters, performance in the nursery, aids to identification, and size and performance of fruit trees worked upon them. The entire series is listed, including some rootstocks that are not dwarfing, some that are not recommended, and some that are even not readily available. This is done to make the record complete and to satisfy questions regarding numbers not commonly mentioned in the literature but concerning which there is some interest. The important rootstocks are marked with an asterisk (*). The scale used in describing the size tree the rootstock makes with the scion variety is (1) dwarf, (2) semidwarf, (3) semistandard, and (4) standard. This corresponds to (1) very dwarf, (2) semidwarf, (3) vigorous, and (4) very vigorous as used by the East Malling Research Station. The latter designation is included in parentheses for comparison.

*EM I. BROAD-LEAVED ENGLISH PARADISE (OF RIVERS)   Of English origin, selected by T. Rivers as a chance seedling about 1860. Vigorous and tall growing in stool beds; shoots stout, rather woody, greenish, zigzagging from bud to bud; leaves large, broadly ovate, "bumpy" or rugose; stipules large; maturing early, rooted shoots can be removed from the mother plant in fall (Figs 38 and 52).

Not very prolific, roots well, relatively hardy.

Makes semidwarf to semistandard tree (EM vigorous) well anchored, early fruiting; susceptible to collar rot, has not done well south of Michigan and New York State, prefers cool, moist, relatively heavy soils, susceptible to drought. Good for weak-growing varieties; well suited to Jonathan and McIntosh.

*EM II. DOUCIN, OF THE BEST FRENCH NURSERIES; OFTEN CALLED ENGLISH PARADISE   At least two centuries old. Shoots stiff and erect in stool beds, short internodes; lenticels on the wood numerous and conspicuous; leaves rather narrowly ovate, often slightly convex, even-shaped, deep bluish-green (Figs. 39 and 52).

Fairly prolific, matures early but roots sparsely in its early years in stools, best left mounded until spring if possible; hardy.

Makes semidwarf to semistandard fruit tree (EM vigorous), usually slightly smaller with most American varieties than EM I but larger than

Fig. 38. (Upper right) **EM I** (Broad-Leaved English). Large, broad, bumpy leaves with large stipules (leaflike appendages attached to the base of the leafstalk), on somewhat woolly, greenish stems zigzagging from bud to bud.

Fig. 39. (Lower right) **EM II** (Doucin). Stiff growth. Even-shaped leaves fairly close together. Very many conspicuous dots (lenticels) on the wood.

EM VII. Does well on light soils. Good for slow-growing varieties; well suited to Jonathan, Delicious, and Stayman.

Highly recommended.

EM III. DUTCH DOUCIN; ALSO CALLED HOLLYLEAF PARADISE AND KÖNIGS SPLITTAPFEL   Widely distributed on the Continent; considered by George Bunyard as typical of the Doucin types, resembling the Codlin apples in fruit characters—smooth, acid, juicy, flavorless. Shoots slender, spreading, flexible, medium in size, uniform; leaves small, narrow, very acutely and deeply doubly serrate (hollyleaf); wood covered with dense grayish pubescence. Very susceptible to apple scab and mildew.

Prolific, roots easily and well.

Makes semidwarf fruit tree (EM semidwarf), but slightly smaller than EM I, fruits early; suckers badly.

Not recommended; not readily available.

*EM IV. HOLSTEIN DOUCIN; ALSO CALLED DUTCH DOUCIN AND YELLOW DOUCIN   Common in Holland and Germany; originally identified as *Malus pumila*. Sturdy, erect, clean grower in stool beds; shoots compact in appearance due to short internodes; wood light brown, faintly grained, with fine cracks like old varnish, "lumps" near buds; leaves thick, with rough surface and wavy edges (Figs. 40 and 52).

Fairly prolific, roots easily, fairly good nursery plant.

Makes semidwarf to semistandard tree (EM semivigorous), early bearer, heavy producer, high top/root ratio, which tends to subject tree to blowing over in exposed situations. Excellent for dwarf pyramids and spindelbush, which do not require staking.

Recommended for limited use.

EM V. DOUCIN AMELIORÉ; ALSO CALLED IMPROVED DOUCIN IN ENGLISH NURSERIES AND RED PARADISE IN DUTCH AND GERMAN NURSERIES   Once a common stock in Europe, superseding EM II in some places. Shoots

Fig. 40. (Upper left) **EM IV (Holstein Doucin).** Thick leaves with rough surface and wavy edges; being close together, they give compact appearance. Wood light brown with fine cracks like old varnish. Lumps near buds are common.

Fig. 41. (Lower left) **EM V (Doucin Amelioré).** Leaf tips are drawn out to a point. Very few but large dots on the wood; buds and leafstalks are red when rubbed.

slender, flexible, medium in size, uniform; leaf tips drawn to a point; very few but large, conspicuous lenticels on wood; buds and leafstalks red when rubbed (Figs. 41 and 53).

Prolific, roots easily.

Makes semidwarf to semistandard fruit tree (EM vigorous), similar to EM II, early fruiting, induces small fruit, shallow-rooted, inclined to lean, suckers easily. Appears unable to absorb K, under conditions where other rootstocks secure adequate amounts.

Not recommended.

EM VI. NONSUCH PARADISE; ALSO KNOWN AS RIVERS' PARADISE   Selected by T. Rivers as a chance seedling in England about 1860. Erect, stiff, compact, sturdy grower in stool beds; shoots medium in size, uniform, distinctly ridged, with short internodes; leaves flat, crinkled, glossy; wood dull greenish-yellow, covered with dense grayish pubescence.

Rather few shoots per stool, roots readily.

Makes semidwarf tree, early fruiting, shallow rooted, leans, suckers badly.

Not recommended; not readily available.

*EM VII. UNNAMED   Long known in English nurseries as a mixture in Doucin stock; called a true Reinette by George Bunyard, the fruit resembling Golden Pippin, but this is questionable because the fruit of EM VII as now grown is not considered edible; sometimes called English Paradise; known in France in the time of La Quintinye (1626–1688). Reselected clone free of rubbery wood, mosaic, and chat fruit virus issued in 1959–1960 as "EM VII (1959)" by East Malling Research Station. Shoots long, slender, flexible, "whippy"; leaves thin, papery, rather shiny, almost circular (circular if tip and petiole are removed), lower leaves sometimes lobed; petiole long and much upturned to give upright appearance (Figs. 42 and 53).

Fairly prolific, roots easily and well, observed to be subject to crown gall in American nurseries, requiring special attention to keep beds free from this disease.

Makes semidwarf to dwarf tree (EM semidwarf), next smallest to

Fig. 42. (Right) EM VII. Whippy, clean growth. Thin, papery, somewhat shiny leaves. Leafstalks are upright. Leaves are circular when tip and stalk are removed. Lower leaves are sometimes lobed, and buds usually are far apart.

EM IX in the series but considerably larger, early fruiting, wide range of adaptation, tendency to produce root suckers. Prefers good soils; good for strong-growing varieties; needs deep planting for anchorage.

Highly recommended.

*EM VIII. FRENCH PARADISE; ALSO CALLED "RED PARADISE"; NOT DISTINGUISHABLE FROM CLARK DWARF   Common on the Continent; used as intermediate stempiece to produce a dwarf tree; considered by George Bunyard to be a representative of the Russian group of apples by virtue of fruit that turns quite white when ripe, with a bloom and a granular or loose-textured flesh and peculiar "strawberry flavor"; reached England in 1696 and was widely grown by 1747.

Weak grower in the stool block; shoots short, spreading, flexible; wood dark purple-brown; leaves smooth, flat, gradually acuminate, with tip drawn out.

Not very prolific, does not root freely, very subject to apple scab, subject to crown gall.

Makes dwarf fruit tree (EM dwarf), nearly as small as EM IX, early fruiting.

Not recommended.

*EM IX. JAUNE DE METZ; ALSO CALLED YELLOW METZ, YELLOW PARADISE OF METZ, AND DIEUDONNÉ   Selected as a chance seedling in France about 1879. Shoots stout, only medium in height and diameter, straight, no trace of zigzagging; wood reddish yellow, burnished or silvery sheen; leaves large, oblong, rather shiny, with "piecrust edges" (Figs. 43 and 53).

Not prolific, but roots fairly well; roots snap and break easily.

Makes dwarf tree (EM very dwarf), early bearing, first or second year planted; likely to lean unless supported, roots easily broken; well suited for dwarf pyramids and spindlebush on good soils.

Highly recommended.

EM X. UNNAMED   Selected and named Doucin U. 1 by Späth, Berlin. Vigorous, shoots erect, long, slender, stiff; leaves small, dark green, waved, frilled.

Fig. 43. (Left) EM IX (Jaune de Metz). Large, oblong leaves, somewhat shiny, with piecrust edges. Shoots straight (no trace of zigzag). Reddish-yellow wood with silvery sheen.

Prolific, roots readily, many small roots, like EM XIII rather than EM XII.

Makes semistandard to standard tree (EM very vigorous).

Not recommended; not readily available.

EM XI. UNNAMED; ALSO KNOWN AS BLACK DOUCIN, GREEN DOUCIN, PRACHT'S DOUCIN, SANDER'S DOUCIN, TORNESCHER SPECIAL, HOHENHORSTER IDEAL, WESSELING'S PARADISE, AND WINTER SUNSHINE. SOMETIMES CONFUSED WITH EM XIII   From Pracht, Germany, 1904. Widely used on Continent of Europe where its winter hardiness makes it valuable, it is the only rootstock in the series that produces edible fruit. Shoots long, slender, vigorous, erect zigzagged; wood deep greenish-brown to green on lower part of shoot; leaves gradually acuminate, tips drawn out.

Prolific, roots well.

Makes semistandard tree (EM vigorous), not unlike EM II.

Recommended for limited use.

EM XII. UNNAMED   Seedling selected from a seedling collection (probably crab stock) in England. Shoots stocky, stiff, upright, medium in height; wood deep, rich, reddish chocolate brown; lenticels orange brown; leaves very shiny, upfolded, concave; petiole kinked at base (Figs. 44 and 54).

Not prolific, does not root easily, and should not be dug until spring so as to encourage rooting.

Makes standard tree (EM very vigorous). Roots are sparse but tough and deeply penetrating.

Not recommended.

*EM XIII. BLACK DOUCIN   Selected and named U. 2 by Späth, Berlin, about 1890. Vigorous; shoots upright, compact, stout, long, smooth; wood blackish-brown with distinctive creamy white lenticels; leaves held erect, deeply cut leaf edges; large stipules, spiny shoots, generally ragged appearance; petiole deep wine-red on underside extending up midrib (Figs. 45 and 54).

Fairly prolific, roots well, holds leaves until spring.

Makes semistandard tree, semidwarf with some varieties (EM vigorous); does well on heavy soils and high water table; well anchored;

Fig. 44. (Upper right) **EM XII.** Very stiff, erect, reddish-chocolate shoots bearing very shiny, upcupped leaves. Leafstalk kinked at base.

especially good for weak-growing varieties; set of fruit poor with some varieties; excellent for Golden Delicious, Cortland, and Jonathan.

Recommended for particular varieties and heavy or wet soil; intolerant of dry soil conditions, as is EM I.

EM XIV. UNNAMED   Selected and named U. 5 by Späth, Berlin. Fairly vigorous; shoots erect, long, slender; leaves rather drooping, margins deeply and shortly waved.

Fairly productive, does not root readily.

Makes semistandard tree (EM vigorous).

Insufficiently tested.

EM XV. UNNAMED   Selected and named U. 6 by Späth, Berlin. Fairly vigorous; shoots stout, zigzagged, considerable feathering; wood covered with dense whitish pubescence.

Not very productive, roots moderately well.

Makes semistandard to standard tree (EM very vigorous), smaller than EM XII.

Insufficiently tested.

*EM XVI. KETZINER IDEAL   Selected and named Doucin U. 3 by Späth, Berlin. Esteemed in East Europe because of winter hardiness. Fairly vigorous; shoots stiff, erect, stout; older wood black; leaves dark glaucous green or blackish-green excepting at tip, which is light green in striking contrast; leaves almost entire (no serrations), especially at base (Figs. 46 and 54).

Fairly prolific, roots slowly but well, best left on mother plant until spring; leaf break very late in spring; matures late and sometimes subject to winter injury; susceptible to woolly aphid in nursery.

Makes semistandard to standard tree (EM very vigorous); well anchored, uniform, productive, but slow to begin cropping unless lightly pruned, has done well for Delicious.

Fig. 45. (Upper left) **EM XIII.** Leaves held erect. Deep cut leaf edges; large leaflets at base of leafstalk and spiny shoots give a general ragged appearance. Older wood is black-brown with prominent white dots (lenticels).

Fig. 46. (Lower left) **EM XVI** (Ketziner Ideal). Growth is very erect. Leaves are dull black-green, except at growing tip, which is light green in striking contrast. Lower leaves almost entire (i.e., no saw edge). Older wood is black.

Highly recommended for semistandard trees in some areas, but largely superseded in England by EM XXV.

EM XVII. UNNAMED   From Sprenger, Wageningen, Holland; considered identical to EM V.

Makes semidwarf tree.

Insufficiently tested.

EM XVIII. UNNAMED   From Sprenger, Wageningen, Holland. Vigorous; shoots erect, long, stout, with considerable feathering; leaves deep green, rather dull, thin; bud break early.

Fairly prolific, roots well.

Makes semistandard tree (EM vigorous).

Insufficiently tested.

EM XIX. UNNAMED   A selection from Späth, Berlin. Originally EM XVII. Sprenger added EM XVII and EM XVIII to the series, not being aware of Hatton's EM XVII, which was then later changed from EM XVII to the EM XIX noted here.

Not very vigorous; spreading; shoots slender, very flexible.

Not very prolific, roots easily.

Makes semistandard tree (EM vigorous).

Not tested.

EM XX. SPURIOUS IX   Distributed erroneously from France as EM IX. Shoots long, erect, slender, flexible, with long internodes; leaves smooth, thick, and fleshy, coarsely serrate; bright grass-green, bronzed.

Prolific, roots easily.

Makes semidwarf to dwarf tree (EM very dwarf).

Insufficiently tested.

EM XXI. CUT-LEAVED IX   Distributed erroneously as EM IX, characteristically deeply incised. Vigorous, spreading; shoots long, stout; leaves coarse, deeply incised, acute, thin, crisp; wood chestnut brown with a metallic luster.

Prolific, roots well.

Makes semidwarf to dwarf tree (EM very dwarf).

Not recommended.

EM XXII. UNNAMED   From Seabrook, Chelmsford, England. Not vigorous; shoots erect, stout, stiff; leaves deep green, rather dull, ovate, tapering, unfolded; late to leaf out in spring.

Not very prolific, roots readily.

Makes semistandard tree (EM vigorous).

Insufficiently tested.

EM XXIII. UNNAMED   From Seabrook, Chelmsford, England. Very vigorous; shoots very erect, long, slender; leaves deep dull green, thin, crisp.

Very prolific, roots well.

Makes semidwarf tree (EM vigorous).

Insufficiently tested.

EM XXIV. PARADIS DE MENTON, NOIR DE MENTON, AND DOUCIN NOIR DE MENTON   Vigorous, spreading; shoots long, stout; wood covered with very dense whitish pubescence; leaves thick, grayish-green.

Fairly prolific, roots well.

Makes semistandard tree (EM vigorous).

Not tested.

*EM XXV. UNNAMED   Seedling from Northern Spy by EM II raised at East Malling, England in 1930–1931. Not very vigorous; shoots stiff, erect, stubby, zigzagged; old wood dull green; leaves usually flat, often convex, much crinkled and crumpled, sea green.

Fairly prolific, roots well.

Makes semistandard to standard tree (EM very vigorous); as large as EM XVI but earlier fruiting, resistant to collar rot (*Phytophthora cactorum*).

Recommended for trial.

EM 26. UNNAMED   Numbering system changed from Roman to Arabic because of complexity of higher Roman numerals. Seedling from cross between EM XVI and EM IX in 1929, and numbered 3436. Introduced from East Malling, England, in 1959.

Fairly vigorous, rather spreading; stiff chocolate-brown shoots with a distinct silvery sheen and few laterals; leaves dark green, broadly wedge-shaped at base, wavy margins, late leaf break in spring, and late maturity in fall (Fig. 47).

Prolific, roots well.

Makes dwarf tree, intermediate between EM IX and EM VII; better

Fig. 47. **EM 26 (3436). Medium oval, bluish green leaves with wedge-shaped bases and small lanceolate, erect stipules.**

anchored than EM IX, but probably needs support except under very sheltered conditions.

Promising. Recommended for trial.

EM CRAB C. INTRODUCED FROM EAST MALLING Very vigorous, shoots woolly throughout their length; two-year wood "burned" red on sunny side, never blackish-brown as in EM XIII; leaves upfolded, broad, dull, with "crosscut" saw edge and large stipules.

Not prolific, roots poorly and with difficulty, best from root cuttings.

Makes semistandard tree (EM vigorous), well anchored, crops early.

THE MALLING–MERTON APPLE ROOTSTOCKS

It is obvious that just as new and improved varieties (cultivars) of fruit are bred and developed, so it is possible to breed and develop improved rootstocks for fruit trees. Recognizing the needs and the possibilities, the John Innes Horticultural Institute then at Merton, England, and the East Malling Research Station at East Malling, England, in 1922 jointly began to raise a series of rootstocks by systematic plant breeding. The work was intensified in 1928. M. B. Crane of the John Innes staff at Merton and H. M. Tydeman of the East Malling staff at East Malling collaborated in the project. The name "Malling-Merton" is derived from the locations of these two institutions. The abbreviation used is "MM."

This breeding project was aimed primarily at improving the range of stocks available, and meeting problems of Australian fruit growers, who were troubled by the woolly aphid (*Eriosoma lanigerum*) and were especially interested in rootstocks resistant to this pest. Northern Spy was their principal aphid-resistant rootstock. Accordingly, Northern Spy was used as one of the parents, and various Malling and other rootstock materials were used as the other parent.

A total of 3,758 seedlings were raised. They were tested for resistance to woolly aphid, and all susceptible ones were eliminated. Strains of aphids were brought in from Australia and Canada and used in the testing. The surviving seedlings were also examined for dwarfing tendencies, propagation qualities, stock/scion compatibility, vigor, precocity, and general field performance on two different soils. Only fifteen of these survived the rigorous tests. The first report on these rootstocks was made in 1952.

It is important to emphasize these facts because it gives an appreciation of what to expect in the MM rootstocks. Resistance to the woolly aphid was an important consideration, but, as has been shown, it was by no means the only consideration. In areas of the world where woolly aphid is important, the MM rootstocks assume added significance, although it must be understood that resistance is not transmitted to the scions worked upon them. Where the woolly aphid is not a serious pest the rootstocks may be compared with existing ones for vigor and cropping. If rootstocks are to be grown as clipped hedges for propagation as hardwood cuttings, this factor of resistance may assume considerable importance to the nurseryman.

Further, since these rootstocks are so new and have had only limited orchard trial, in comparison to the many years and extensive trials with the Malling rootstocks, they should be judged conservatively. On the other hand, the Malling-Merton clonal rootstocks have been subjected to unusually thorough critical, scientific testing. The initial reports are very encouraging. It is hoped that they will prove superior to some of the older rootstocks that have been grown for centuries. Some have been found at East Malling to be better anchored and more fruitful than the older rootstocks within the same vigor group. But only time will tell how they will do in orchards in different parts of the world.

From this breeding work, then, fifteen rootstocks were introduced for testing. They were numbered 101 to 115, inclusive. The EM XXV rootstock might well have been included with these, since it is from a cross made at the same time between Northern Spy and EM II. Yet it lacks the aphid resistance of the other MM rootstocks, and for this reason was omitted and placed in the EM series.

Following is a list of the MM apple rootstocks, giving a brief description, principal identifying characters, and general performance both in the nursery and up to fifteen years in the orchard. Only MM 104, 106, 109, and 111 have been released and recommended for trial; the others are included here in order to complete the record and satisfy inquiry.

MM 101. Parentage Northern Spy x EM I (John Innes Horticultural Institute). Vigorous; shoots stiff, erect; lenticels slightly raised, deep brownish-orange; leaves somewhat upturned, with large serrate stipules.

Fairly productive in stool beds, roots very well.

Makes semidwarf tree, not significantly different from EM VII, Spy, and MM 106, but better anchored than EM VII and Spy; does not sucker.

Not recommended.

MM 102. Parentage Northern Spy x EM I (East Malling Research Station). Vigorous; shoots stiff, erect; rather zigzagged; leaves downturned, with twisted tips; serrations downturned, often brown-tipped.

Fairly productive in stool beds; roots well.

Makes semidwarf to dwarf tree; smaller than EM VII; does not sucker.

Not recommended.

MM 103. Parentage Northern Spy x Ben Davis (East Malling Research Station). Vigorous; shoots rather spreading, flexible, slender, with long ridges and ashen-gray buds; leaves flat, often convex toward the base.

Fairly productive in stool beds, roots only fairly well.

Makes standard tree, as large as EM XVI (EM very vigorous); suckers somewhat.

Not recommended.

MM 104. Parentage EM II x Northern Spy (East Malling Research Station). Vigorous; shoots rather spreading; wood with numerous conspicuous lenticels and bright red buds; leaves much outturned; petiole tinged brilliant wine-red.

Very productive in stool beds; roots very well.

Makes semidwarf to semistandard tree (EM vigorous), larger than EM II, better anchored and heavier yielding, may not require staking, prefers well-drained soil, resistant to collar rot (*Phytophthora cactorum*); does not sucker, may be too vigorous on some soils, should not be

Fig. 48. **MM 104.** Shoots with numerous, conspicuous lenticels and bright red buds; upfolded concave leaves have upturned margins and brilliant wine-red, much out-turned leafstalks.

planted on poorly drained soils or soils liable to waterlogging (Figs. 48 and 55.)

MM 105. Parentage EM II x Northern Spy (John Innes Horticultural Institute). Not very vigorous; wood with faint pewter-like sheen and red buds; leaves wedge-shaped at base, flat, often unequal; tips much drawn out, very acute.

Very productive in stool bed, roots very well.

Makes semidwarf tree, slightly larger than EM II, suckers rather freely; not superior to EM VII.

MM 106. Parentage Northern Spy x EM I (East Malling Research Station). Vigorous; shoots somewhat spreading; wood covered with fairly profuse whitish pubescence, somewhat swollen at the node, indistinct lenticels and ashen-gray buds; leaves flattish, rather glossy.

Fairly prolific in stool bed, roots well, susceptible to mildew.

Makes dwarf to semidwarf tree (EM semidwarf) similar to EM VII in size and fruiting, smaller on light soils; nearly as early fruiting as EM IX with some varieties; best suited to strong varieties; fairly well anchored; does not sucker (Figs. 49 and 55).

Recommended for trial.

MM 107. Parentage Northern Spy x EM XV (East Malling Research Station). Vigorous; shoots somewhat spreading, stout, flexible; wood covered with fairly profuse whitish pubescence; leaves deep green, very glossy.

Fairly prolific in stool bed, roots well, sheds leaves late in the season; poor performance in nursery.

Makes semidwarf to dwarf tree (EM semidwarf), smaller than EM VII or nearly the same size, slightly larger than MM 106; suckers somewhat; variable in performance, with different scion varieties.

Discarded.

MM 108. Parentage Northern Spy x EM XV (East Malling Research Station). Weak; shoots stiff, wiry; wood with faint pewter-like sheen; leaves upfolded, broad.

Fig. 49. **MM 106.** Pubescent shoots, somewhat swollen at the node, with indistinct lenticels and ashen gray buds; flattish, horizontal, rather glossy leaves with long, much out-turned stalks.

Fairly productive in stool bed, roots well; poor general performance in nursery.

Makes semidwarf tree similar to EM II; does not sucker; variable in performance with some scion varieties.

Discarded.

MM 109. Parentage EM II x Northern Spy (John Innes Horticultural Institute). Not very vigorous; shoots stiff, erect; wood with slightly raised lenticels and reddish buds; leaves acutely serrate, deeply incised, margins raised and frilled in short waves (Figs. 50 and 55).

Fairly prolific in stool bed, roots well.

Makes semistandard tree (EM very vigorous); nearly as large as EM XVI, well anchored, but tends to lean after summer gales when bearing heavy crops under wet soil conditions; very productive; prefers well-drained soils; suited to dry soils; does not sucker.

Recommended for limited trial.

MM 110. Parentage Northern Spy x EM I (East Malling Research Station). Vigorous; shoots spreading, stiff, with fairly long internodes, zigzagged; leaves finely and acutely serrate, flat or slightly convex, pointing downward; stipules large, spreading.

Very prolific in stool bed, roots very well.

Makes semidwarf tree (EM semidwarf), similar to EM II; does not sucker.

Not recommended.

MM 111. Parentage Northern Spy x Merton 793 (John Innes Horticultural Institute). Vigorous; shoots erect, stiff, stout, somewhat swollen at the nodes; leaves coarsely serrate, tipped with red, often faintly lobed, upcupped.

Productive in stool bed, more profuse than EM II (Figs. 51 and 55).

Makes semidwarf tree, similar to EM II but fruits earlier and more heavily; smaller than MM 109; good for Golden Delicious; suckers only slightly; has withstood drought better than EM II.

Recommended for trial.

Fig. 50. **MM 109.** Shoots with fairly numerous, slightly raised lenticels and reddish buds; upturned and upcupped leaves with acutely serrate margins, raised and frilled in short waves, long upturned stalks, and small stipules.

MM 112. Parentage Northern Spy x Winter Majetin (East Malling Research Station). Vigorous; shoots spreading, stoutish, flexible, leaves flat, abruptly acuminate, downturned and twisted; stipules large.

Fairly productive in stool bed, roots fairly well.

Makes semidwarf tree, similar to EM II; does not sucker.

Not recommended.

MM 113. Parentage Northern Spy x EM XII (East Malling Research Station). Very vigorous; shoots spreading, slender; somewhat swollen at nodes; leaves finely rounded, almost crenate, upcupped.

Not productive, roots poorly.

Makes standard tree (EM very vigorous), larger than EM XVI; does not sucker.

Not recommended.

MM 114. Parentage Northern Spy x EM XII (East Malling Research Station). Fairly vigorous; shoots erect, flexible, distinctly ridged, with short internodes; leaves downturned, curved over; petiole distinctly kinked at base.

Not productive, roots poorly.

Makes standard tree (EM very vigorous), larger than EM XVI; suckers badly.

Not recommended.

MM 115. Parentage Northern Spy x Ben Davis (East Malling Research Station). Very vigorous; shoots stout, flexible, with long internodes, much swollen at the nodes; leaves large, oblong, upturned, shortly waved, giving a ragged appearance.

Fairly productive, roots well.

Makes standard tree (EM very vigorous), larger than EM XVI but fruits earlier and more heavily; does not sucker.

Not recommended.

Fig. 51. **MM 111.** Stout, stiff, erect shoots, considerably swollen at the nodes with rather ragged, ashen gray buds; upcupped leaves with raised, wavy margins, coarse serrations often tipped with red, and stalks often lumpy at the base.

MISCELLANEOUS VEGETATIVELY PROPAGATED APPLE ROOTSTOCKS

Northern Spy

The Northern Spy variety (cultivar) of apple has long been known for its resistance to the woolly aphid (*Eriosoma lanigerum*). This variety originated at East Bloomfield in western New York in 1828, and is prized for the excellence and long-keeping quality of its fruit.

In Australia and New Zealand, where this aphid is serious in its attack upon the roots of apple trees, Northern Spy rootstocks have been rather generally used. The variety propagates with difficulty and is not a good nursery plant. Nevertheless, in the absence of something better it is still recommended. Introduction of a chalcid wasp (*Aphelinus mali*), which is parasitic upon the woolly aphid, and the introduction of new spray materials have greatly reduced the ravages of this pest and have made resistance less important than it once was.

In addition to its resistance to woolly aphid, the Northern Spy when used as a rootstock makes a semidwarf tree not unlike EM V. It is a poor rooter in the nursery. The stock requires very good soil conditions, is difficult to replant, often develops a one-sided root system, is shallow-rooted and short-lived.

The Merton 778, 779, 789, and 793 Rootstocks

Prior to the introduction of the Malling-Merton series of rootstocks discussed in the preceding paragraphs, four apple rootstocks were produced from crosses between EM II and Northen Spy by the John Innes Horticultural Institute at Merton, England. They were designated Merton 778, 779, 789, and 793, and were introduced into New Zealand in 1941 for preliminary trial. There they were compared principally to rootstocks of Northern Spy, EM XII, and EM XVI, as possible improvements over these well-known rootstocks. Special attention was paid to resistance to woolly aphid and to collar rot (*Phytophthera cactorum*), which were serious problems. The originators of these rootstocks dis-

Fig. 52. **(Left to right)** EM I, II, IV.

carded them for use in England as comparable to EM II but not equal to the new MM series. Observations reported in 1958 from New Zealand are, however, of some interest.

Merton 778 was found not appreciably different in performance from the common, widely used Northern Spy, and was discarded.

Merton 789 was found to be an excellent rootstock, producing trees slightly smaller than standard (EM vigorous) and about the size of EM XVI. But the stock proved susceptible to collar rot and was therefore discarded.

Merton 779 was found resistant to collar rot, but less so than 793. Further, it possessed brittle wood, and was less long-lived in the stool bed, although it propagates very readily. It produced a vigorous tree somewhat smaller than standard and about the same size as EM XVI.

Merton 793 proved an excellent propagating stock, producing large numbers of uniform, well-rooted shoots. It appeared to produce a semi-standard to semidwarf tree (EM vigorous), smaller than EM XVI and larger than Northern Spy, but more resistant to woolly aphid, adapted to a wider range of soil, and inclined to earlier and heavier cropping than that stock. It was also resistant to collar rot and has appeared worthy of further testing.

EM 3426, 3428, 3430, 3431, 3436, 3438, and 3461

From 1008 seedlings produced by H. M. Tydeman of the East Malling Research Station in 1929, six have been carried further for trial, and a seventh (3461) has been mentioned in the literature as having special merit for trees in pots.

EM 3426 (parentage EM VII x IX): medium ovate leaves, rather flat, slightly crinkled, glossy, with drawn-out, downward tips; produces tree smaller than EM IX (EM very dwarf).

EM 3428 (parentage EM IX x VII): bright grass-green leaves, flattish, upturned with much upturned petioles and small, faintly serrate, crescent-shaped stipules; produces tree similar to EM II (EM vigorous).

EM 3430 (parentage EM XII x IX): dark green, glossy, horizontal leaves; few, fairly large reddish-orange lenticels; produces tree as large or larger than EM XVI (very vigorous).

Fig. 53. (**Lower left, left to right**) **EM V, VII, IX.**

EM 3431 (parentage EM XIII x IX): smallish, double-upfolded and upturned leaves; produces tree smaller than EM IX (EM very dwarf); useful in experimental work and in fruit breeding where early fruiting is desired.

EM 3436 (parentage EM XVI x IX), now named EM 26, intermediate between EM IX and EM VII.

EM 3438 (parentage EM IX x XII): large broad leaves with frilled edges, and large broad stipules; produces tree nearly as large as EM XVI, better fruiting (EM very vigorous).

EM 3461: mentioned as being superior for pot plants.

Alnarp 2(A2)

This apple rootstock was selected in the nurseries of the Alnarp Gardens in southern Sweden in 1920 from among mixed imported rootstock material. In 1927 it was taken to the nurseries of the State Horticultural Research Department at Alnarp for propagation and testing.

It is a strong growing stock, comparable to EM XVI in vigor. Shoots are upright, with zigzag tendency. Bark is reddish-brown or grayish-brown with yellow, round, conspicuous lenticels. Leaves are oval, broad, dark green, falling in early autumn.

One of the features of A2 is the ease with which it propagates. It propagates readily in stool beds, from hardwood cuttings, and from leaf-bud cuttings in a tight propagation frame or under mist. It has been found hardy to winter cold and compatible with all varieties so far worked upon it.

It makes a tree that is semistandard (EM vigorous) in size, slightly smaller than EM XVI and not unlike EM XIII. It has proved to be a heavier and earlier bearer than either EM XIII or EM XVI, being not unlike EM IV in fruiting but firmly anchored. It would seem to possess the same characteristics of early fruiting and vigor as does EM XXV.

This rootstock has proved adapted to conditions in Sweden but has not been tested extensively outside that country.

Malus Robusta 5

The Malling series of rootstocks are generally unsuited to propagation in stool beds where winter cold is severe, as in all but the more

Fig. 54. (Below, left to right) EM XII, XIII, XVI.

favored sections of Canada. To meet this need, a series of rootstocks has been developed in Canada from the Cherry Crab apple (*Malus baccata* x *M. prunifolia*). Among these is a clone designated Malus Robusta 5 or Robusta 5, originated by the Canadian Department of Agriculture in 1928 from seed obtained from Siberia in 1927 through the courtesy of the Arnold Arboretum. It was a selection from a number of seedlings. It is resistant to blight and hardy to winter cold as far north as Ottawa, Canada.

It is a vigorous grower, and sends up a great number of medium-size, uniform shoots in the stool bed, which root readily. It has yielded 70,000 rooted shoots per acre. In addition, it has been rooted commercially by softwood cuttings taken from plants in the juvenile condition. Plants in the adult condition do not provide shoots that root well. Commercial varieties of apples have been worked upon it successfully with no incompatibilities reported. The resulting trees are only somewhat smaller than standard if at all, perhaps classified as semistandard (EM vigorous).

Where winter hardiness is a problem and where a uniform, clonal rootstock is desired that makes a slightly smaller than standard tree, Robusta 5 merits consideration.

K-41

K-41 is a rootstock selected from among seedlings grown at Kansas State University, under conditions of low winter temperatures and high summer temperatures. It may be propagated vegetatively by mound layering. It has also been used as a dwarfing intermediate stempiece between a vigorous seedling rootstock and a scion variety. It seems compatible with most varieties and is hardier than either EM VIII or EM IX when used in this way.

Praecox

To complete the record, mention should be made of Praecox, which has also been called *Pyrus malus praecox*. It is mentioned by George Bunyard, prominent nurseryman and pomologist of Maidstone, England, as the form of dwarf apple discovered by Pallas in the Volga district and in Samara and Tainim. It is extremely dwarf in habit, perhaps

Fig. 55. (Left, top to bottom) MM 104, MM 106, MM 109, MM 111.

the most dwarf of all, with small oval leaves that are densely hairy. This may have been the "Pigmy" Paradise of Thomas Rivers. It was used by English nurserymen during the middle of the last century but was considered by George Bunyard in 1919 to have been entirely abandoned by that time. There is still another rootstock, also called Praecox, which has been observed as similar in behavior to EM II.

DWARFING ROOTSTOCKS PROPAGATED FROM SEED

All the apple rootstocks discussed so far in this chapter are clones and are propagated vegetatively, as by layers and cuttings. This leads, of course, to great uniformity in performance, since the tree is accordingly composed of a clonal scion variety and a clonal rootstock. On the other hand, most apple trees grown in America are worked onto seedling rootstocks raised from seed.

This introduces a factor of variability, since seedlings from seed are not identical. Some sources and some lots of apple seed, however, are notable for the relative uniformity in the seedlings they produce. That is, they tend to stamp their offspring with certain dominant characteristics. The common wild French Crab of Normandy in France is one of these. The Delicious, Rome, and McIntosh varieties and the Whitney Crab behave similarly. And so there has always been hope that some seed parent might be found that would provide uniform rootstocks with dwarfing properties.

The results have not been very successful. The closest seem to be seedlings from some of the various species of apple, such as the Siberian Crab apple (*Malus baccata*). With some varieties, the general tendency is to reduce the size of tree slightly—varying from semistandard to semidwarf, but they are very variable. Nothing has as yet been developed that is generally reliable and useful. And the Sargent Crab apple (*M. sargenti*), which it has been thought would be dwarfing, has been shown in one trial to be vigorous.

In limited tests, K. D. Brase and R. D. Way of the New York State Agricultural Experiment Station have used the Manchurian Crab (*Malus baccata* var. *mandshurica*) with some degree of success. The

Delicious and Cortland varieties of apple on this rootstock developed into trees about the size of EM II or EM VII. After thirteen years in the orchard these size differences were still apparent. Yields of fruit were considered less than on EM II and EM VII with which they were compared. Two other varieties, McIntosh and Virginia Crab, made weak unions and seemed not congenial.

An interesting seed source that has been from time to time suggested is that from various species or forms of apple that develop apomictic seed from apogamic embryos. This means the embryos are not formed by union of gametes from the male and female parents, but arise from the "body tissue" of the plant and are therefore identical to the parent plant, just as are cuttings and layers. This process is more fully discussed in connection with paragraphs on citrus rootstocks. In the case of citrus, apomictic production of seeds and seedlings by proliferation of the nucellar tissue is very high—frequently 90 per cent or more. The seedlings that may be produced by normal processes of fertilization are quite unlike the uniform nucellar seed, and are easily observed and rogued out. This results in what is virtually a clonal line of rootstocks, and this is what has been sought in the apple.

Among species experimented with in this regard are *Malus hupehensis*, *M. toringoides*, and Sikkim Crab (*M. sikkimensis*). Unfortunately the seed content of the fruit is very low, as is the viability of the seed. If the mother tree is near cultivated varieties of apple, the number of embryos that develop from normal processes of cross-fertilization is increased, but the resulting seedlings are variable.

Seedlings of these three species have been tried as rootstocks for several varieties of English apples at the Long Ashton Research Station in England. *M. hupehensis* was entirely incompatible with the varieties used. The other two species were compatible. The performance of the trees in the nursery and in the orchard was considered satisfactory, but the relative vigor of the varieties tested on these two rootstocks varied appreciably, and could be neither classified nor predicted.

In general, however, the trees were smaller than on standard seedling rootstocks, being not unlike trees on the EM II or EM XXV. They were vigorous in the nursery, accepted a wide range of varieties without signs of incompatibilities, and were well rooted and early-cropping in the orchard. The variation between trees was more than on some clonal rootstocks but no more so than on some others, such as EM XVI.

K. D. Brase and R. D. Way found that several varieties of apples succeeded on apomictic seedlings of *Malus sikkimensis* and attained the size of trees on the EM VII rootstock (semidwarf). On the other hand, the performance of other varieties was inferior, suggesting the type of incompatibility often associated with the presence of a virus disorder.

Nevertheless the possibility of developing dwarfing lines of uniform rootstocks from such material is entirely possible and likely in time.

OTHER MISCELLANEOUS ROOTSTOCKS FOR THE APPLE

A number of other materials may be used upon which to work the apple, and which are dwarfing to varying degree. Among them are the chokeberry (*Aronia* spp.), the pear (*Pyrus* spp.), the serviceberry (*Amelanchier* spp.), the hawthorn (*Crataegus* spp.), and the mountain ash (*Sorbus* spp.). Such combinations are of value only as novelties.

VIRUSES IN CLONAL APPLE ROOTSTOCKS

The virus problem with clonal rootstocks is potentially very serious. It is under close study in a number of research institutions throughout the world. Attempts are being made to identify viruses and to introduce virus-free materials for local areas. Orchardists should be alert to the possibilities and should take advantage of virus-free rootstocks whenever possible.

The East Malling Research Station has taken active steps to ensure rootstock materials free from virus diseases, principally rubbery wood, mosaic, and chat fruit. The following list, prepared by T. N. Hoblyn, formerly Assistant Director of the East Malling Research Station, indicates the progress that has been made:

EM I   Clones available free of rubbery wood and mosaic, but not free from chat fruit.

EM II   Clone A free of rubbery wood, mosaic, and chat fruit: all sub-clones infected with mild form of latent virus.

EM III   Clones free of rubbery wood and mosaic.

EM IV   Clone B free of rubbery wood, mosaic, and probably chat fruit.

EM V   Clone A free of rubbery wood and mosaic.

EM VI   No virus-free clone available.

EM VII   Clone free of rubbery wood, mosaic, and chat fruit; all sub-clones infected with latent virus.

EM VIII   Clones free of rubbery wood and mosaic only.

EM IX   Clone A free of rubbery wood, mosaic, and chat fruit; but infected with latent virus.

EM X   No virus-free clone available.

EM XI   No virus-free clone available.

EM XII   Clone free of rubbery wood and mosaic, clone A under test for chat fruit; free of latent virus.

EM XIII   Clone free of rubbery wood and mosaic.

EM XIV   No virus-free clone available.

EM XV   No virus-free clone available.

EM XVI   Clones free of rubbery wood, mosaic, and chat fruit.

EM XVII   Clones free of rubbery wood and mosaic.

EM XVIII   Clones free of rubbery wood and mosaic.

EM XIX   One clone free of rubbery wood and mosaic.

EM XX   Clones free of rubbery wood and mosaic.

EM XXI   One clone free of rubbery wood and mosaic.

EM XXII   No virus-free clone available.

EM XXIII   One clone free of rubbery wood and mosaic.

EM XXIV   One clone free of rubbery wood and mosaic.

EM XXV   Clones free of rubbery wood and mosaic; under test for chat fruit.

EM 26   Clones free of rubbery wood and mosaic; under test for chat fruit; free of latent virus.

EM Crab C   Clones free of rubbery wood and mosaic.

MM 101, 102, 103, 105, 110, 112, 113, 114, and 115   Clones free of rubbery wood and mosaic.

MM 104, 106, 109, 111   Clones free of rubbery wood and mosaic; under test for chat fruit; free of latent virus.

MM 106   Clones free of rubbery wood and mosaic; under test for chat fruit; clones 1, 2, 4 and 5 free of latent virus; clone 3 infected with latent virus and discarded.

MM 109 Clones free of rubbery wood and mosaic; under test for chat fruit; infected with latent virus.

MM 111 Clones free of rubbery wood and mosaic; under test for chat fruit; free of latent virus.

EM 341, 3414, 3426, 3428, 3430, 3431, and 3438 Clones free of rubbery wood and mosaic; under test for chat fruit.

M 778, 779, 789, and 793 Clones free of rubbery wood and mosaic.

COMPATIBILITY OF APPLE ROOTSTOCKS AND SCION VARIETIES

The apple rootstocks found in commerce have a remarkable record of congeniality with a wide range of commercial varieties worked upon them. Over fifty varieties of apples, including the ones grown commercially in the United States and Canada, have been found to unite satisfactorily with the major apple rootstocks commerce.

However, difficulties have appeared that have been associated with virus disorders, as discussed in the chapter on stock-scion relations. The introduction of viruses into either rootstock or scion by budding and grafting is a very real possibility. The Virginia Crab and the U.S.D.A. 227 rootstocks are examples. Combinations of these rootstocks with budwood from one source of a given variety have been uncongenial, whereas combinations with budwood from another source of the same variety have been congenial.

Finally, the Bechtels' Flowering Crab apple has proved incompatible with apple EM IX. But the general concept with apple stock-scion combinations is one of congeniality in the great majority of instances.

SUMMARY AND CONCLUSION

For very dwarf apple trees there is no rootstock commercially available that rivals EM IX. Trees on EM 26 (3436) are slightly larger than EM IX, better anchored, and in many ways superior. EM VII is a cosmopolitan rootstock, widely adapted, and very useful for a semidwarf tree. MM 106 is similar to EM VII, with which it competes. EM II is the long-time favorite for semidwarf to semistandard (EM vigorous) trees. EM IV and XI are still much used on the Continent. EM IV is a heavy cropper but has a high top/root ratio and is poorly anchored. EM XI is very resistant to winter cold and useful where winter injury is a problem. Both MM 104 and MM 111 are competitiors to EM II, being better anchored and heavier fruiting, with MM 111 producing trees the same size as EM II, and MM 104 slightly larger.

EM XVI is widely adapted as a semistandard to standard (EM very vigorous) rootstock. Where collar rot is a problem, EM XXV and MM 109 are superior to EM XVI, which is very susceptible to this disease. Crab C has many desirable characteristics for standard trees but is not easily propagated.

As regards virus, commercial sources of EM VII, II, and XVI are relatively free. Many sources of EM IX, I, and XI carry the rubbery wood virus. The MM series is virtually free of virus.

Cultural methods are important considerations in the selection of a rootstock. Under the modern system of light pruning and sod-mulch or mulch, trees may be brought into production earlier than under the older system of heavy pruning and clean cultivation.

Under these conditions, the more vigorous rootstocks, such as EM XXV, MM 104, MM 109, and MM 111, may be found advantageous. They will quickly develop well-anchored trees with good framework, capable of carrying heavy crops, and may possibly need less nitrogenous fertilizer than will weaker rootstocks, though this is a debatable point.

8

DWARFING APPLE INTERSTOCKS

A FRUIT tree can also be dwarfed by the interposition of an appropriate piece of stem between the rootstock and the scion variety. This piece of stem is often called an "interstock" or an "intermediate stempiece." Such a tree is composed of three separate parts—a rootstock, a stempiece, and the scion variety. It is said to have been "double-worked."

Although this method is often thought of as a development of the 1930's and 1940's, it is really very old. The use of an interstock or intermediate stempiece for dwarfing was described by John Rea in *Flora*, published in England in 1665, as follows: "I have found out another expedient to help them forward [promote earlier fruiting], that is by grafting the Cyen of the Paradise apple in a Crab, or other apple-stock, close to the ground, with one graft, and when that is grown to the bigness of a finger, graft thercon about eight inches higher, the fruit desired, which will stop the luxurious growth of the tree, almost as well as if it had been immediately grafted on the fore-mentioned layers, and will cause the trees to bear sooner, more and better fruits."

In 1830, a "new" method of grafting was described by M. Martin de Bressolles of France in the *Bon Jardinier* and reported in *New England Farmer* in the same year. It was essentially the method of John Rea; namely, the insertion of a stempiece of Paradise apple between a vigorous seedling root and the chosen scion variety.

Numerous other references appear in the literature pertaining to the

155

effect of an interstock on the performance of a fruit tree. The double-working of pears on the quince is an example, but is discussed in the section dealing with pears and need not be reviewed further here. John Parkinson in *Paradisi in Sole Paradisus Terrestris*, published in 1629, tells of the improvement of nectarines by working upon apricot inter-stocks that have in turn been worked upon plum rootstocks. More recently, N. H. Grubb, of the East Malling Research Station in England, noted in 1923 that an intermediate stempiece of pear placed between a quince rootstock and a pear scion not only influenced compatibility with the quince but also affected the vigor of the scion and the time of its coming into bearing. R. C. Knight reported in 1927 and 1934 similarly for apples.

INTERSTOCK TESTS IN NEW YORK STATE

In 1933 Tukey and Brase of the New York State Agricultural Experiment Station at Geneva, New York, published results of their studies on the effect of scion and of an intermediate stempiece on the development of roots of young apple trees. The work was begun on the basic problem of stock/scion relationship. It was known that the rootstock affected the scion, as in dwarfing rootstocks. The question now arose as to whether the scion, or even an intermediate stempiece, could in turn affect the root of the tree. They proceeded to make 1,800 grafts, involving 43 combinations of single- and double-worked apple trees, employing 12 varieties of apple and EM IX and XIII on French Crab seedling roots (Fig. 56). Graft combinations were made, for example, involving French Crab as the rootstock, EM IX as a 2-inch intermediate stempiece and Jonathan as the scion variety (Jonathan/EM IX/French Crab seedling). The reciprocal was also made, in which a 2-inch stempiece of Jonathan was interposed between the French Crab rootstock and EM IX used as the scion (EM IX/Jonathan/French Crab seedling).

The results were conclusive: "The question is no longer which *one* part dominates the entire plant (that is, whether root, stem-piece, or scion), since *all* have been shown to have marked influence at one time or another. The rootstock may be the limiting factor one time, the union (stem-piece) another, and the top scion another."

The studies brought out still another interesting fact: namely, "Where a dwarf stock has been used as either rootstock, intermediate stem-piece or top cion, the effect has been to dwarf the entire plant." This was the beginning of renewed interest in an old method that has sometimes been erroneously called "new."

In the light of this possibility of producing a dwarfed fruit tree by the use of a dwarfing interstock, double-worked trees of Baldwin, Delicious, Early McIntosh, and McIntosh were planted in the orchard, in which a 3-inch interstock of EM IX was inserted as an intermediate stempiece.

The results showed that a 3-inch stempiece of EM IX very definitely dwarfed the entire tree. For example, trees of McIntosh/EM IX/French Crab were 67 per cent the size of McIntosh/French Crab, and they came into fruiting a year earlier (Fig. 57). On the other hand, McIntosh trees budded directly on EM IX rootstocks were only 51 per cent the size of standard trees, and began to fruit the second year. In other words, although the EM IX stempiece did dwarf the tree somewhat, the effect was nowhere near so great as when the EM IX was used as the rootstock (Fig. 58).

They concluded that in the more favored horticultural sections, when high winds and severe winter cold were not limiting factors, trees single-worked upon the EM IX rootstock gave more uniform and more reliable performance, coupled with a smaller tree and earlier fruiting, than did double-worked trees involving EM IX as an interstock. There the matter rested.

Fig. 56. Double working by grafting, employing a rootpiece, an intermediate stempiece, and the scion variety: (A) Simple whip graft of scion (above) and rootpiece (below), before and after tying, for comparison with double grafting. (B) Double whip graft of rootpiece (below), intermediate stempiece (middle), and scion variety (top). (C) Longitudinal section of double-grafted, two-year-old tree showing unions of scion, intermediate stempiece, and rootpiece. (D) Portion of two-year-old, double-grafted tree, showing the rootstock (below), intermediate stempiece (middle), and scion variety (top), similar to section shown in (C).

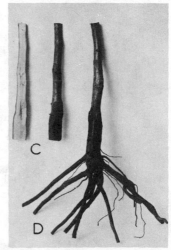

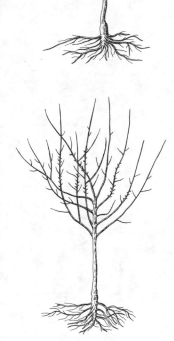

INTERSTOCK TESTS WITH "CLARK DWARF" IN IOWA

In 1943, Professor T. J. Maney of the Iowa State University at Ames, Iowa, reported before a meeting of the Iowa State Horticultural Society on some interesting results he had had with intermediate stempieces in double-working experiments with apples. This is the beginning of interest in the so-called "Clark Dwarf" and the "Clark Dwarf Method" of propagating dwarfed fruit trees. Because of the historical significance and because of the relative inaccessibility of the original publication, a full review of the development seems worthwhile.

Dwarf trees had not been recommended widely in Iowa because of the severity of the winters there and because the dwarfing rootstocks used "have not been reliably hardy." Maney was referring to the Malling rootstocks (EM I to XVI inclusive), and continued: "Experience indicates that the English stocks have a hardiness equal to some of the Eastern apple varieties which are not well adapted to our climatic conditions. It is questionable how these stocks might react to our test winters where only reliably hardy varieties come through without injury."

He then discussed hardy apple rootstocks, and said that chief interest in apple stocks had been centered on vigorous, hardy standard stocks, such as Virginia Crab and Hibernal, and that studies on dwarf stocks had received only minor attention.

About 1924 he acquired an "interesting hardy dwarf stock which was introduced to us by H. Walton Clark, a biologist, who was connected with the United States Bureau of Fisheries Biological Station at

Fig. 57. **Six-year-old double-worked tree of McIntosh on French Crab seedling roots, showing comparative dwarfing effect of dwarfing intermediate stempiece: (A) (Upper left) With intermediate stempiece of McIntosh scionwood (McIntosh/McIntosh/French Crab seedling). (B) (Lower left) With intermediate stempiece of EM IX scionwood (McIntosh/EM IX/French Crab seedling). Compare with Fig. 58.**

Fig. 58. **(Below) McIntosh tree directly on EM IX rootstock (McIntosh/EM IX), decidedly smaller and earlier fruiting than double-worked tree with intermediate stempiece of EM IX. Compare with Fig. 57.**

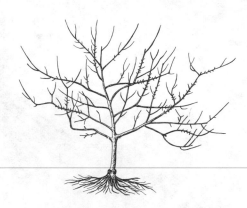

Fairport, Iowa. Mr. Clark was an ardent amateur horticulturist and he had discovered a dwarf type of tree growing in the garden of an old lady living in Muscatine, Iowa. The tree at the time was probably 20 to 25 years old and during its life had experienced temperatures as low as −25° F without showing any signs of injury. The history of the variety as given by the lady was that her son was a sailor and that he had brought the stock back from South America. The characteristics of the leaves and type of growth indicate that it undoubtedly is a type which might have originated as a seedling from the original Paradise, but it is distinct from the new standardized English stocks.

"This stock was carried along in the nursery for a number of years and was found to be a very slow-grower. Finally we set some buds on a line of seedlings which was derived from a *Pyrus baccata* cross. The buds seemed to be well adapted to these particular seedlings and made excellent growth. The year after budding we set into the dwarf shoots buds of a number of standard varieties, so when the shoots of the dwarf type were cut back to the bud, a section of the dwarf, to the extent of two or three inches, remained in the stem of the tree. In this make-up we had a strong vigorous root, a stem-piece insert of the dwarf and the top of the standard variety. . . .

"We later found that the tops were dwarfed and the trees fruited quite regularly through several seasons. . . . With this technique it may be possible to produce hardy, dwarfed trees with good root foundations at a cost not excessive to the nurseryman. Apparently the intermediate section in the trunk exerts a full dwarfing effect."

At ten years of age the trees were reported as not more than 5 to 7 feet in height. And they survived without injury the severe drought periods of 1934 and 1936, the low temperature of −30 degrees F. in 1935–1936, and the unseasonal blizzard of November 11–15, 1940, when 90 per cent of the apple trees in Iowa were either destroyed or severely injured.

The next step was the introduction of this method into the apple industry. This was done by the Stark Brothers Nursery and Orchard Company of Louisiana, Missouri. Mr. Paul Stark, Sr., a member of the firm, was in close touch with Professor Maney and his work. He had seen the studies with intermediate stocks being carried on at the New York State Agricultural Experiment Station at Geneva, New York, by Tukey and Brase. He was sufficiently close to the winter-hardiness problem to grasp the significance of Professor Maney's findings. Finally, the double-working operation involving the grafting of dormant wood was well suited to the operations of his organization, where bench grafting was a common practice.

It may be observed again in passing that the practice of double-grafting in the production of nursery trees is especially suited to regions of relatively long growing season, strong soils, and warm summer temperatures, such as the Ozark regions of Missouri and Kansas. Trees that are budded and grown under such conditions often grow too large to be easily handled and transplanted.

On the other hand, bench grafting does not result in sufficiently large trees in regions of short growing seasons and cool summers, as in western New York and New England. Budding is the preferred practice there.

At all events, the point is that a set of circumstances arose that fitted together nicely to cause the double-working of fruit trees with Clark Dwarf to assume considerable practical value.

The Clark Dwarf makes a tree that is semidwarf to dwarf, depending upon the variety and several other factors. Unfortunately, some of the details had not yet been thoroughly worked out when the Clark Dwarf fruit trees were introduced into the trade, and some unpredictable complications arose that slowed the process. Since hardiness to winter cold was a problem in the Central States area, recourse had been taken to using hardy rootstock material in the double-grafts. Virginia Crab was the prime favorite. It was grafted onto a seedling root and planted deep to induce scion-rooting of the Virginia Crab. Then onto this hardy body stock the Clark Dwarf was worked, followed in turn by the desired scion variety.

But the hardy body stock often failed to scion-root, leading to variation in tree performance in the orchard. Further, the three grafting operations (scion variety/Clark Dwarf hardy body stock/seedling nurse root) were complicating. They accentuated any difficulties in grafting technique and added to variability in orchard performance. In addition, the length of the intermediate stempiece was found to alter the dwarfing effect, and this length was not always constant. Generally speaking, the longer the stempiece of a dwarfing interstock, the greater the dwarfing effect. And finally, some Virginia Crab was found to be very susceptible to the stem-pitting virus, often present in the scion variety worked upon it, so that congeniality and performance of the scion was adversely affected. The result was lack of uniformity and some uncertainty.

As these problems are surmounted, the double-working technique may be expected to find a useful place in the propagation of dwarf fruit trees. This is especially so in less-favored horticultural sections, where winter cold is severe, where strong winds may demand good anchorage, and where summer heat and drought may find some of the European rootstocks, such as the Malling and Malling-Merton stocks, less at home.

Yet the Clark Dwarf when used as an intermediate may not always prove as winter-hardy as the Clark Dwarf tree itself on its own roots. Reports of winter killing of the interstem have been received from South Dakota. Similarly the writer has observed severe winter injury on young trees at the point of union of the interstem. In this instance, a 12-inch interstem of Clark Dwarf was used, both unions being well above the ground. Minimum winter temperature of —25 degrees F. was experienced. Severe blackening and winter injury of the tissues occurred both at the region of union between the scion and the interstem as well as between the rootstock and the interstem. Neither the scion (Yellow Transparent) nor the rootstock (Virginia Crab) was severely injured; and the Clark Dwarf wood itself was not severely injured. But the damage to the region of the unions was so severe that the trees died the following summer, although they leafed out in spring. The behavior resembled immaturity at the union such as is often observed in stock-scion combinations.

Identity of the Clark Dwarf

There had been some suspicion that the Clark Dwarf might be either an old European stock material or derived from such a source. In 1953 Brase confirmed this suspicion by studies he made, critically comparing Clark Dwarf with EM VIII. He fruited both rootstock materials and found them to be seemingly identical in leaf, blossom, and fruit, as well as in general growth and fruiting habits.

Some doubt has been cast upon this by the fact that EM VIII was not introduced into this country as such until 1922, whereas the original Clark Dwarf tree was twenty to twenty-five years old when it was discovered in 1924. But EM VIII has sometimes been considered the original "Paradise" stock, and it is not at all unlikely that it may have reached South America from Europe many years ago, and long before the Malling rootstocks were standardized and introduced by number. Further, EM VIII was not a recommended rootstock, and it was present in America under that number to a very limited degree only. Opportunity for comparison was, therefore, not often at hand, and any similarities could have been easily overlooked.

Nevertheless there is always the shadow of a doubt when trying to identify material as complex as a plant, involving a multitude of fine differences. Perhaps the Clark Dwarf may eventually be found to differ in some degree from EM VIII, but at least the two are so similar that they can readily pass one for the other.

OTHER INTERSTOCK COMBINATIONS

Materials other than Clark Dwarf may be used as interstock material for dwarfing. The greatest degree of dwarfing has occurred with EM VIII, followed closely by EM IX. EM VII makes a somewhat less dwarfed tree, approaching semistandard in size, and larger than the EM VII used directly as a rootstock. A. N. Roberts of Oregon reports success with a 12-inch stempiece of Clark Dwarf worked upon the EM XVI rootstock. Such combinations have produced trees nearly as dwarf as those propagated directly on EM IX as a rootstock. The trees have seemed much better anchored without breakage than trees worked directly on EM IX rootstocks. An intermediate stempiece shorter than 12 inches did not give as great a degree of dwarfing.

EM I and EM II have been tried as interstocks. Again, preliminary observations are that the degree of dwarfing is not as great as when these two materials are used directly as rootstocks.

K 41 is a rootstock developed at Kansas State University from a series of seedling rootstocks. When used as an intermediate stempiece it has dwarfed the tree somewhat and induced somewhat early bearing. It overgrows both the stock and the scion but seems compatible with most varieties. It is hardier than either EM VIII or EM IX, which are often used in this way. Trees are substantial and well anchored.

Robusta 5, developed in Canada as a hardy rootstock, may be used as an intermediate stempiece. It has brought McIntosh into early bearing. It is hardy and it will tolerate wet, heavy soils.

Ottawa 524, also developed in Canada, induces some degree of dwarfing when used as an interstock but is very susceptible to the stem-pitting virus. In fact, it is so susceptible that it is used as an indicator plant for the virus.

Robin is a variety of apple originated in Canada by W. T. Saunders, of the Dominion Experimental Farm, from a cross between *M. baccata* and Simbirsh No. 9 apple made in 1904 and named in 1911. Robin is resistant to fire blight and is hardy to extremely low temperatures. It does not show stem pitting, but when used as an interstock it under-grows both the stock and the scion so severely that faulty unions and severe constrictions eventually develop. In early years it is only slightly dwarfing.

SUMMARY AND CONCLUSIONS

Double-worked trees employing a dwarfing intermediate stempiece have proved useful in forming substantial, early fruiting, well-anchored trees with strong scaffold branches. EM XVI has proved a desirable root-stock upon which EM VIII, A 2, and K 47 have been successfully worked as intermediates to produce trees of this type. For very dwarf trees, the Clark Dwarf and the EM IX have been similarly used. Dwarf-ing is never as great when the dwarfing material is used as an inter-mediate stempiece as when it is used directly as a rootstock with its own root system; and short sections of intermediate stempiece are less dwarfing than longer ones.

9

DWARFING ROOTSTOCKS
FOR THE CHERRY

THERE is no really good dwarfing rootstock for the cherry in common use. Of the several rootstocks available, some, to be sure, produce a smaller-growing tree than others, yet none produces a tree so characteristically dwarf as does the EM IX apple rootstock for the apple or the quince rootstock for the pear, unless it be by unhealthy stunting.

Most sour cherries and some sweet cherries are in themselves fairly early fruiting. In commercial orchards in America, trees have been kept fairly low by proper heading and good management. Accordingly, the need for a dwarfing rootstock has not been so urgent as with the apple and the pear. Further, the cherry is a plant that is exacting as to soil and climate, and does not thrive under any but the most favorable conditions. Incompatibilities with rootstocks, an unsuited environment, and insect and disease attacks all seem accentuated in the cherry. A rootstock that dwarfs the plant appreciably may result in its failure under other than a very favorable environment.

The Mazzard and Mahaleb Rootstocks

Rootstocks commonly used for the cherry are seedlings of the mazzard, or wild sweet cherry (*Prunus avium*), and the mahaleb, or perfume cherry, of Europe (*P. mahaleb*).

164

The mazzard rootstock is characterized by large leaves, by fruit borne singly or in twos or threes, and by a strong, vigorous, spreading root system that carries many fibrous roots and is dark brown in color. The bark of both the root and of the stem has a distinctly bitter taste, and it gives a black color reaction when either stained with or immersed in an iron salt solution such as ferric chloride. It is a vigorous grower and produces large long-lived trees with both the sweet and most of the sour varieties worked upon it. Selections of clonal material of the sweet cherry have been made by the East Malling Research Station in England. Certain of these (F 5/4 and F 1/1) are somewhat dwarfing in comparison with the popular virus-free F 12/1, suggesting the possibility of securing improved clonal dwarfing rootstocks for the cherry through selections of this kind. Yet the degree of dwarfing so far revealed is comparatively slight; the most resulting only in a reduction to about two-thirds the size of the most vigorous rootstocks.

The mahaleb rootstock, which is quite distinct but, nevertheless frequently confused with the mazzard, is characterized by small, glossy leaves, by fruits that are borne in "racemes" or "bunches," and by a downward-spreading root system that is almost devoid of any fine fibers and is yellowish brown in color. In contrast to the mazzard, bark of both the root and the stem of the mahaleb cherry lacks a distinctly bitter flavor, and fails to give a black color reaction with solutions of iron salts. It is hardier than the mazzard, easier to raise from seed and to grow in the nursery. It makes a tree somewhat smaller, a year or two earlier in bearing, and shorter-lived than does the mazzard. It is relatively less dwarfing with the sour cherry than with the sweet cherry. In fact, commercial sour cherry orchards in America are almost exclusively on mahaleb roots, but the mahaleb is seldom considered dwarfing, though doubtless it should be, since Montmorency trees are often as close as 16 x 16 feet on the mahaleb rootstock, whereas they may be planted 25 x 25 on the mazzard rootstock. It is recognized, however, that Montmorency sour cherry trees come into fruiting a year or two earlier on the mahaleb rootstock and that they begin to fail at seventeen to twenty years, whereas they may live to thirty-five years on the mazzard rootstock. In Europe, by contrast, the mahaleb rootstock is now out of favor, trees upon it having shown signs of incompatibility and early death.

It is interesting to note that the length of the stempiece of mahaleb affects the degree of dwarfing. Most cherry trees are propagated by budding just above the ground level, so that the stempiece of mahaleb is commonly very short. On the other hand, when the rootstock is run up to a height of 3 feet and then budded or grafted to the desired variety, the effect is to dwarf the top markedly. In the case of some sweet cherries the dwarfing may be so severe at this height as to be classed as an uncongeniality. The method has possibilities, in which the more vigorous varieties, such as Schmidt, should be grown on a longer stempiece than the less vigorous Black Tartarian. From preliminary observations that have been made, it seems a likely suggestion to grow the mahaleb rootstock to a height of 26 to 30 inches and to bud or to graft at this height with a long scion—8 inches.

The Sour Cherry as a Rootstock

The sour cherry (*Prunus cerasus*) has been tried as a dwarfing rootstock for the cherry. Sour cherry seedlings produce smaller trees when used as rootstocks than do mazzard seedlings. Yet the seedlings are difficult to grow in the nursery, partly because of susceptibility to leaf spot, and varieties of sweet cherry do not take well on them. They are reported, however, to produce a satisfactory dwarf tree for the common Morello and Duke varieties, but unsatisfactory for sweet varieties. The scion variety Kentish Red (*P. cerasus*) has been used in Australia for bush trees of sweet varieties. This combination has given trees of moderate growth, but of somewhat poor anchorage and a tendency to sucker. At East Malling, selected and vegetatively propagated *P. cerasus* has proved unhealthy, and short-lived compared with *P. avium*. Here again, as with the mahaleb, a long stempiece of sour cherry, specifically the Morello-type Northstar variety, has been suggested as useful in producing a dwarfed cherry tree.

A natural-occurring selection of sour cherry that propagates from suckers and softwood cuttings under mist, and is known as the Stockton Morello, has been used with some success for both the sweet and sour cherry. The plant belongs to the Morello group of sour cherries, small growing and weeping. It has the definite advantage over seedlings in that it is propagated vegetatively, so that the type is really a clone and each plant of the type is identical in genetic makeup to the others of the type. However, the sweet cherry does not take well on it, and it is infected with virus disorders. Yet it does offer possibilities as a clonal slightly dwarfing rootstock for the cherry, provided virus-free material can be found. If interest warranted, it could undoubtedly be propagated by leaf cuttings much more rapidly and easily than from suckers from orchard trees, as is often done.

Miscellaneous Cherry Rootstocks

Another possibility as a dwarfing rootstock for cherries is the ground cherry (*Prunus fruticosa* Pall). It is a native of eastern Europe and Siberia, growing as a small spreading bush three to four feet tall. It is propagated from seed. Both sweet and sour cherries are reported to have been budded successfully on this rootstock and to have fruited the first year in the orchard. The seedlings are very susceptible in the nursery row to the leaf-spot fungus (*Cocomyces hiemalis*), and the rootstocks tend to sucker in the orchard. It might, however, be used as an interstock, as suggested for both the mahaleb and the sour cherry.

The Western Sand cherry (*Prunus besseyi*), which is in itself a dwarf form, has been suggested as a dwarfing stock for the cherry, but it has proved generally unsatisfactory. Varieties of both the sweet cherry (*P. avium*) and the sour cherry (*P. cerasus*) unite with it with difficulty, perhaps 20 to 50 per cent stand of buds. Further, the growth of budlings from successful bud takes is poor, and most trees die at the end of the third growing season.

The Breeding of Dwarfing Cherry Rootstocks

Horticulturists at the Institute of Horticultural Plant Breeding, Wageningen, Netherlands, have suggested the possibility of developing a series of clonal dwarfing rootstocks for the sweet cherry from crosses between various species of *Prunus*. Among the dwarf types are *P. incisa compacta* and two types of *P. nipponica Kurilensis*.

At the John Innes Institute at Bayfordbury, England, attempts have been made at developing dwarfing rootstocks for the sweet cherry (*Prunus avium*) from genetic dwarfs that typically segregate from certain crosses between varieties of sweet cherry. Certain crosses in particular (No. 8123 x Kassin's Frühe Herz) have produced a high proportion of genetic dwarfs, and may be useful in developing dwarfing rootstocks for the cherry.

The degree of dwarfing secured from these dwarfing cherry rootstocks has proved substantial. At four years of age, trees of the Merton Glory variety worked onto these rootstocks were 4.2 inches in circumference at the point of bud union, compared to 6.8 for the same variety on vigorous cherry rootstocks (F 12/1). In addition to their dwarfing effect, they induce earlier fall defoliation and earlier maturity of the fruit. Anchorage has been reported as good. The problem now is to find a means of propagating these materials vegetatively.

In such a breeding program, an early method of identifying dwarfing rootstocks without recourse to grafting tests would be of great assistance. For example, the dwarfing effect of apple rootstocks has been shown to be correlated with the presence of a greater proportion of living tissues in the roots, which is reflected in a higher bark/wood ratio, a great amount of medullary ray tissue, and a reduction in the amount of wood fiber. Unfortunately, this relationship does not carry over into species of *Prunus*, and this criterion cannot be used reliably for identification of potential dwarfing rootstocks in this genus.

The Future

Much remains to be done with rootstocks for the cherry, both by selection of seedling lines and by development of clonal lines. There is much variation in mahaleb and sour cherry seedlings, and limited vigorous range in the mazzard. It would seem worthwhile to attempt to pick out some of these different forms for development as rootstocks. From them, improved dwarfing rootstocks might reasonably be expected. There is ample evidence that dwarfing rootstocks and natural dwarf varieties of cherry are already present in America but have not been sought and have been overlooked. For example, at the New York State Agricultural Experiment Station at Geneva, a progeny was developed in hybridization that was dwarf in habit. Yet, because attention was focused at that time upon large and vigorous forms for commercial orcharding, the plants were merely commented upon and then discarded. Now that attention is being directed at dwarf forms, it seems entirely reasonable to expect they may be forthcoming.

SUMMARY AND CONCLUSION

The mahaleb cherry (*Prunus mahaleb*) is the standard rootstock with which to dwarf both the sweet and the sour cherry; but it dwarfs the sour cherry only slightly (semistandard) and is variable with the sweet cherry, producing trees that are poor and short-lived with some varieties and that are semistandard with others. The dwarfing effect is increased by lengthening the stempiece of mahaleb (3 feet). There are a number of related species and forms of cherry from which dwarfing rootstocks might be developed, including the ground cherry (*Prunus fruticosa*).

10

DWARF FORMS OF PEACHES, AND DWARFING ROOTSTOCKS FOR THE PEACH

SINCE the nectarine is merely a bud sport of peach that has glabrous fruit (free from pubescence), all that is said regarding peaches applies equally to the nectarine. Both of these fruits, especially the nectarine, are among the most beautiful and richest of fruits. They are nicely suited to growing against a wall or under glass, where they may be brought to great perfection.

The apricot (*Prunus armeniaca*) responds very much like the peach, and can be considered along with the peach in the general discussion of dwarfing rootstocks.

Some General Considerations

Unfortunately, there are no thoroughly satisfactory dwarfing rootstock materials in common use for dwarfing the peach (*Prunus persica*). This may be due in large measure to the fact that peach trees are neither large nor long-lived, and the problem has been more one of inducing sufficient vigor and providing proper care to develop a tree of good size for heavy cropping and for long life. It is surprising to realize that the average life of peach trees in some eastern peach-producing areas in the United States is not over six years. Under such conditions there has not been much attention paid to finding methods and rootstocks for dwarfing the peach.

The rootstocks in most common use for the peach are seedlings of the Lovell variety secured from California where Lovell is grown as a canning peach. The peach is also used for the apricot. Other materials are wild natural peaches growing in the southern Appalachian mountains of the United States, and red-leaved forms that are useful because they can be quickly identified from the scion variety that is budded upon them and that reduces errors in the nursery. But these all produce standard trees.

There is, however, constant though modest demand for a dwarf form of peach by home gardeners and amateur fruit growers. The interest is for ornamental purposes, for a point of interest in the landscape planting, and for a dual-purpose tree—both ornamental and fruit-bearing.

In Japan the flowering apricot (*Prunus mume*) serves as this kind of plant. It makes an interesting and substantial ornamental tree with attractive flowers and good foliage. Though the fruit is not comparable to improved varieties of peach, it is nonetheless useful, and is widely employed by housewives for pickling and drying. Even the seeds within the pits are used, as one uses almonds, in the preparation of esteemed delicacies. A Mume tree in the yard is considered a good omen. Unfortunately, *Prunus mume* does not do well in the United States, but it illustrates the type of plant desired.

DWARF FORMS OF PEACH

There are plant materials available that may be adapted or developed for dwarfing the peach. For example, there are a number of dwarf forms of peach that come fairly true from seed and either belong to *Prunus persica* or are very closely related to it. They develop into bushlike plants 3 to 5 feet tall. Their fruit is peachlike but not comparable to that of cultivated varieties of peach. While they are often called "dwarf peach," this is an improper designation. Nevertheless, one wonders whether these dwarf forms could not be used for the purpose of developing a suitable dwarfing rootstock for the peach.

Pomological literature describes several small or dwarf forms of peach.

Dwarf Aubinel is one of these. It reproduces from seed as a small, bushy plant. The fruit is said to be yellow, freestone, good in quality, and ripening in late September. Then there is a Dwarf Champion, described from New Mexico in 1899, and a Dwarf Cuba, mentioned from Michigan in 1895. Dwarf Orleans is a variety that originated in Orléans, France, early in the eighteenth century. The tree is said to reach a height of only 2 or 3 feet. Flowers are large and showy. Fruit is 2 inches in diameter, roundish, white-fleshed, with bitter juice, freestone, ripening early in October.

In 1846 A. Poiteau of France described a very dwarf form of peach. He designated it *Pecher nain* (*Persica nana*. Poit. et Turp.). *Persica nana* (Stokes) is the scientific name commonly used for the Dwarf Russian almond, yet the description Poiteau gives does not seem to fit precisely plants of this species as known today. The Dwarf Russian almond is a shrub 3 to 5 feet tall, whereas Poiteau says: "This peach tree does not grow bigger than the common stock gillyflower (*Mathiola incana*); so that it is frequently grown in a vase in order to serve it with its fruit on the table.

"The shoots are thick, very short, and loaded with a great number of buds located close to one another. Its leaves are numerous, pendant, of large size, edged with sharp indentations, and without glands on the petiole.

"The flowers are large, it is true, but not as large nor as colorful as in Madeleine blanche, as Duhamel says. I have never seen this fruit color or ripen in Paris."

He then quotes Duhamel du Monceau as saying: "It is round, rather thick, and relatively abundant in proportion to the size of the tree. One of these little peach trees whose top has only a 9- to 10-inch spread sometimes bears 8 or 10 fruits with a diameter of two inches. A deep groove (suture) divides it and ends at the stem end in a shallow cavity. The color is bright red. The stone is small and white. The fruit, very mediocre, is grown only for curiosity, and ripens toward mid-October. It is produced from seed, and there is a variety with double flowers."

The United States Department of Agriculture introduced an ornamental double-flower dwarf form of peach from China under Plant Introduction Number 41395 in 1915. It was called Swatow, from the place of its origin in Yunnan. It was described from China as "a curious little tree grown here only in pots as a house plant. . . . This particular tree is now just 15 inches high and had five full-sized peaches. They are borne on the main trunk on stems about a quarter of an inch long and make one think of Papayas. . . . The fruit is of good color and tastes better than any I have tasted in China. The flesh is white and it clings to the pit. . . . It hangs on the tree a very long time and is quite ornamental. The blossom is quite showy, too. The Chinese say it comes true from seed."

Swatow has been used in California in an attempt to breed dwarf varieties adapted to the United States, and especially to California, which are useful not only as ornamentals but as dwarf fruit-bearing trees. Two varieties developed from this program are of special merit and are available from nurserymen; these varieties are Flory and Elbertita. The former is described as a densely foliaged, shrublike tree that seldom reaches more than 5 feet in height. Blossoms are profuse, rosy-red. The fruit is small, white-fleshed, and freestone. Elbertita is similarly a densely foliaged, shrubby tree that is said seldom to reach over half the size of a typical peach tree. The fruit is said to be a large, yellow-fleshed clingstone, ripening in August.

W. E. Lammerts of southern California began a breeding program in 1936 in which he hybridized what he has called "The Chinese Dwarf Mandarin Peach (a double-flowering evergreen dwarf variety)" with the large-fruited commercial Rio Oso Gem variety. He also used "Early Double Flowering Pink" and "Early Double Flowering Red" in his breeding program.

In 1948 four varieties were introduced from this breeding program. They were known as the "Daily News Star Peaches," named for the *Los Angeles Daily News* and combining ornamental flowering characteristics with fair fruit quality. Of these, the Daily News Four Star and the Daily News Three Star proved most popular. The former is characterized by large, very double, two-inch light-pink flowers and by highly colored white freestone fruits, 2¼ to 2½ inches in diameter, ripening in late June. The Daily News Three Star variety carries two-inch, deep-pink double flowers and large freestone yellow fruits that ripen in August.

H. C. Swim, working with the same material between 1940 and 1954 similarly developed a large light-pink double-flowered variety named Altair. The fruit is large, 2¼ to 2¾ inches in size, highly colored, freestone, white-fleshed, ripening in mid-August. Another variety, Saturn, is characterized by possessing large, double rose-pink flowers, and bearing large, yellow-fleshed freestone fruits in midsummer.

Still another variety, Bonanza, has been introduced by the Armstrong Nursery Company from crosses made by David L. Armstrong beginning in 1954, involving the Swatow Dwarf Peach and several large-fruited commercial varieties. The tree of Bonanza is small, roundish, compact, with dense foliage, reaching a height of five to six feet, which has been used successfully as an ornamental in pots and tubs (Fig. 59). The compact almost rosetted type of growth is due to very short internodes (approximately one-fourth inch). The tree is precocious blossoming and setting fruit the second year planted. Thinning is required for the attainment of good-size fruit.

The flowers are pink, semi-double, large, and showy. The fruit is medium in size, moderately red-blushed, yellow-fleshed, freestone, ripening early—about a week to 10 days before Redhaven.

Chilling requirements of the tree are low (550 to 600 hours below 45 degrees F.). All of the varieties introduced from this program were developed especially for southern California, and their performance in other climatic regions is undetermined.

It is possible, also, to grow the peach as a miniature bonsai plant in the home. It can be grown in pots in greenhouses, as is done in Europe, with proper manipulation and attention. Further, among standard cultivated varieties of peaches there are considerable differences in size and vigor. Some are relatively small and some are very strong-growing.

DWARFING ROOTSTOCKS
FOR THE PEACH

Peach Rootstocks

No successful dwarfing rootstock for the cultivated peach has yet been developed from the species (*Prunus persica*) to which it belongs.

The several dwarf forms of peach that have been mentioned in the preceding paragraphs are at least suggestive. It would seem that if enough
time and energy were devoted to the problem, a really satisfactory dwarfing rootstock might be forthcoming from among these and other forms
of this species. There is much to be done in this area.

Plum Rootstocks

The statement is commonly made that the peach can be dwarfed by
working it upon the plum. While this may be technically correct, there
are so many species and forms of plum that one must be more specific.
In fact, with many plums used as rootstocks the degree of dwarfing is
slight, although the variety Redhaven has been reported from Canada as
being satisfactorily dwarfed upon the St. Julien A plum. This and other
plums used as rootstocks are described in Chapter 12.

It is noticeable that many favorable reports with the peach on the
plum are from areas where the soil is heavy and cool, as in southern
England, in Holland, and in northern Germany. The plum is well suited
to such soils, whereas the peach is not. One is led to wonder whether
the recommended use of plum stocks is not mostly because of soil adaptation, and whether the slight dwarfing effect is not merely a secondary
observation. Indeed, the plum is said to be used in France for the peach
on heavy soils, and the hard-shelled almond for lighter, drier soils. In
Canada, St. Julien A has been reported as successful on heavy soils but
inferior on light soils.

The Narrow-Leaved or Bastard Shiny Mussel is mentioned on the
Continent as being the most dwarfing for the peach among the Mussel
group. The Yellow Kroos (Yellow Kroojespruin) plum is used considerably in Holland for peaches and is said to dwarf the tree slightly.
Among the St. Julien plum stocks, the St. Julien A is considered the
most successful slightly dwarfing peach stock.

In tests at the East Malling Research Station with the Hale's Early

Fig. 59. **Bonanza Dwarf Peach of the ornamental fruit-flowering type. Note
the rosetted appearance of the trees, which is due to the very short internodes that give the dwarf character.** *Courtesy of David L. Armstrong*

peach on several plum rootstocks, there have been wide differences both in compatibility and in dwarfing. Brompton and Common Mussel have proved compatible and have produced vigorous trees. Damas C has also proved valuable, and is recommended further by easier propagation. Pershore, St. Julien C, and Kroosjespruin have seemed of value as dwarfing rootstocks. Some idea of the range of dwarfing may be seen from the following figures of trunk diameter after five seasons' growth:

Narrow-Leaved Shining Mussel	194 mm.
Damas D	185 mm.
Broad-Leaved Shining Mussel	184 mm.
Common Mussel	178 mm.
Brompton	178 mm.
Damas C	174 mm.
Damas B	159 mm.
Kroosjespruin	158 mm.
Pershore	150 mm.

At all events, the Mussel plum stocks, most of the St. Julien plums, and the Black Damas plum stocks are said to dwarf the peach somewhat, and to induce early fruiting. It must be again pointed out, however, that seedlings are variable and that clonal stocks should be used if any degree of certainty in performance is to be obtained. The stocks mentioned belong to the species *Prunus insititia*, characterized by the damsons and Mirabelles (not to be confused with Myrobalan), and are slower-growing as a group than the common European plum (*P. domestica*) and the Myrobalan plum (*P. cerasifera*). These and other plum rootstocks are described fully in Chapter 12.

Miscellaneous Dwarfing Rootstocks for the Peach

Prunus davidiana, or Father David's peach, has been tried as a stock for the peach sporadically at least since 1865. It dwarfs slightly if at all, and is noted more for its hardiness to winter cold than for any dwarfing properties. The fruit of this species is in itself small and of little value.

Prunus subcordata, or Pacific plum, is said to dwarf the peach somewhat, but records are fragmentary and uncertain.

Prunus maritima, or Beach plum, will accept the peach and will dwarf it to some degree. There is a dwarf form of this species that has also been tried, but the reports are not consistent.

Prunus tomentosa, or Nanking cherry, has shown considerable promise as a dwarfing rootstock for the peach. The plant is itself a bush or small tree, propagated from seed. Peach buds are said generally to succeed fairly well upon it, although some uncertainties exist about compatibilities. Trees grow well in the nursery as budlings and young trees. They fruit early; in fact, they may bear a few fruits even in the nursery row. The Redhaven peach has fruited the second year on this rootstock. Trees have remained shoulder-high for several years, although there has been lack of uniformity of the trees as they became older. Further, magnesium deficiency has appeared. Accordingly, although the Nanking cherry may prove successful as a dwarfing rootstock for the peach, there is still much to learn before it can be enthusiastically recommended for this purpose. It would seem that clonal lines must eventually be developed.

The Sand cherries, which are discussed more in detail under the section on plum rootstocks, have been used successfully from time to time for dwarfing the peach. A report appearing in *Revue Horticole* in 1847 tells of producing peach trees no taller than 3 feet on rootstocks of *Prunus pumila*. The trees were said to be full of flower buds the first year of growth from the graft, and to have produced large, beautiful fruit. A dozen different varieties were used, and no incompatibilities were reported among them. They were exhibited in Paris on different occasions. Furthermore, an acquaintance of the author's has written that he has grown the Greensboro peach successfully on seedlings of *Prunus pumila* near Bar Harbor, Maine.

Prunus besseyi, or Western Sand cherry, is among the more promising plant materials with which to dwarf the peach. It propagates readily both by seed and by layers. It is in itself a bush or small tree to about 4 feet in height. The fact that it can be propagated by mound layering suggests that clones may be selected and developed, with resulting greater satisfaction in uniformity and dependability of performance. Peach buds unite well with it, although care must be taken to bud the stock before too late in the season. It matures early, and unless budded in late July, rather than late August or early September as is done for peaches on peach rootstocks, success is likely to be poor. Budlings grow well in the nursery, and trees fruit the first or second year in the orchard. For comparison, it may be noted that trees of the Redhaven peach have been slightly larger and sturdier on *P. besseyi* than on *P. tomentosa*. There are insufficient records to establish to what age trees will live on this rootstock, but in view of the severe dwarfing it would not be surprising if they were not long-lived.

The late Professor Niels E. Hansen, of South Dakota, who worked extensively with the Western Sand cherry, wrote to the author in 1941 as follows: "I was the first to bud peaches on sand cherry (*Prunus besseyi*). One of my early bulletins shows a photograph of a dwarf tree of this kind bearing fruit in a box about the size of a sixty-pound butter tub. I remember a nursery in Tennessee that followed this idea; they found the tree had to be cut back twice the first season owing to the bushy growth. I think this stock will cause earlier bearing and more dwarf habit of growth, but I do not know if it will be of any real value in your region [western New York]."

Prunus glandulosa, one of the flowering almonds, has been reported from Massachusetts as producing a very dwarf peach tree, and to be compatible with a wide range of peach varieties. Similarly, in 1860 dwarf peach trees were shown in France worked successfully on *P. sinensis*, and characterized as carrying very beautiful large fruits in abundance. On the other hand, records from 1896 in Wisconsin with flowering almond stocks showed complete failure, not a single peach bud uniting with the rootstocks. There are, of course, several species of flowering almond, and it is impossible to determine whether the Wisconsin trials were with the same species used in Massachusetts. In addition, *P. glandulosa* is said have the fault of suckering badly as a rootstock.

Additional miscellaneous rootstock materials that have been tried for dwarfing the peach include the Mediterranean Ground cherry (*Prunus prostrata*), the Sloe (*P. spinosa*), the Marianna plum (*P. munsoniana* x *P. cerasifera*), and the American plum (*P. americana*). All four have been reported as successful, but no details of substantial tree performance have been found—though several varieties of peaches have been worked upon them successfully. Favorable results have been reported also for the wild Oregon Pacific or Western plum (*P. subcordata*) when it is used as a dwarfing intermediate stempiece between a standard peach rootstock and the peach scion variety.

ROOTSTOCKS FOR THE APRICOT

As has been said, the peach is commonly used as a rootstock for the apricot. As used in California, it is said to check the tree slightly in comparison with apricot seedlings, and is somewhat tolerant of alkali and of drought. The St. Julien plum (EM A) has been used for dwarfing purposes in Europe, but it is only slightly dwarfing. Brussels is no longer recommended because of the fault of suckering. The Common Mussel is mentioned as being esteemed on the Continent for the apricot. In New Zealand the Marianna plum has been found slightly dwarfing, and especially useful in wet soils, although one of its faults is that it suckers badly. Seedlings from commercial varieties of apricots have been suggested as dwarfing material, but any that are known make magnificent, vigorous, and late-fruiting trees instead. The almond has been tried off and on, but generally with unsatisfactory results. The union with the apricot is poor, and trees frequently break off. Double-working has been tried with some success, using an intermediate stempiece of peach or of plum between the almond rootstock and the apricot scion.

Some appropriate remarks about the apricot written to the author by the late Professor Niels E. Hansen in 1941 may be of special interest: "*Prunus besseyi* is being used successfully as a budding stock for very new hardy apricots from the Harbin region. They make a fine tree as far as the roots are concerned. California has had good results with the standard apricot on *Prunus besseyi*. Apricots on *Prunus americana* (a native American plum) stock break off easily. There is a large swelling at the union. I use wide cloth bands and old steel fence posts for staking to take the strain off the point of union.

"Wherever the peach root is hardy, it will be the favorite stock for the apricot. Soon we will be able to get seedlings of the Manchurian apricots (*Prunus armeniaca* var. *Mandshurica*) themselves, which will solve the problem where hardiness is a factor.

"I have had excellent success with budding the apricot on *Prunus sibirica* (*Prunus armeniaca* var. *sibirica*; Siberian apricot) which I brought over myself from Mendo and Shilka (Manchuria). They make a perfect union in the nursery. There is a tendency to early bearing, and the trees are much smaller than the real apricot."

ROOTSTOCKS FOR THE ALMOND

Almonds are commonly grown on seedling almond rootstocks. The peach has also been used. No dwarfing effect has been suggested from either of these materials. Seedlings of commercial varieties of apricot do not make a good union, and are to be avoided. Plum roots are also disappointing.

EXPLANATION OF CONFLICTING REPORTS WITH DWARFING ROOTSTOCKS

The conflicting nature of reports on dwarfing rootstocks for the peach, apricot, and nectarine can be explained in several ways, and serve to emphasize the difficulty of the problem. First is the matter of the identity of the rootstock material itself. There are many species and subspecies of *Prunus*, and some of them have been confused. In addition, there are strains and clones that are also distinct and that behave quite differently even though closely related botanically. One is not always sure that what is reported in one place is the same material reported in another. In other words, standardization is badly needed before comparisons can be made with any degree of reliability.

Second, the time of budding is an important consideration. As has been previously said, peach rootstocks are commonly budded in the North Temperate Zone some time in late August or early September. On the other hand, some of the species of *Prunus* that are contemplated as rootstocks for peaches have a shorter growing season. Unless budded while they are still actively growing, as in late July or early August, a poor stand of buds may result, which might be interpreted erroneously as an incompatibility between peach and the particular rootstock.

Third, virus disorders seem especially frequent and severe with *Prunus*, and this may also prove misleading insofar as compatibility is concerned. Thus, *Prunus besseyi* is suspected of carrying virus troubles, which reduce bud take and also result in poor unions. If the foliage of a peach budling on the *P. besseyi* stock shows pale-green foliage in mid-summer, and if the leaves tend to roll upward toward the midrib, it may contain virus. Such trees usually defoliate prematurely, develop an abnormal overgrowth by the scion, and die in a year or two. Until virus disorders are properly screened, and until virus-free rootstocks are available, much confusion is bound to occur in the entire area of dwarfing rootstocks for peaches—in fact, for all species of *Prunus*.

SUMMARY AND CONCLUSION

There is no thoroughly reliable rootstock now in use that can be recommended for producing dwarf peach, nectarine, almond, or apricot trees. The St. Juliens and the Black Damas plums have been used to considerable degree for this purpose for the peach, producing semi-standard trees at best. Since seedlings are variable and uncertain as to both congeniality and tree performance, only clonal lines should be used. Of these St. Julien A is most suitable for a tree of intermediate vigor. Common Mussel is the traditional stock for semidwarf peach trees.

For more severe dwarfing, the Western Sand cherry (*Prunus besseyi*) has been used with limited success, followed by the Eastern Sand cherry (*P. pumila*), the Nanking cherry (*P. tomentosa*), and the flowering almond (*P. glandulosa*). All these, however, have shown incompatibilities from time to time, and are far from being really satisfactory.

Ornamental dwarf forms of the peach have been found that possess showy blossoms and edible fruits, thus combining attractive spring blossoms with edible fruit in summer.

For the apricot, both the St. Julien plum (EM A) and the Marianna plum have been found slightly dwarfing as well as adapted to wet soils. The Western Sand cherry (*Prunus besseyi*) and the Manchurian apricot (*P. armeniaca* var. *mandshurica*) offer some promise.

No good dwarfing rootstock for the almond has as yet been reported.

11

DWARFING ROOTSTOCKS
FOR THE PEAR

T H E R E are as yet no forms or species of pear (*Pyrus* spp.) suitable as rootstocks upon which to dwarf the pear, comparable to what is found among species of apple (*Malus* spp.) for dwarfing the apple. To be sure, there are differences in vigor between pear seedlings and between species of pears. And these differences are sufficient to suggest that satisfactory dwarfing pear rootstocks might be found or developed from among them. But there are none at the moment.

On the other hand, the pear is closely related to a number of genera in the family *Rosaceae*, to which it belongs, and upon which it can be budded or grafted with varying degrees of success for dwarfing purposes. Among these are the apple (*Malus domestica*), quince (*Cydonia oblonga*), hawthorn (*Crataegus oxyacantha*), mountain ash and wild service-tree (*Sorbus* spp.), serviceberry (*Amelanchier* spp.), Oriental photinia (*Photinia villosa*), and cotoneaster (*Cotoneaster* spp.). In fact, the apple, mountain ash, pear, and quince are so closely related that some botanists place them together in the same genus, *Pyrus*.

THE QUINCE AS A
DWARFING ROOTSTOCK

Yet insofar as practical fruit production is concerned, the quince is the only one of those mentioned that provides a satisfactory dwarfing rootstock for the pear. It is as ancient a plant horticulturally as the apple.

Olivier de Serres called attention to the quince as a useful rootstock for pears in France as early as 1600, and suggested it even for the apple. In 1662 Le Gendre published *La Manière de cultiver les arbres fruitiers* in which he recommended working the pear on the quince for the growing of espalier forms in France; and English writers describe its use in England in 1685. The exact nature of the quince material used at this time is not known, but it is described as propagating easily by cuttings and layers.

In 1667 Merlet advised a quince he called Portugal quince, and which was apparently continued widely in use until at least the early nineteenth century. Later in the nineteenth century, another quince came into use in the vicinity of Paris; it was called the Fontenay quince. In 1880 Baltet called attention to the Angers quince, which was apparently quite different from the Fontenay. There has been considerable speculation as to the origin of these types, but nothing positive has been determined.

For several centuries France was the great nursery for quince rootstocks, notably in the region around Angers and Orléans. From there, quantities of rootstocks were supplied to nurseries in England and America. In fact, until the United States embargo against importation of rootstocks in 1930, the American supply of quince rootstocks came each year largely from French nurseries, and some American nursery companies maintained nurseries of their own in France to supply themselves with quince rootstocks.

Early Groupings of Quince Rootstocks

The quince propagates readily by such vegetative means as hardwood cuttings and mound layering, so that distinctive commercial strains or types of quince have been quite easily and naturally developed, and are more properly called clones. In the different nursery areas of France, certain types of quince that were observed to grow especially well there gradually became characteristic of the different regions and came to be identified by the name of the locality. Thus, the type commonly grown in the vicinity of Angers became known as the "Angers quince," the type propagated near Fontenay became known as the "Fontenay quince," and the type that has been identified with no particular locality yet is frequently met with in England has been broadly dubbed there as the

"Common quince." Although each is different from the others in appearance, in growth, and in effect upon the pear worked upon it, all belong to the same species (*Cydonia oblonga*), and all might properly be given the group name of common quince.

In botanical literature there is occasionally distinction made between quinces that bear roundish or apple-shaped fruit and those that produce pyriform or pear-shaped fruit. Thus, there are listed *Cydonia oblonga piriformis* (pear quince) and *C. oblonga maliformis* (apple quince), and *C. oblonga lusitanica* (Portugal quince).

Confusion Among Quince Rootstocks

In commercial nurseries, over a period of years, the various types from different regions became mixed, so that when quince rootstocks were purchased on the open market, without critical attention, almost any type of quince and mixtures of several might be received. The type received in America from France a hundred years ago was thought to be the Portugal quince, but later proved to be a variety of apple quince; it proved slow and feeble in growth and entirely unsuited to the pear, quite in contrast to the recognized vigor of the true Portugal quince. In fact, because of the ease with which the quince is propagated, stool blocks may be found scattered here and there in nursery sections of the Continent, England, and America—frequently containing mixtures and often going by some local designation. As has been suggested, the several different types are decidedly different in their effect upon the pear worked upon them. Undoubtedly, the differing experiences with dwarf pear trees in America reported in horticultural literature trace back in part to mixtures, misnaming, and differences in rootstock.

THE MALLING QUINCE ROOTSTOCKS

Because of the apparent confusion among quince rootstocks, R. G. Hatton and his associates at the East Malling Research Station began a study of quinces in 1914, which was reported in 1920. Collections of quince rootstocks were secured from fourteen different nursery sources in England and abroad. Because of the historical significance of the work, it is reviewed here in some detail.

It at once became apparent that there were a number of varieties or types of quince in common use and that the common names by which they were known were of little value in accurate identification. Varieties differed considerably in ease of propagation, in vigor, in time of maturity, and in general habit of growth and botanical characteristics.

Since the nomenclature was so confused, five main groups or types were selected; these were then designated with letters of the alphabet as Types A, B, C, D, and E. Subsequently the number was increased to seven by the addition of Types F and G. The word "Type" has been dropped, and the prefix "EM" has been substituted, as with the apple rootstocks from East Malling. In common practice the EM is often dropped and the designation becomes "Quince A," "Quince B," and so on.

Of these seven quince rootstocks, only A, B, and C are of commercial importance, and A is the most important of these. D, E, F, and G are useful for their fruit rather than as rootstocks. By some, A, B, C, and D are classed as weak growers in contrast to the strong-growing E, F, and G. A more critical classification classifies C as a weak grower and strongly dwarfing, B, D, and E as intermediate in vigor and dwarfing, and A, F, and G as strong growers and the least dwarfing of the lot. A more detailed description will follow in later paragraphs.

Quince EM A is the true Angers quince, the most widely used type, and the one most commonly found in stool blocks (Fig. 60). It has been identified by Maurer of Germany as similar to "Pillnitzer Klon R 20b." Mixtures of B, D, and C are sometimes found in A, and by some it is erroneously called Fontenay quince. It is characterized by very vigorous, upright-growing plants, with shoots that spread in all directions, leaves large, often abruptly accuminate, with some soft puckering, and with reddish tinge the entire length of the midrib. In size the leaves are intermediate between the larger Portugal (E), apple-shaped (F), and pear-shaped (G) types and the B, C, and D. The fruit is of medium size, approaching apple-shape, fair to good in quality. It is both a good orchard plant and a good nursery plant. It layers readily, comes rather easily from cuttings, is vigorous and healthy, is relatively free from the leaf-spot disease, and develops a good fibrous root system. Trees worked upon it have good anchorage and are generally liked.

Quince EM B is called "common" quince, considered by Hatton to be the old common quince of England (Fig. 61). But it also sometimes

Fig. 60. Quince rootstocks, East Malling selections: (Left to right) Quince A, Quince B, Quince C.

goes under the name of simply "Angers." It is somewhat similar in plant characteristics to Quince A, but is less vigorous, with shoots slightly more erect. The leaves are smaller than for A, longer in relation to width, downy on underside and can be distinguished by the smaller, acutely-tipped drooping leaf, with some soft puckering as in A, but with less red in the midrib. The fruit is small, hard, gritty, downy, quite different from Quince A in shape, and poor in quality. In the nursery it roots well from both layers and cuttings and develops a slightly less fibrous root system. It is a slow grower. Trees worked upon it are somewhat smaller in size than those on Quince A, so that it is useful and preferred for the smaller bush-type dwarf pear tree. It is also not overly hardy to winter cold.

Quince EM C is rarely met with in commerce, being almost extinct.

Fig. 61. (Below) Close-up of leaves of quince rootstocks: (Left to right) Quince A, Quince B, Quince C.

It has not been identified with certainty as to origin. Maurer considers it identical to Stroops quince and Stroops Ideal VII, often mistakenly called Angers quince. Hatton has suggested that the "New Upright" or "Paris de Fontenay" introduced to America by Ellwanger and Barry of Rochester, New York, under those names a century ago, may have been of this type. Quince EM D has been designated by Hatton to receive the name Fontenay, yet he concedes that the original Fontenay of old horticultural literature may perhaps have been Quince EM C. Quince C is characterized by very erect yet dwarf habit of growth, bunchy "candelabra" in general appearance, with evenly thin shoots that are faintly brown compared with A and B. The leaves are roundish, blunt-tipped, small, with much rigid puckering (as though starched), the leaves appearing very yellow in early spring. In the nursery it lacks vigor, is not as winter hardy as Quince EM A, and is very subject to the leaf-spot disease, so that it is frequently entirely defoliated in midsummer. It roots readily both from layers and from hardwood cuttings. The root system is small and shallow. Trees worked upon it are smaller and more precocious in fruiting than those on either Quince A or Quince B.

Quince EM D is now the type name given to what has been designated the true Fontenay, or de Fontenay, quince. It is characterized by a distinctive drooping habit of growth and dark green elongated leaf. The fruit is small, hard, pear-shaped, and very downy. In the nursery it is not so vigorous as Quince A and roots poorly from layers but better from cuttings. It has shown incompatibilities with a number of varieties, has produced weak plants, and is not recommended.

Quince EM E is the Portugal quince, now grown mostly for its fruit. It was recommended as a rootstock for the pear by writers nearly three hundred years ago, but has not been used in recent years. It is said to be especially hardy, as quinces go. It is characterized by very large, downy leaves and stout, vigorous growth. The fruit is somewhat pear-shaped, loses its down, and becomes smooth. In the nursery it roots poorly and is incompatible with many varieties. Clones of certain of its seedlings are used as indicators of viruses. It is not recommended as a rootstock.

Quince EM F is a cultivated, apple-shaped quince, and Quince EM G is a cultivated, pear-shaped quince. They are both old in horticultural use, and are grown primarily for their fruit but not for use as rootstocks for the pear. Several varieties of these types are in cultivation, as Rea,

Van Deman, Orange, and Champion, and it is possible that from among them additional types might be developed that would be of use as rootstocks. Both F and G are vigorous growers. The leaves of F are larger than those of G. In the nursery, F and G root poorly and show many incompatibilities. They are not recommended.

Fruiting characteristics of pear trees on EM A, B, and C

Some conception of the fruiting characteristics of pears worked on EM A, B, and C may be noted in the report of Conference trees made by Hatton in 1928. On EM A, up to the seventh year, the production was 547 pounds per acre; on EM B, 415 pounds; and on Quince EM C, 2,397 pounds. The number of fruits per tree was, respectively, 4.5, 3.9 and 24.1.

ATTEMPTS TO GROUP QUINCE ROOTSTOCKS

Subsequent to the reports by Hatton and his co-workers at the East Malling Research Station, Tydeman in 1947 reported additional studies with quince rootstocks. It had been recognized that there were a number of varieties or types of quince in use as rootstocks for pears on the Continent that had not been represented in the fourteen collections studied by Hatton.

Accordingly, several varieties of so-called "Provence quince" were secured from French nurseries, three varieties of "Angers quince," five types of "Pillnitz quince," and one lot of "de Fontenay quince."

It was again reaffirmed that although the general gross characteristics of all quinces were similar, they differed widely in their behavior as rootstocks. Also, it became again apparent that the common names frequently employed as a matter of convenience, such as "Angers quince" and "Provence quince," were names that applied to geographic regions from which the quince came rather than to specific botanical forms. Nevertheless, Tydeman was able to make a general grouping of varieties he had collected that is both interesting and useful, and is summarized as follows:

1. *The Quinces of Angers.* This group includes both EM A and EM B. Although EM B is slightly less vigorous than EM A, it is nevertheless closely related to it botanically. Two other varieties (C 28 and C 30) also fall into this general group. The former was described as "à port érigé" (erect carriage), and the latter as "à petites feuilles" (with small leaves).

As a class, the true Angers is distinguished by slender, wiry, rather zigzagged, deep-brown, almost glabrous shoots and by vigorous, erect, rather dense bushes. They are slow to leaf out in spring, and are susceptible to damage from winter cold.

2. *EM C and Related Quinces.* EM C, as has been noted, was received at the East Malling Research Station mixed in a lot designated "Angers," from which it was readily separated and renamed "EM C."

Similarly, a variety (C 29) was received by Tydeman under the name "Angers" and with the description "à rameaux buissonnants" (with bushy branches), which very much resembled EM C. Still another variety, known as "Kwee," was received from Holland. It also appeared similar to EM C.

While not identical, these three resembled each other sufficiently to be grouped together. The group is recognized by the extremely dwarfed, compact, and erect growth. The leaves are small, yellowish-crinkled, appearing very yellow in early spring. Plants have been flowerless in England over a period of thirty-four years, and have remained very small.

It should be noted that this may have been the Fontenay of old horticultural literature, and the stock introduced into America by Ellwanger and Barry of Rochester, New York, as "Paris de Fontenay."

3. *The Quinces of Pillnitz.* This group had its origin in Germany, where Schindler, of Pillnitz, selected them for seedlings of Angers quince because of their resistance to winter cold, among other good qualities.

Five varieties were received by Tydeman, as R 1, R 2, R 3, R 4, and R 5. All five proved resistant to cold injury, but the R 4 was less so than the other four. It differed also in other characteristics, and in many ways seemed more related to the Provence group than the Pillnitz group.

All five, however, very much resembled the Angers quince (EM A), with leaves that appeared late in spring and with the rather dense, bushy habit and slender, somewhat wiry shoots characteristic of the Angers. The plants matured earlier in the fall than the Angers, a fact that may help to explain the observed resistance to cold. The Pillnitz quinces seemed less vigorous than either the Provence or the Angers.

4. *The Quinces of Provence.* The so-called "Cognassier de Provence," or Provence quinces, were called to attention in France about 1925 or 1930. They grow wild in the hedgerows of the Provence region of France, which lies along the Mediterranean Sea east of the Rhone River. Materials are said to be collected from these hedgerows and sold in commerce as rootstocks for the pear. The more prominent nurserymen have made their own special selections from among the native materials, and have propagated them vegetatively.

The Provence quinces as a group are characterized by their vigor and their reported resistance to drought and to disease. They are more tolerant of alkaline soils than is the Angers quince. They are preferred to Quince A in southern France and are said, also, to be compatible with a wider range of varieties of pears. While this may be so, it should also be pointed out that there is a great number of Provence quinces, with considerable range and variation in performance. Thus, at the Station de Recherches d'Arboriculture Fruitière at Belle Beille, Angers, France, there are more than eighty types of this quince under test. Some of the reports of congeniality are probably associated with the compatibility of a specific pear variety with a specific quince rootstock rather than the compatibility of pears in general with the Provence quinces in general.

Four selections of the Provence quince were received by Tydeman and designated C 51, C 52, C 53, and C 54. Although differing markedly in performance from one another, they appeared sufficiently similar to be grouped. An outstanding characteristic of the group was the profuse whitish pubescence that covered the young wood in early winter immediately after leaf-fall. Shoots were stouter and straighter than the Angers, the wood was rough-textured, and the leaves were somewhat longer and narrower, with pointed, accuminate tips. They bore flowers that were large and showy.

5. *The Quinces of Fontenay.* The Cognassiers de Fontenay are used extensively as a pear rootstock in certain parts of France, particularly in the vicinity of Paris. EM D was designated by Hatton as the true Fontenay quince, but the designation is by no means certain. In fact, as noted, there is some suggestion that EM C may have been the Fontenay of old French literature.

Tydeman received a variety (C 39) under the name "de Fontenay," and found it quite unlike EM D. It resembled the Provence quince, and he included it in that grouping. At all events, although the Fontenay quince is recognized in French nurseries it appears to be a confused group, with considerable variation.

6. *Quince EM D*. Although EM D was designated by Hatton as the true Fontenay, Tydeman has preferred to put it in a class of its own, and separate from the Fontenay group. It is characterized by a low-drooping habit of growth that makes a large spreading bush. Young shoots are susceptible to cold injury. The plants are also characterized by early and profuse production of large white flowers.

7. *The Quinces of Orléans*. The Cognassiers d'Orléans are also found in French nurseries and in French literature. By some it has been called the Fontenay quince. Tydeman received plants from Denmark under the name "Quince of Orleans" that bore a close resemblance to the Provence quince. In short, this group, too, is confused and uncertain.

THE QUINCE AS A ROOTSTOCK FOR THE PEAR IN FRANCE

The importance of the quince to pear culture in France is shown by the fact that 90 per cent of the pears are grown on quince roots. As has been noted, the use of the quince as a rootstock for the pear goes back well into French horticultural literature. The identity of this early material is not known. About the middle of the seventeenth century, as already noted, the Portugal quince was introduced and widely used well into the nineteenth century. Later in the nineteenth century, largely in the vicinity of Paris, another quince gained prominence, known as the Fontenay. About 1880 still another quince, the Angers, came into use. Finally, between 1925 and 1930 the Provence quince was called to attention.

As reported by Brossier, these three principal types of quince are now recognized in France for dwarfing the pear; namely, the Angers, the Provence, and the Fontenay. And of these three, the Angers and the Provence are by far the most used. Unfortunately, each of these three main types is not uniform; instead, there are several clonal types to be found in each group. Nevertheless, the three main types or groups have been deemed sufficiently different from one another to be distinguishable. The color of the leaves in early spring and general growth characteristics have been found useful for identification, especially for the Angers and the Provence quinces.

The Angers Quinces

As studied in France, the Angers quinces seem to have originated in the northern or middle part of France, possibly in the vicinity of Angers. At least they are restricted largely to these regions and have never been found spontaneously in the southeastern or southern part of France.

The Angers quince is characterized by yellowish-green leaves in early spring, which are only slightly pubescent; by slender, zigzag branchy twigs that show only slight pubescence; by susceptibility to leaf spot; and by plants that are very branching and moderately spreading. It propagates freely both by layers and stem cuttings. Figures of 75 per cent or more have been secured by both methods as compared to 60 per cent for the Provence quinces.

Compatibility has been found to vary. For example, one clone of Angers has been found entirely incompatible with the Passé Crassane variety. The union appeared perfect but the trees died after a few inches of growth. Virus troubles have been suspected.

The Pillnitz quinces are considered as belonging to the Angers group, probably having been selected as seedlings in northeastern France. Four of five clones in this group have been found relatively resistant to cold.

The Provence Quinces

The Provence quinces are found over a much wider range than the Angers. They are common in the region of Avignon in the Provence District, and are found commonly in the nurseries of southern France. Most of the quince stocks found in commerce today are said to belong to this group, as subgroups and special clonal lines.

The Provence group is characterized by leaves that are deep green in early spring, with rather heavy pubescence; by sturdy, erect twigs that are very pubescent at the tips and on the small twigs; by resistance to leaf spot; and by plants that are in general more vigorous, less spreading, and more erect than the Angers.

In general, they propagate less rapidly than do the Angers, but are good growers and are apparently congenial with a wider range of varieties. They are more tolerant of alkaline soils and of droughty conditions than the Angers quince, but they do not tolerate wet soils as well.

The Fontenay Quince

The Fontenay quince is not widely distributed. It is used principally in the vicinity of Paris. It is valued for its excellent root system and its resistance to cold. Under this name, quinces have been observed that have resembled not only EM D but also EM A and the Provence quince. The greater number of Fontenay quinces that have been examined in France have seemed to belong to the Provence group. In fact, stocks secured from nurserymen of Fontenay that are reportedly of the type grown in that region since 1850 have been characterized as belonging to the Provence group. The question is raised whether the Fontenay quince may not be a clonal selection of the Provence quinces.

Other Quinces

From an examination of quinces from three hundred different sources, including some from Israel, Italy, Portugal, Egypt, and Tunisia, some seven groupings have been made by Brossier and his associates at Angers. Some of these quinces are not generally found among quinces used as rootstocks. In addition there are apparently several subgroups and special clones within each group, with considerable difference in growth characteristics and performance as rootstocks. With the intensive studies being undertaken with quince rootstocks, it would seem that considerable improvements should be forthcoming.

MISCELLANEOUS DWARFING ROOTSTOCKS FOR THE PEAR

As has been mentioned, a number of other closely related materials have been used as rootstocks for dwarfing the pear. Patrick Barry, in the middle of the nineteenth century in western New York, reported success in working certain varieties of pears on both the mountain ash (*Sorbus* spp.) and native hawthorn (*Craetegus* spp.). He preferred to grow the mountain ash seedlings for two years before budding them to the pear, and for the hawthorn he suggested three years from seed before budding. He considered the mountain ash useful for sandy soils where pears and quince frequently fail. On the other hand, he recommended the hawthorn for wet, cold soils. Although suggesting these materials in this way, he nevertheless concluded that it was better to grow pears on soils that were better adapted to the pear than to resort to these rootstocks in an attempt to adapt the pear to otherwise unsuited soils.

Occasionally fruiting pear trees may be met with, or may be found reported in the literature, which have been worked not only on English hawthorn (*Craetegus oxyacantha*) and various other species of hawthorn (*C.* spp.) but also on the mountain ash (*Sorbus aucuparia* and *S.* spp.), on *Cotoneaster multiflora*, on English cotoneaster (*C. integevrima*), and even on the common apple (*Malus domestica*). In fact, in 1860 a wide assortment of pear varieties was reported from France as having been successfully worked upon the English hawthorn. Included in the list of varieties were Bon Chrêtien d'Espagne, Napoléon, Beurré Royal, Beurré Bretonneau, and Louise Bonne d'Avranches.

Here again the suggestion is often noted that hawthorn will withstand wet, cold soils and that it may be useful as a rootstock for these conditions. But, as has already been observed, one might rather improve such soil or not plant upon it. The combinations with hawthorn, mountain ash, cotoneaster, and apple are not long-lived, and are to be considered essentially as novelties.

Attempts have been made to secure pear rootstocks from various species and forms of the pear to serve as dwarfing rootstocks, but there has been little success. Yet it would seem that possibilities might exist. Thus, Vavilov of Russia studied the wild progenitors of fruit trees in Turkestan and the Caucasus. He found the quince growing in great forests in eastern Transcaucasia along the west banks of the Caspian Sea at about latitude 40 to 42 degrees North. He described the great variation that exists in this material. H. M. Tydeman of the East Malling Research Station received quince seed from this region and grew seedlings from it. He found them strikingly similar botanically to quince EM C.

There are also hybrids between the quince and the pear that have been suggested for possible use as rootstocks for the pear. These are given the generic name *Pyronia*. However, attempts to hybridize the pear and the quince have not been overly successful.

There are also chimeras of pear and quince that are of passing interest and that have similarly been suggested as rootstocks for the pear but have not been tested. Among these are *Pyronia danieli*, also called *Pirocydonia*, and *Pyronia luxemburgiana*, which is a form of *Pyronia veitchii*.

SOME ADDITIONAL CONSIDERATIONS WITH QUINCE ROOTSTOCKS

The quince root prefers fairly heavy and even moist soil, much as the pear prefers. Light, sandy, and droughty soils are not good. The stock is very successful for trained trees and for pyramids. A limiting factor in its use is its susceptibility to fire blight and its tenderness to severe winter cold. Where temperatures reach —10 degrees F. and there is not good snow cover, the quince should not be used. In western New York during the severe winter of 1933–1934, when a minimum temperature of —31 degrees F. was experienced for a few hours, all pear trees on the quince root were destroyed by virtue of the killing of the rootstocks on which the pears were growing. The pear tops themselves were not killed; they leafed out in the spring, but wilted and died as summer approached. There have been many dwarf pear trees grown in eastern America during the last 175 years, but there are few left. One wonders whether occasional severe winter cold may not have been an important factor in their disappearance. The discovery or development of a strain of quince hardy to winter cold would be a valuable acquisition.

It is often recommended to plant dwarf trees deep so as to encourage them to scion root. Some of the most productive and longest-lived commercial "dwarf pear" orchards have been developed this way, but they are obviously not true dwarf fruit trees when scion-rooted. A successful planting of 14,000 Bartlett trees on quince roots, with Beurré Hardy as interstock, is reported from California.

Quince roots have been used in California for pears since at least 1900. Comice/quince was one of the early combinations. Later, orchardists used Comice/Beurré Hardy/quince. However, California orchardists found it necessary to plant the tree with the Hardy bud below ground level so as to reduce quince suckering and the attendant hazard of fire blight. Fire blight is severe in parts of California. But when the Hardy stempiece was planted deep in this way, it eventually struck root and the desired dwarfing effect was lost. By using Old Home as the intermediate in place of Beurré Hardy, fire blight was much reduced, since Old Home is resistant to this disease.

Quinces EM A, B, C, D, E, F, and G have all been tested commercially in California. Of these, EM A has the best record of performance. It has been found that the quince rootstock is relatively shallow-rooted, extending not deeper than 2 to 2½ feet. Although tolerant of wet soil, it has proved susceptible to lime-induced chlorosis.

The varieties most commonly grown on the quince in America are Bartlett (Williams' Bon Chrêtien), Beurré d'Anjou, Duchess d'Angoulême, Easter Beurré, and Louise Bonne de Jersey. Others which may be grown successfully on the quince are Beurré Giffard, Clapp Jargonelle, Belle Lucrative, Beurré d'Amanlis, Beurré Clairgeau, Triomphe de Vienne, Beurré Hardy, Comte de Lamy, Beurré Superfin, Bloodgood, Buffum, Elizabeth, Flemish Beauty, Howell, Pitmaston Duchess, Doyenné Bousock, Doyenné d'Alençon, Glou Morceau, Beurré Diel, Olivier de Serres, Conference, Doyenné du Comice, Passé Crassane, Josephine de Malines, Madeleine, Osband, Pound, Tyson, White Doyenné, and Vicar of Winkfield.

Double Working

A special consideration in the use of the quince as a rootstock for dwarfing purposes is the fact that not all varieties of pear do equally well upon it. That is, some varieties either do not make good unions or lack vigor when worked directly on the quince.

To meet this situation, the practice of double-working has been employed, in which a variety of pear that is congenial with the quince is first worked upon the quince and then the pear stock worked on the desired pear variety that lacks congeniality with the quince (see Chapter

15). The intermediate pear variety is known as an "intermediate stock." Good intermediate stocks are Beurré d'Amanlis, Beurré Hardy, Fertility, Nouveau Poiteau, Old Home, Pitmaston Duchess, and Uvedale's St. Germain. Of these, Beurré Hardy has by far the best record of performance over a wide area of the world. It is hardy to winter cold, vigorous-growing, good to handle in the nursery, and reported to improve cropping of the variety worked upon it.

Over a period of years, through trial and error, pear varieties have been grouped into three classes according to their ability to prosper on the quince rootstock, as listed below. The list is not exact; rather it gives only general indications of tendencies. Thus a variety grown north of its climatic range may seem to do well on the quince by virtue of the fact that the quince root matures the fruit earlier than does the standard pear root. In a more favorable climate, such a variety would show no such improvement on the quince root.

Similarly, it has been noted that varieties with granular and gritty flesh are often made more so on incompatible rootstocks, whereas those varieties with melting or buttery flesh are less affected. This is due to the fact that factors that check the optimum growth and development of a pear fruit tend also to make it more gritty, less well matured, and lacking in flavor and quality.

Group 1. Varieties that make good unions with Quince A and succeed upon it (those marked with an asterisk * have proved especially suited to the quince):

Alexander Lucas	Doyenné du Comice	Old Home
Bergamotte Poiteau	* Duchess de Angoulême	Osband
* Beurré d'Anjou	Duroudeau	Passé Crassane
Beurré Dumont	Early Seckel	Phelps
Beurré d'Esperen	Easter Beurré	Pitmaston Duchess
Beurré d' Amanlis	Elizabeth	Pound
Beurré Fougueray	Flemish Beauty	President Barabe
Beurré Gifford	Gorham	Pulteney
Beurré Hardy	* Glou Morceau	Rosteizer
Beurré Lebrun	* Howell	* Tyson
Beurré Superfin	Kieffer	Urbaniste
Bloodgood	* Louise Bonne de Jersey	Uvedale's St. Germain
Buffum	Madeleine	* Vicar of Winkfield
Colmar	Maxine	* White Doyenné
Covert		Winter Bartlett

Group 2. Varieties that will unite with the quince and grow upon it but in many cases do not make a lasting union and are improved by double-working:

Bartlett (William's Bon Chrêtien)	Gray Doyenné
Belle Lucrative	Guyot (Early Bartlett)
Beurré d'Arenberg	Jargonelle
Beurré Clairgeau	Josephine de Malines
Beurré Rance	Kings
Beurré St. Nicholas (Duchess d'Orléans)	Lawrence
Bristol Cross	Laxton's Progress
Cayuga	Lemon
Chapin	Madeleine
Clapp	Max Red Bartlett
Comte de Lamy	Merton Pride
Doyenné Boussock	Monarch
Doyenné d'Eté	Nec Plus Meuris
Dr. Hogg	Onondaga
Dr. Jules Guyot	Packham's Triumph
Emile d'Heyst	Reine des Poires
Ewart	Souvenir du Congress
Forelle (Trout)	Willard
Gansel Bergamotte	Winter Nelis

Group 3. Varieties that either fail to unite directly with the quince or make a feeble growth upon it:

Beurré Bosc	Foukouba
Beurré Six	Hoblit
Burbank Holiday	Illinois Bartlett
Columbia	Madame Treyve
Dana Hovey	Marie Louise
Dix	Seckel
Doyenné Goubault	Sheldon
Duchess Bérard	Thompson's
Dunmore	Waite
Dumont	Washington
Early Seckel	Worden Seckel
Farmingdale	

SUMMARY AND CONCLUSIONS

Quinces in use as rootstocks are often designated by group names, as Angers, Fontenay, Orléans, Provence, and Pillnitz. While some of the

groupings, notably Angers, Pillnitz, and Provence, are justified by similarities in botanical characters, yet others are badly mixed and cannot
be designated as groups on botanical grounds. Most of the groupings
that appear in commerce represent the local races of seedling quinces or
vegetatively propagated strains that are indigenous to certain geographical locations or originated there, from which they have derived their
names. It is necessary to separate each of the many varieties of quince
in each group and study each separately before any real evaluation can
be made as rootstocks for dwarfing the pear.

The two most important groups are the Angers and the Provence,
the former having come from northeastern France and the latter from
southern France. These two are separated by characteristic appearance
and growth habits.

Among the Angers, three clones have been standardized, namely,
EM A, B, and C. EM A is the true Angers quince, the preferred material, and the one most commonly and widely used. It is substantially
free from virus. EM B is slightly more dwarfing, and EM C is the most
dwarfing and the earliest to promote cropping. Most sources of quince
C contain quince stunt virus. The Pillnitz quinces are selections of
Angers, recognized mostly for their relative hardiness against winter cold.

Less is known about the Provence quinces, but they are of considerable promise and are receiving considerable study and attention. The
true Fontenay quince, EM D, may belong here. There are a great number of clonal lines in the group, from which substantial progress may be
expected in providing a range of dwarfing rootstocks useful for the pear.

Many other dwarfing materials of pear and related species and genera
are available and have been used sporadically—mostly as novelties, and
with only limited success.

Varieties of pear that do not succeed well on the quince may be improved by double-working, using a body stock, such as Beurré Hardy,
and Old Home, which are compatible with both the quince rootstock
and the pear scion variety.

12

DWARFING ROOTSTOCKS FOR THE PLUM

T H E R E is no rootstock in common use for the plum that provides as severe a degree of dwarfing as does the EM IX apple rootstock for the apple or the Malling C quince rootstock for the pear. In consequence the statement is frequently made that there is no rootstock that dwarfs the plum. But here again the difficulty rests in part with the interpretation of the word "dwarf." For a very dwarf type of plant, it is true that no satisfactory rootstock is now commonly met with for the plum, yet at the same time there are rootstocks available that do show different degrees of vigor and fruitfulness and that affect the vigor and productiveness of the scion varieties of plums worked upon them. In this sense, then, certain rootstocks are more dwarfing than others, and may be so employed where some reduction in size of tree is desired. Unfortunately, although stature may be affected, fruiting is not greatly altered. There does not seem to be the close correlation between small trees and early fruiting that there is for the apple and the pear.

Complex Factors Associated with Plum Varieties and Rootstocks

The rootstock problem with the plum is complicated by the wide assortment of both scion varieties and rootstock materials available. The rootstocks are distributed among the following species: *Prunus domestica*, represented by the European plum, including the prune types; *P.*

insititia, to which the damsons belong; *P. myrobalan*, the cherry plum; *P. americana*, a native American species; *Marianna*, an American hybrid between *P. munsoniana* and *P. cerasifera*; *P. pumila*, the Sand cherry; *P. besseyi*, the Western Sand cherry; *P. persica*, the peach; and *P. armeniaca*, the apricot; *P. tomentosa*, the Nanking cherry; and several related species. And even in the cases that seem like distinct species and subspecies, there is considerable uncertainty as to their identity and their probable hybrid origin.

Further, several of the well-known plum rootstocks, such as the so-called Common plum, refuse to be placed in any recognized botanical species. They are best thought of as horticultural forms or as natural hybrids involving two or more species. They merely add to the difficulty and complexity of the rootstock problem with plums. Added to this is the factor of virus diseases, which are common in scion varieties of plum as well as in plum rootstocks and which affect compatibility.

Similarly the varieties of plums commonly cultivated and worked upon one or several of these rootstocks are from a wide range of species of similar uncertainty. They are, however, commonly placed in three groupings for convenience: (1) *Prunus domestica*, the common European plum, including the prune types, and *P. insititia*, the damsons types; (2) *P. salicina*, the Japanese types; and (3) *P. americana*, *P. hortulana*, *P. munsoniana*, and related species of native American plum.

From this it can be seen why it becomes difficult, if not impossible, to generalize freely as to stock-scion relations with plums. In fact, incompatibilities occur much more frequently among plums than is commonly realized. One exception seems to be the Myrobalan plum, which produces a vigorous commercial tree with a wide assortment of plum varieties worked upon it, regardless of species. But in practice it is necessary to treat each variety and rootstock combination separately in order to be accurate and to avoid difficulties with incompatabilities.

RESEARCH AT THE EAST MALLING RESEARCH STATION

About 1914, Hatton and his associates at the East Malling Research Station, Kent, England, began a study of plum rootstocks. They collected representative rootstocks from a number of sources and attempted to classify them and standardize them in a manner similar to their studies with apple rootstocks. In his first report, Hatton commented upon the lack of guiding principles in selection of plum rootstocks by growers and nurserymen. Unlike dwarfing rootstocks for the apple and the pear, where there is a substantial historical background and various classifications, none seemed evident for plums.

Horticulturists of the eighteenth and nineteenth centuries, for example, recorded no principles of selection and little knowledge of the effects of plum rootstocks. The interest appeared to be mostly directed at vigor and productivity. Between 1724 and 1835, the names of a number of plum rootstocks now in use were mentioned. The Shining Mussel plum, the Black Damas, and the St. Julien are examples. Rootstocks were raised both from seed and from layers, with considerable dependence upon suckers from the roots of plum trees. In this way, many local plum rootstocks arose, which added materially to the confusion.

Enumeration of Plum Rootstocks

In order of preferential use in Europe, the collection made at the East Malling Research Station included the following principal plum rootstocks: Mussel, Brussel, Common Plum, St. Julien, Black Damas, Myrobalan, Marianna, Brompton, *Prunus domestica*, Kroogespruin and *P. spinosa*. Of these, the Myrobalan, St. Julien, Black Damas, *P. domestica* and *P. spinosa* were raised from seed. The others were propagated vegetatively. They have all been fully described and discussed in the publications from the East Malling Research Station. Following is a listing of plum rootstocks, with some of the major characteristics of each, taken largely from this source, but including several others as well.

Ackermann plum (*Marunke*) (*Prunus domestica*) came from the Ackermann-Torgau nursery in Thüringen, Germany, where it is found in the wild. Leaves are small to medium, oval or obovate, thin, soft; fruit is medium, roundish ovate, dull purple. It produces a medium or intermediate-size tree.

Aylesbury prune (*Prunus domestica*), also known as Michaelmas prune, originated as a variety in Buckinghamshire, England, where it is grown on its own roots and also used as a rootstock for other varieties.

Leaves are medium to large, oval, thick, somewhat rough; fruit is medium, oval, dull blue. It produces an intermediate to semivigorous tree.

The Black Damas stocks (*Prunus insititia*) resemble the St. Julien. The name is also found as Black Damascena. There are scores of varieties of Damas plums, including a white Damas. They are propagated by seed as well as by layers and cuttings. The name dates back several centuries before Christ and is associated with the ancient city of Damascus, in Syria, where this type is supposed to have originated. A Petit Damas Noir, a Gros Damas Noir, and a Damas d'Italie have long been used on the Continent as rootstocks for the plum. Clonal selections have been made from this group, including four by the East Malling Research Station; namely, Black Damas A (large-leaved), B (small narrow-leaved), C (ragged-leaved), D (flat-leaved), E, and a White Damas (WD4). Black Damas C is said to be identical to Pillnitzer Klon R26C. It is the only one of the group still judged useful. Leaves are small to medium in size, broadly ovate, almost circular, slightly rough, resembling a Domestica; fruit is medium, roundish oblate, purplish red. It produces a medium or intermediate-size tree.

Brompton (*Prunus domestica*) is of English origin. Leaves are large, oval to almost circular, with irregular waved margins and rough surface, thick, fleshy. The fruit is medium in size, oblate, dark purple. Shoots root readily from hardwood cuttings. It is still recommended for the production of semivigorous trees, not so large as those on Myrobalan B.

Brussel (*Prunus domestica*) came from Holland. It has been widely used on the Continent. Leaves are very large, broadly elliptical, thick, fleshy, somewhat rugose, with hairy undersurface. Shoots are vigorous and somewhat spiny. The fruit is roundish, dark purple, flat-sided, medium to small in size. It propagates easily by layering, and makes a large tree. Because of disease susceptibility and other undesirable characters it has finally been recommended for discard as a rootstock for the plum. It is included here only because it is so often mentioned in the literature.

Common plum is from England, where it has been long used. It has not been identified botanically as belonging to any single species, but is found fairly true to type. Leaves are small, oval to oblong, thick, rugose, curled, drooping. Fruit is small, round-oblate, and dark purple-black. It propagates readily from both layers and hardwood cuttings. It is a clean stock, easy to work. Though it is recommended for production of semidwarf trees, it is not congenial with all varieties. Nevertheless, it is widely and successfully used as a semidwarfing plum rootstock.

EM 478 (*Prunus domestica* x *P.* sp.) is a selection by the East Malling Research Station, England, from a cross between Brompton and Common plum, made in 1933. Leaves are small, medium oval, thin, slightly rough. It has been suggested as a semidwarfing rootstock but needs further trial.

Grosse grüne Reineclaude (*Large Green Reine Claude*) (*Prunus domestica*) is found in the trade on the Continent. Leaves are large, elliptical, rugose; fruit medium, round, green. It makes a medium-sized to vigorous tree.

Hüttner IV (*Prunus domestica*) is from Germany as a selection from Ackermann. The leaves are medium in size, broadly ovate, fairly smooth. It makes a medium-sized tree.

Kroogespruin (*Kroosjes gelb, Yellow Krujespruin, Gelekroos*) is from Holland, where it is used mostly for peaches. Leaves are medium to large, medium oval, thick, soft. It is one of the few plum rootstocks that does not bear a dark-purple fruit. It is propagated by layering, and is medium in vigor. Various clonal selections have been made of this rootstock, including K_3 and K_5 by the East Malling Research Station, but none has proved outstanding.

Marianna (*Prunus cerasifera* x *P. munsoniana*) is from America. It originated in Marianna, Texas, in the 1870's, probably from a cross between De Caradeuc (*P. cerasifera*) and Wild Goose (*P. munsoniana*). Leaves are medium to large, long, narrow, lanceolate, thin, smooth, shiny, with drawn-out tips. The fruit is small, round oval, bright carmine. It propagates readily from hardwood cuttings, is widely congenial, and is useful for a medium or intermediate-sized tree.

The Mussel stocks (*Prunus insititia, P. domestica?*) are a mixed group of considerable variability, of English origin. Three or four main types, and several minor ones, have been recognized. They are described below:

Broad-Leaved Shiny Mussel (*P. insititia*) has variable, medium-sized leaves, broadly ovate, thin, smooth, flattish, with shiny surface; fruit is medium to small, oval, necked, blackish-purple. Shoots are less thorny than Common Mussel, but propagation is more difficult. It produces a medium to intermediate-sized tree.

Common Mussel (*P. insititia*) has medium to small ovate leaves, thin, very curled, rugose, with finely serrate, wavy margins; fruit spherical, smoky black, suggestive of Damson. Shoots are generally spiny. Propagation is rapid by root cuttings. It makes a desirable medium-sized tree.

Flat-Leaved Mussel (*P. insititia*) has leaves of medium size, broadly ovate, thin, smooth. This stock is not often met with.

French Mussel (*P. insititia*) has small leaves, medium to broadly ovate, thin, rather rough. This stock is not often seen.

Narrow-Leaved Shiny Mussel (*Bastard Shiny Mussel*) (*Prunus domestica*) resembles the Broad-Leaved Shiny Mussel. Leaves are small to medium, narrowly ovate, thin, soft; fruit is medium, oblong oval, with no neck, blackish-purple. It produces a medium or intermediate-sized tree.

Myrobalan (*Prunus cerasifera*) is the Cherry plum, characterized by cherrylike leaves and fruit. It should be noted that the Myrobalan is sometimes erroneously called "Mirabelle." The latter, however, belongs to *P. insititia*, the same species as the Damsons and the St. Juliens. Leaves are thin, elliptical, soft, often tattered. Fruits are small, round, yellow to purplish-red. It is propagated readily from seed as well as by vegetative means. It is compatible with a wide range of varieties, and generally produces a large tree. The fact that at one time it was considered a dwarfing rootstock leads to the belief that there are strains or selections that might be used in this way. At present, however, the types used generally produce large trees. Myrobalan is the most widely used plum rootstock for vigorous orchard trees.

From seedlings of the Myrobalan, a number of selections have been made that have been propagated vegetatively. Selections by the East Malling Research Station are Myrobalan A (narrow-leaved), B (green-wooded), C (broad-leaved), D (yellow-leaved), and E. Of these, Myrobalan B has proved outstanding and has become the recognized standard clonal rootstock for vigorous plum trees. Leaves of this rootstock are medium in size, elliptical, thin, soft, often tattered; fruits are small, round, purplish-red.

Other selections are Myrobalan Blanc (Myrobalan alba) with fairly large leaves, narrowly ovate, thin, soft; Belgian Myrobalan, with medium to large leaves, narrowly ovate, thin, soft; and Myrobalan 29, with large, narrowly ovate leaves, thin, smooth. All three of these are said to produce vigorous trees.

Pershore (*Prunus domestica*) is the Yellow Egg variety of plum, common in the great plum-growing area of Evesham in Worcestershire, England. Leaves are large, broadly elliptical, wavy, thick, fleshy, with finely serrate margins. The fruit is yellow and of good size. Propagation is by layers and suckers. It is still a recommended rootstock and produces a medium or intermediate-sized tree.

Prunus divaricata is similar to Myrobalan (*P. cerasifera*). In fact, it has been at times grouped with this species. Extensive tests with seedlings at the East Malling Research Station have found them not superior to rootstocks already available.

Prunus domestica stocks as found in the trade include a wide variety of seedlings from this species, dominated by the German Prune or Quetsche types, which are the oldest of the prune types and are grown widely throughout the world. These stocks have been used mostly on the Continent. Although few improved clones have been developed from this source, the wide variation from vigorous to dwarfing forms that may be found among them suggests further exploration.

Prunus spinosa (European Sloe) is the common wild plum of western Europe. Plants are small, spreading, much branched, thorny; leaves small, oval, finely serrate; fruits small, round, black.

The St. Julien stocks (*Prunus insititia*) are the St. Julien plum of old horticultural literature, closely related to the damsons, and constituting a division within *P. insititia*. The St. Juliens have been widely known and used for several centuries as rootstock material that dwarfs the plum to some degree. Among recognized forms are a Petite St. Julien, a Gros St. Julien, and a St. Julien de Toulouse. Plants are characteristically stiff, stocky, often spined, with medium-sized, thick rugose leaves and small, roundish, oblong dark fruit. They are propagated from seed as well as by layers and cuttings. Several clonal selections have been made, including six by the East Malling Research Station, England. The more important clonal lines in this group are enumerated below:

S. Julien A (dark curly-leaved) has leaves of medium size, oval, fairly thick, rough; and fruit that is round oblate, olive green. It has been found valuable as a semidwarfing rootstock for the plum, and is highly recommended for this purpose. It is replacing Common Plum as a semidwarfing rootstock.

St. Julien B (glaucous flattish-leaved) is thought to be a hybrid between *P. insititia* and a tetraploid form of *P. spinosa*.

St. Julien C (large curly-leaved) has broadly ovate leaves, rather thick and fleshy, resembling a Domestica plum. Fruit is medium in size, round, dark blue.

St. Julien J has small leaves, oval thin; its fruit is very small, globular, blue-black.

St. Julien K has small to medium leaves, oval, rather thick and fleshy, resembling a Domestica plum; and the fruit is small, round, purplish-red. This is the most dwarfing of the plum rootstocks, producing a full dwarf tree, which merits more extensive trial.

CLASSIFICATION OF PLUM ROOTSTOCKS

Classification According to Summer Characteristics

Tydeman, at the East Malling Research Station, has attempted to group this array of widely diverse rootstocks for the plum according to their summer characteristics as follows:

A. *Pershore group*—vigorous, stout, stiff reddish or purplish shoots with large, deep-green leaves: Brussel, Brompton, Pershore.

B. *Black Damas group*—shoots usually rather less vigorous than in the Pershore group, flexible, leaves of medium size: Ackermann (Marunke), Narrow-Leaved Shiny Mussel, Gelekroos (Yellow Krujespruin), Broad-Leaved Shiny Mussel, Black Damas C, Flat-Leaved Mussel, French Mussel.

C. *St. Julien group*—only moderately tall, stiff stocky shoots, often considerably spined, with rugose, thick or fleshy, often rather drooping leaves: St. Julien A, St. Julien K, Aylesbury Prune, St. Julien C.

D. *Common plum group*—shoots not usually very vigorous, often rather spreading, slender, flexible, often fairly clean but occasionally spiny, with smallish dark-green, rough leaves: St. Julien J, Common Mussel, Common plum, EME 478.

E. *Myrobalan group*—vigorous, shoots long, very flexible, leaves fairly large, narrow, thin, smooth, almost hairless, with very indistinct glands and finely serrate teeth: Belgian Myrobalan, Myrobalan Blanc, Marianna, Myrobalan 29, Myrobalan B.

European Classification According to Size

It may be of some interest, especially for those who wish to delve into the rootstock problem with plums more deeply, to understand the general grouping of plum stocks used in England and on the Continent according to the vigor of the tree that is produced when plum varieties (cultivars) are worked upon them. While disappointing in the fact that there is less dwarfing among the groups than one might like, yet they do settle the question of which ones are not dwarfing as well as those that are.

Based on the vegetative performance in England of budded or grafted trees, they have been grouped by one writer as follows:

Semidwarfing: Brussel (not a recommended stock), Common plum (shows considerable incompatibility), St. Julien C (not readily available), St. Julien A.

Intermediate: Common Mussel.

Vigorous: Myrobalan B, Pershore, Brompton, and perhaps Marianna and Black Damas C.

A comparable grouping from the Continent of Europe based on the growth of the rootstock itself, not budded or grafted trees, is as follows:

Medium strong growing: Kroosjes gelb, Pershore, Black Damas C, and St. Julien A, C, and D.

Strong growing: Brussel, Brompton, Grosse grüne Reine Claude.

Very strong growing: Myrobalan B, Marianna.

A second grouping from the Continent, and one that agrees closely with the preceding, is:

Medium strong growing: Ackermann, Hütner IV, Kroosjes gelb, Brussel, Common Mussel, Narrow-Leaved Shiny Mussel, Broad-Leaved Shiny Mussel, Common plum, Black Damas C, St. Julien A, B, C, and D, St. Julien de Toulouse, Grosse grüne Reine Claude.

Strong growing: Myrobalan B, Myrobalan alba, Marianna, Pershore, Brompton.

The majority experience seems in England and on the Continent, then, to place the Common plum (England) as the most dwarfing of the lot, followed by St. Julien C, St. Julien A, and Common Mussel, in that order. Further substantiation comes from tests with the Czar variety of plum on these rootstocks. Trees at nine years of age were 2 meters high on St. Julien C, 2.4 meters on Common Mussel and on St. Julien A, compared with 3.2 meters on Myrobalan B. But it must be again pointed out that St. Julien C is not readily available and that incompatibilities are observed with Common plum.

Finally from the long-time studies with plum rootstocks at the East Malling Research Station, the following rootstocks have been recommended. Each represents a selected clonal line:

Vigorous—Myrobalan B
Semivigorous—Brompton
Intermediate—Marianna, Pershore (Yellow Egg)
Semidwarfing—Common plum, St. Julien A
Dwarfing—St. Julien K

EXPERIENCES IN AMERICA WITH PLUM ROOTSTOCKS

Rootstocks for dwarfing the plum have received very little attention in America, owing partly to decline in interest in the plum itself, and partly to the fact that many varieties of plums are characteristically early-bearing and not overly large in tree. In fact there has seemed to be more need to improve the vigor and stature of many plum trees in America than to repress them. Interest has for the most part been directed at producing large, vigorous, high-yielding plum trees rather than dwarfed trees. Most mention of dwarfing in American literature has been by way of warning against rootstocks of this nature for the plum. The Myrobalan plum (*Prunus cerasifera*) has proved desirable as the rootstock for vigorous, productive trees, and there the matter has largely rested.

Nevertheless there is a potential interest in plum trees, both standard and dwarfed. The plum is a fruit that runs the gamut of color, size, shape, flavor, and use. In some ways it is the most versatile and varied of all the deciduous tree fruits. If a reliable dwarfing rootstock could be developed, it is a fair guess that the plum as a fruit might stage a considerable resurgence of interest, especially for home garden and ornamental planting.

Patrick Barry, of Rochester, New York, in 1859 mentioned several rootstock materials as useful where a small tree was desired. Among these was the "Myrobalan," which he said produced a small-sized garden plant; but it is more than likely that he referred to the "Mirabelle" plum, which is an Insititia plum (*Prunus insititia*). The names were often confused in literature of that time. Also, he described the Sloe as being valuable for a prolific small tree, although it is not clear whether he referred to the European Sloe (*Prunus spinosa*) or to some American plum. The Beach plum, the Canadian plum, and the Chickasaw plum were also mentioned for dwarfing. The Beach plum was probably *Prunus maritima*. The Canada plum is uncertain but may have been *Prunus nigra*. The Chickasaw plum may have been *Prunus angustifolia* or *P. munsoniana*, or it may have been a damson type (*Prunus insititia*) from the noted damson district around Chillicothe, Ohio. It is of some interest to note that this type of damson was brought from Europe by early colonists and found its way to Chillicothe by way of Virginia and North Carolina.

Waugh found in Massachusetts that St. Julien seedlings produced trees slightly smaller than standard but did not bear fruit until the fourth year set. On the other hand, some American species of *Prunus* gave a much greater degree of dwarfing, such as *P. pumila* (Sand cherry), *P. besseyi* (Western Sand cherry), and *P. americana* (American plum). Yet incompatibilities with these species are much more frequent than one would like.

Hedrick, in New York State, conducted tests with fifteen varieties of plums on six rootstocks. The scion varieties were selected to represent a wide range of plums, including five Domestica plums (*P. domestica*), two Insititia plums (*P. insititia*), two Hortulana plums (*P. hortulana*), two types of *P. munsoniana*, and one representative of *P. americana*. The rootstocks were seedlings of *Prunus americana*, St. Julien, and peach (*Prunus persica*); and hardwood cuttings of Marianna.

After ten years in the orchard, the largest trees were produced by the Myrobalan rootstocks, the smallest were on Americana, and the others were intermediate. A representative rating in terms of trunk diameter (inches) of the scion varieties is as follows: Myrobalan 5.9; St. Julien 5.4; own-rooted 4.5; Marianna 4.4; Peach 4.1; and Americana 4.0. These results agree very closely with general observations in Europe.

Native American Plum Rootstocks

In general terms, the American plum (*P. americana*) propagated from seed has proved somewhat successful as a moderately dwarfing rootstock for a wide range of varieties of plum, including European prune types, damson types, Japanese types, and native American types. The scions tend to overgrow the rootstocks, and with strong winds the trees may break at the union, as reported from Iowa. However, it must be again noted that the greatest interest in this stock is in regions where winter hardiness is a problem. It would seem that the hardiness of Americana rootstocks may be the more important factor, and that dwarfing may be secondary. Nevertheless, this material deserves more attention as a dwarfing material for the plum.

The Sand cherries appear to be the most dwarfing of all the stocks reported. They are promising for both the plum and the peach. There is some confusion about the species involved in this group. *Prunus pumila* is found on sandy shores in the Great Lakes region of the United States. It is a narrow-leaf, upright form growing three to five feet tall. *P. besseyi* (or *P. pumila besseyi*, as it is sometimes called) is a broad-leaf form, bushy, and more dwarfed. It is found in the Great Plains area of the United States from Manitoba to Kansas and Colorado. A rarer form of Sand cherry is one known as *P. cuneata, P. pumila cuneata, P. susquehanae,* or *P. pumila susquehanae*. Erect to three or four feet, it is found in woods, hills, and shallow bogs from Maine to Manitoba and south to Pennsylvania. Then there is a prostrate dwarf cherry found on beaches and shores from Quebec to Massachusetts and Ontario which is called *P. pumila depressa*.

Of these, *Prunus pumila* has been known to be successful as a dwarfing rootstock for plums and peaches since at least 1847, when the satisfactory results of Dr. Bretonneau, of Tours, France, were reported. Peach trees reached a height of three feet and plum trees were three to six feet. The trees fruited very early (the first year after budding), produced heavily, and were considered extremely satisfying.

In some respects, the Western Sand cherry (*Prunus besseyi*) is the best of the Sand cherries for dwarfing. It produces a tree which is more dwarfed than that from the eastern form of the Sand cherry (*P. pumila*). On the other hand, there are records that the stands of plums, and peaches as well, in the nursery are better on *P. pumila* than on *P. besseyi*. However, this difference may be due to propagation techniques, because reports from other sources are just the reverse. As has been noted in the discussion on dwarfing stocks for peaches, the Sand cherries tend to mature early, so that unless the buds are inserted in late July or early August, before growth has ceased, the buds of plums and peaches may not take well and a poor stand may result.

The Nanking cherry (*Prunus tomentosa*) has been reported from Canada as being more dwarfing than the Sand cherries but not thoroughly reliable because of uncertain incompatibilities. Brase and Way report from New York State that Stanley prune (*Prunus domestica*) was successful on *P. besseyi*. Trees began to bear at two years after planting in the orchard, and did not fail to bear during a twelve-year test period. The Beauty variety (*P. salicina*) also behaved well on this stock, but some other varieties such as Fellenburg and Pacific (both *P. domestica*), did not do so well. The American Mirabelle and Sweet Damson (*P. insititia*) did very poorly. This again emphasizes the specificity of stock/scion combinations with plums, and why so much dissatisfaction occurs. Apparently only by trial and error will the best combinations be revealed.

The late Professor N. E. Hansen, of South Dakota, an authority on the Western Sand cherry, wrote to the author in 1941: "I have budded many plums on Sand cherry stock (*Prunus besseyi*). The trees fruit freely, but as they got into heavy bearing the trees sagged over and the root was not strong enough to sustain the top. I think the root will be strong enough to sustain varieties of small habit."

This communication would seem to emphasize the fact that no rootstock alone will give the desired dwarfed tree. Rather, there must be an appropriate system of culture adopted that is adapted to the dwarfed tree. This means height of heading, type of training and pruning, and general cultural requirements, as will be discussed elsewhere.

SUMMARY AND CONCLUSION

The rootstock most widely used in America upon which to work the plum is the Myrobalan (*Prunus cerasifera*). In older horticultural literature it was considered to be a somewhat dwarfing stock. Today, raised from seed, Myrobalan rootstocks are considered the standard for vigorous orchard performance. Neither do the clonal selection Myrobalan A (narrow-leaved), Myrobalan B (green-wooded), Myrobalan C (broad-leaved), or Myrobalan D (yellow-leaved) have potentialities as dwarfing rootstocks of the kind desired, although when used with weak-growing varieties some of them do produce a semidwarf tree. On the other hand, Myrobalan B is the standard clonal rootstock for vigorous, productive orchard trees.

Among the Domestica group (*Prunus domestica*) there are some possibilities for dwarfing purposes, but these have not been fully explored. Domestica seedlings in themselves show variations and produce some very dwarfish types among the more numerous very vigorous forms. But seedlings are too variable to be used with any certainty. The large-leaved Pershore (Yellow Egg) produces a medium or intermediate-sized tree, and the Brompton, a semivigorous tree—both lines being propagated vegetatively but not readily. The Broad-Leaved, or Shiny-Leaved, Mussel, which is propagated with difficulty from root cuttings, is said to have some dwarfing tendency.

Among the Insititia groups, familiar as damson types with generally small, stiff, ovate leaves, there is more promise. In general, they produce smaller growing trees in comparison to the Myrobalan and Domestica stocks. To this group belong the Common Mussel, the Narrow-Leaved or Bastard Shiny Mussel, the Black Damas stocks, and the St. Juliens. The Common plum (England) is related to this group, and has been freely used to produce a desirable semidwarf tree. The Common Mussel, which is propagated from root cuttings, has a dwarfing effect; in fact, some European authorities contend that for fan-shaped and bush-shaped trees of early-bearing characteristics it is more like EM IX among apple rootstocks than any other plum stock. The Black Damas stocks raised from seed, and including types formerly known as Damas d'Italie, Petit

Damas Noir, and Damas Noir, are also said to be slightly dwarfing but variable. Clonal lines of this group have been developed that are more uniform than seed lines, but they are said to be only slightly dwarfing, if at all. Black Damas C is the most preferred of the lot.

The St. Julien stocks as a group are not so hardy against winter cold as might be desired, but they are sufficiently hardy for the major horticultural centers. They are raised largely from seed but can be raised vegetatively, and several types have been known for some time, such as St. Julien de Toulouse, Petit St. Julien, and Gros St. Julien. Clonal selections have been made that have proved semidwarfing to dwarfing. St. Julien A is considered the best for producing a semidwarf tree, and may eventually replace Common plum. It is further recommended by the fact that virus-free plants are available and that it is readily propagated by both layers and cuttings. St. Julien K is promising for a semidwarf to dwarf tree.

Many other species have been used, with limited success and to varying degrees, for dwarfing the plum. The most promising of these are the Nanking cherry (*Prunus tomentosa*) and the Western Sand cherry (*P. besseyi*), which have produced dwarf to semidwarf tree with a number of varieties of plums worked upon them.

13

DWARFING ROOTSTOCKS FOR CITRUS AND MISCELLANEOUS FRUIT PLANTS

CITRUS (*Citrus* spp.) rootstocks that have a dwarfing effect on the scion have not been used to any extent in commercial citrus production in the United States. They are, however, so used in China, Japan, and Palestine. In general, as with production of deciduous fruits, the industry has identified size and vigor with high yields, and has shunned anything dwarfed. There are indications, however, that citrus trees of smaller than standard stature may have a place in commercial citrus production. More than passing interest has been evidenced in various parts of the world.

In the United States, chief interest in dwarfed citrus is from those who desire an ornamental plant and from those who have limited space in which to grow citrus trees and who have limited facilities for their care. Of these two uses, the ornamental aspect has the greater immediate appeal, although the use in commercial plantings may have the greater potential.

Only within the last twenty years has there been much research with dwarfed citrus in the United States. It has come mostly from the Citrus Experiment Station, Riverside, California, begun there by H. J. Webber in 1927, and carried along by L. D. Batchelor and W. P. Bitters, and from the University of California at Los Angeles by R. W. Hodgson and his associates. As more interest is shown in dwarfed citrus trees, it is not too much to expect that considerably more information will be forthcoming, which may tend, in turn, to generate more interest.

Nucellar Embryos

In most varieties of Citrus, as well as Fortunella and Poncirus, extra embryos are developed adjacent to the ovule that are derived, not from fecundation but from somatic tissue. This means that such embryos have the same genetic makeup as the body cells of the plant, and are therefore identical to it. When they germinate, they produce plants identical to the mother plant. They are the same just as though they had been produced from cuttings. This is in sharp contrast to embryos derived by the more typical process of fertilization, in which there is union of male and female reproductive cells that combine characteristics from both parents and result in progeny that are dissimilar.

Nucellar embryos are very important to the propagation of citrus rootstocks. By roguing out the variable hybrid seedlings (they can be identified by their different growth habits) from the seedling population, a clonal lot of seedlings is available all of which are identical one to another. This lends for ease of propagation and for uniformity of performance.

CITRUS ROOTSTOCKS

The dwarfed citrus trees discussed in this chapter are produced by budding upon certain dwarfing rootstocks. In general, citrus trees may be grouped into three categories based on size; namely, large, standard, and dwarfed. The results of several rootstocks have been summarized as follows:

Large trees: Sweet orange (*Citrus sinensis*) and Sampson tangelo (*Citrus reticulata* x *C. paradisi*) rootstocks;

Standard trees: Sour orange (*Citrus Aurantium*), Cleopatra mandarin (*Citrus reticulata*), and grapefruit (*Citrus paradisi*);

Dwarfed trees: Trifoliate orange (*Poncirus trifoliata*) (with some exceptions), the citranges (hybrids between *Citrus sinensis* and *P. trifoliata*), and minor miscellaneous materials.

None of these categories is absolute, being dependent upon the particular cultivar used, variations in soil and climatic factors, and the particular strain of rootstock. Nevertheless, the information is of value as a general point of reference with which to begin.

The Trifoliate Orange

Trifoliate orange (*Poncirus trifoliata*) has been used as a rootstock throughout the citrus areas of the world. It is a small, much-branched tree, very thorny, with palmately trifoliate leaves (tri-foliate, cloverlike). The trees are deciduous in winter. Seedlings grow readily from seeds and produce about 70 per cent nucellar (vegetative) embryos. All stock/scion combinations generally show an overgrowth of the stock at the union.

In the orchard, budded trees have been precocious in fruiting and have yielded well in proportion to their size. Fruit has matured early, and is of high quality. Trees are decidedly dwarf, are disease-resistant, and have shown considerable hardiness. A disease characterized by shelling or scaling of the bark (exocortis) is still a problem, but if it can be avoided or controlled, this rootstock is of prime usefulness. There are, however, strains of Trifoliate, and these vary in their performance. Among these are Rubidoux C.E.S. 833, Swingle C.E.S. 1498, Pomeroy C.E.S. 1717, Webber-Fawcett C.E.S. 2552, Barnes C.E.S. 7554, the Savage from Australia, the Sartori from Argentina, and various selections from old California orange groves. The strain known as Rubidoux C.E.S. 838 has proved satisfactory as a dwarfing rootstock, and has been used for this purpose to some degree. Eventually these rootstocks must be standardized if they are to become important for dwarfing citrus.

The Marsh grapefruit (*Citrus paradisi*) on the Rubidoux C.E.S. 838 strain of Trifoliate orange rootstock used at Riverside, California, has been dwarfed to a greater degree than the orange. Trees have averaged 9 feet tall at twenty years, compared to 16 feet for grapefruit trees on the sweet orange (*Citrus sinensis*) rootstock. They have reached only 28 per cent the size of trees on sweet orange and yet have yielded an average of 58 per cent as much fruit. Of several rootstocks that have been tried, Rubidoux S.E.S. 838 has proved satisfactory for dwarfing the Marsh grapefruit. Again emphasizing the importance of selected strains, other Trifoliate strains have been observed that produced trees standard in size.

Washington navel orange trees twenty years of age on Rubidoux Trifoliate rootstocks were 54 per cent the size of trees on sweet orange and produced 94 per cent as much fruit. Valencia orange trees averaged 58 per cent the size of trees on sweet orange and produced 94 per cent as much fruit.

Eureka lemon (*Citrus limon*) gave unsatisfactory results on Rubidoux, another Trifoliate orange. They all developed the exocortis disease, exhibited bud-union problems, and were very dwarfed and unhealthy. It is suggested that this disease was introduced into the Trifoliate rootstock in the buds of Eureka lemon with which they were budded.

The mandarin group (*Citrus reticulata*) of oranges does especially well on Rubidoux Trifoliate orange. At eight years of age, trees were 43 per cent as large as trees on sweet orange and produced 79 per cent as much fruit.

The kumquat (*Fortunella* spp.) also does well on Rubidoux Trifoliate orange. Trees at twenty-one years of age averaged about 11 feet in height, and were very fruitful.

The Box Orange

The box orange (*Severinia buxifolia*) is a small spiny shrub that could easily pass as a substitute for the common box (*Buxus sempervirens*) for hedge planting. It propagates well from both seeds and cuttings. It is said to be resistant to high boron and high salt, and to be immune to many citrus diseases. It markedly dwarfs the scion tops worked upon it. Since there are different strains of Severinia, this stock will need to be standardized.

The Marsh grapefruit is especially successful on this rootstock. Trees at twenty years of age had reached a height of 9 feet, compared with 14 feet for grapefruit on sweet orange. Yields were 54 per cent of those on sweet orange.

Results with oranges in California were not satisfactory, but budwood sources and virus troubles may have contributed. Valencia orange trees at twenty years of age were 12 feet in height as compared with 16 feet on sweet orange roots and produced only 39 per cent as much fruit. Washington navel orange trees were 9 feet tall at twenty years, were chlorotic, and yielded only about 27 per cent as much fruit as on sweet orange rootstocks.

The Palestine Sweet Lime (Citrus limettioides)

The Palestine sweet lime, frequently referred to as a sweet lemon, is the principal rootstock of Palestine. It is a large thorny tree. The seeds germinate readily and produce 95 to 100 per cent nucellar embryos. Nursery seedlings are very vigorous. Trees budded on this rootstock are markedly dwarfed (virus xyloporosis), are very sensitive to cold, and the stock is susceptible to gummosis. Fruit grown on this rootstock is low in acid and is somewhat insipid. Because of these shortcomings it is questionable whether this rootstock will find a place in the citrus areas of the United States, even though it does dwarf the scion top substantially.

The Cuban Shaddock

Cuban Shaddock (Citrus grandis hybrid) is highly recommended in Cuba as a rootstock for the Washington navel orange although it is susceptible to the Tristeza virus. It is a large, vigorous, thorny tree resembling the Shaddock in appearance, closely related to the lemon, and perhaps a hybrid. The seeds germinate readily, producing a high percentage of nucellar embryos. The quality of fruit from trees on this rootstock is not so high as on several others, and the trees are inclined to be tender to winter cold. It produces a substandard tree.

Marsh grapefruit trees on this rootstock were 70 per cent as large as those on sweet orange and yielded 101 per cent as much fruit during the first ten years. Washington navel orange trees were approximately 45 per cent as large on this stock as on sweet orange, and produced 105 per cent as much fruit during the first ten years. Yields of both grapefruit and orange declined during the next four years. Eureka lemon trees at sixteen years of age were 48 per cent as large on Cuban Shaddock stock as on sweet orange, yielded 66 per cent as much fruit the first ten years, and increased in yield for the next few years.

The Citrange

The Savage citrange (a hybrid between *P. trifoliata* and *C. sinensis*) has been recommended for grapefruit but not for other combinations, except perhaps mandarin oranges. Grapefruit trees at nineteen years of age were only 58 per cent as large as trees on sweet orange, and Satsuma orange trees were 55 per cent as large.

The Morton citrange (susceptible to Tristeza virus) has proved less dwarfing as a rootstock than some other stocks for both the Washington navel orange (63 per cent) and Satsuma oranges (72 per cent), but is highly productive. It is considered especially desirable for home plantings. The Troyer citrange, although vigorous by some standards, is also considered worthy of consideration for further trial. The Rusk, Coleman, and Cunningham citranges have proved sickly and unproductive.

The Eureka Lemon

Cuttings taken from the Eureka lemon (*Citrus limon*) and worked to the Valencia orange have produced trees that are considerably dwarfed (8½ feet in height at eighteen years). But the fruit quality is inferior, and the tree is subject to several diseases (shell bark and psorosis) as well as winter injury.

Various citrons (*Citrus medica*), lemons, and lemon-citron hybrids have been tried as dwarfing rootstocks here and there. Results have been either unsatisfactory or insufficient to warrant further attention.

SUMMARY AND CONCLUSIONS

Citrus trees on vigorous rootstocks may typically reach a height of 20 feet or more and require a planting distance of at least 20 x 20 feet. Trees on suitable dwarfing rootstocks should not reach beyond 8 feet.

The Trifoliate orange (*Poncirus trifoliata*) has proved to dwarf the grapefruit to considerable degree (78 per cent), and to dwarf both the

Washington navel orange and the Valencia orange effectively but to a lesser degree (54 and 58 per cent, respectively). It is unsatisfactory for the lemon.

The box orange (*Severinia buxifolia*) dwarfs grapefruit and Washington navel and Valencia oranges, producing trees about 9 feet tall in comparison to 14-foot trees on standard sweet orange stock.

The Palestine sweet lime (*Citrus limettiodes*) markedly dwarfs the Washington navel orange (37 per cent), the Valencia orange (58 per cent), and the grapefruit (75 per cent). It is handicapped by lack of winter hardiness and tendency to diseases.

The Cuban Shaddock (*Citrus grandis* hybrid) does not dwarf the Marsh grapefruit (70 per cent) as much as some of the preceding rootstocks, but does successfully dwarf the Washington navel orange (45 per cent) and the Eureka lemon (48 per cent). It is somewhat tender to winter cold.

The citranges (hybrids between *P. trifoliate* and *C. sinensis*) are not so dwarfing as some of the preceding rootstocks, but they are hardy and healthy. Marsh grapefruit is dwarfed 58 per cent, and Washington navel oranges 63 per cent.

In all of this discussion, viruses play an important part. The situation is constantly changing, requiring thorough re-evaluation as the virus problem becomes better understood.

THE MEDLAR

The medlar (*Mespilus germanica*) is an interesting though uncommon fruit, closely related to the apple and the pear. It is an attractive shrub or small tree to a height of 20 feet, relatively hardy. It is propagated by budding on medlar seedlings raised from seed. For dwarfing purposes it is budded and grafted upon the pear, quince, and hawthorn (*Crataegus oxyacantha*).

The fruit is a pome, 1 to 2 inches in diameter, which is conspicuous by being open at the apex in contrast to apples and pears, which are closed. The fruit matures in early fall, and should be allowed to hang until fully mature. It should then be ripened at warm temperatures for two weeks, when it can be used for flavoring and for jelly.

There are a number of cultivated varieties (cultivars) as follows: Dutch or Monstrous—to 20 feet in height; Royal—to 15 feet; and Nottingham—10 to 15 feet. Their chief value is as a point of interest in the landscape planting, and as a conversation item.

THE OLIVE

A wild olive shrub (*Forestiera durangens* Standll.), native to the State of Michoacán in Mexico, has been shown by Mario Calvino to dwarf the olive successfully. It is known in Mexico as "Acebuche." The wild olive of Spain is *Olea europa*, but this is a different species, as indicated above, although belonging to the olive family (*Oleaceae*). The Mission olive, worked upon this wild Mexican olive, is said to have grown and fruited well as a dwarf form.

Part Four

PROPAGATION OF DWARFING ROOTSTOCKS AND DWARFED FRUIT TREES

14

PROPAGATION OF ROOTSTOCKS

ROOTSTOCKS for fruit trees are of two distinctly different types, namely, (1) seedling rootstocks and (2) clonal rootstocks. Seedling rootstocks are raised from seed. Clonal rootstocks are raised from layers and cuttings of some vegetative portion of a plant. The term "clone" and "clonal" are derived from the Greek, meaning a shoot or twig. "Clonal" and "vegetative" are frequently used interchangeably, but clonal is the better term.

SEEDLING ROOTSTOCKS

Seedling rootstocks constitute the bulk of the rootstock material used for fruit trees. Seed supplies are abundant and inexpensive, the cost of production is relatively low, and seedling rootstocks have been by tradition "good enough" for the general run of nursery and fruit operations. Further, most rootstock materials do not propagate readily by means other than seedage. And although seedling rootstocks have the disadvantage that no two individuals are identical in genetic makeup, yet by cut-and-try methods, certain sources and types of seed have been found to give a practical answer to the commercial rootstock supply, and are used for that purpose (Fig. 62).

As far as dwarfing rootstocks are concerned, those produced from seed are not overly satisfactory. Yet, since some are nevertheless used at present, notably for the peach, the sweet cherry, the sour cherry, the apricot, and the plum, no treatise on production of rootstocks would be

225

complete without a brief mention of their source and how they are raised.

Seed may be purchased from seed houses, seed collectors, and importers of seeds who make a specialty of the service. In the selection of domestic sources of cherry, peach, and plum seed, only late-ripening kinds should be taken, inasmuch as early-ripening sorts seldom have viable seed.

After-Ripening

Fruit tree seeds will not germinate immediately they are mature. They must first complete a period of "after-ripening" in which internal changes are brought about necessary for germination. These processes go on most rapidly at temperatures just above freezing (41 degrees F. is optimum) and under moist conditions such as are provided in nature. Contrary to popular belief, freezing is not necessary and in some cases is actually harmful.

For seeds that require a period of after-ripening, good results may be secured by (1) fall planting, (2) stratification, and (3) controlled-temperature storage. In practice, fall planting is confined mostly to large seed, such as peach, plum, and cherry. Rodent attack and injury from soil washing, limit fall planting of smaller seeds.

With other seeds, stratification is best, a practice in which the seeds are packed in a pit, box, or other container in alternate layers of sand or soil and put out of doors in a sheltered position over winter. It is from this placing in strata or layers that the word "stratification" comes. Leaves, peat moss, sphagnum moss, or anything of similar water-holding nature may also be used. Small lots of seed may be wrapped in non-corrosive wire screening for rodent protection, and packed away in boxes or buried in the ground.

For storage under controlled conditions seeds should be packed in boxes with sand or moss, wet down thoroughly, and held at a temperature of about 41 degrees F. If the storage does not keep the seed moist, it must be watered occasionally. The ice compartment of a household

Fig. 62. **Seedling rootstocks in common use in American nurseries:** (Left to right) Myrobalan, Mahaleb, Mazzard, Apple, Pear. These typically produce vigorous trees when budded or grafted to the scion variety, although seed lines are available that dwarf to some degree.

refrigerator, or an icehouse, or a cellar of the right temperature may be used. Under these conditions cherry seed requires 14 to 16 weeks for after-ripening; peach seed, 12 weeks; and apple and pear seed, 8 to 10 weeks.

Fruit seed will germinate at low temperature. They must be watched carefully in early spring for possible starting before the ground is ready to receive them. Packing with ice or lowering the temperature of the storage may be resorted to in order to delay development.

Planting and Care

The after-ripened seeds are planted out of doors in early spring in rows 12 to 24 inches apart. A light coverage of peat moss over the seed-bed or over the top of the row is helpful in heavy soils, preventing the soil from drying and baking on top and thus interfering with the emergence of the seedlings.

The seedlings of cherry and plum are dug in the fall of the year and sorted and graded during the winter months. They are graded according to size at the collar into Extra, above ¼ inch; No. 1, ³⁄₁₆ inch and up; No. 2, ²⁄₁₆ inch and up; No. 3, below ²⁄₁₆ inch. Either the No. 1 or the No. 2 grade is satisfactory for the plum and cherry. In general the smaller sizes are preferable for conditions where a vigorous growth is to be expected and for strong-growing kinds, such as the Myrobalan plum. The larger sizes are best for slowing growing conditions and for slower-growing kinds (Fig. 63).

PROPAGATION OF CLONAL ROOTSTOCKS

Clonal rootstocks are the chief reliance for satisfactory rootstocks for dwarf fruit trees. This is especially true for apple and quince rootstocks,

Fig. 63. Two common types of seedling rootstocks used in the nursery trade: (Left) Straight-root seedling for piece-root grafting; (Right) Branched-root seedling for lining-out and budding. Branched-root seedlings are induced by undercutting the taproot of the seedling when it is two to three inches tall. Straight-root seedlings develop from undisturbed seedlings.

and promises to be of greater concern with other classes of fruits in the future.

They may be propagated vegetatively in several different ways, as from (1) hardwood stem cuttings (a portion of a dormant shoot), (2) root cuttings (a portion of a dormant root), (3) softwood stem cuttings (a portion of a leafy shoot), (4) layers, (5) suckers, and (6) nurse-root grafts.

The particular method to be used depends upon a number of factors, such as availability of wood, expense of the material, supply of labor and land, and the nature of the rootstock itself. Just as each variety of fruit has individual characters of flavor and keeping quality peculiar to itself, so each rootstock has individual characteristics of rooting and of response to different rooting treatments. For example, the EM I and XIII apple rootstocks tend to produce a relatively high proportion of variable, large-sized grades of shoots when layered, whereas the EM VII produces a high proportion of medium-sized uniform grade. Quince EM A and apple EM IX propagate well from hardwood cuttings under some conditions, whereas apple EM XII propagates with great difficulty by this means. There are apparently no differences in the characteristics of the plants produced by the different methods.

The general principles of plant propagation apply to clonal rootstocks; and since circumstances may arise that may make propagation by one means more desirable or advantageous than another, the more important principles and methods are here discussed.

How and Where Roots Form

Roots arise in the tissue of the plant that lie just outside the woody portion of the root and stem. This is in the inner bark of the root, technically known as "pericycle." (See Chapter 4.)

The first indication that a lateral root is being formed, as seen under the microscope, is in the enlargement and elongation of two or more cells in the pericyclic region. They subsequently divide and develop to form a conical growing point, which is called a root primordium and which is buried beneath the "bark" of the root or stem in which it is formed.

Myriads of root primordia are formed throughout the root system of a plant in this way. Some of them may continue enlargement and differentiation, displacing and rupturing the tissues under which they lie and emerging as lateral roots. Root primordia may develop upon them, in turn. In this way a dividing and redividing root system is formed that penetrates and invades the soil or the growing medium. But not all these root primordia emerge as roots. They may lie dormant or quiescent until stimulated to activity by some external environment or internal change, thus remaining as potential roots.

Roots may arise similarly from stems. When they arise from a plant part other than a root they are called "adventitious," although in popular usage the word "adventitious" has come often to be applied erroneously to any root that rises late in the development of the plant, as from a wounded region of a large root.

The significant feature of root formation is that roots arise from tissues that are still "young" or active and that are not yet differentiated into some mature plant part. Accordingly, it is from the active growing region of a plant that roots may be most easily induced, as from tissues near axils of leaves, from the base of nodes, and from cambium, pericycle, and ray parenchyma.

Plants vary in their capacity to form roots and to respond to treatments intended to induce root formation. Certain apple rootstocks, for example, form root primordia so freely as to give shoots and roots a "pimply" appearance. Cherries, on the other hand, form roots only under the most favorable conditions, and even then with difficulty. On this basis plants are designated as rooting "easily," "freely," "with difficulty," and the like.

Plants that are in the "juvenile" condition form roots most freely. They can be aided by etiolation, that is, by preventing light from reaching the parts from which roots are desired. A piece of electrician's tape tied around a stem has been found effective in this manner. And the practice of covering young shoots with friable soil so that they emerge through it is used for this reason in layering of plants.

Use of Growth Regulators to Aid Rooting

It is possible to speed up the rooting of cuttings by the use of certain chemicals known variously as plant regulators, hormones, auxins, and growth substances. The method is best suited to softwood cuttings, but has been found useful with hardwood cuttings in a few instances.

For all practical purposes, these materials do not induce the formation of roots as discussed earlier in this chapter, but they do hasten the rooting process so that the critical period in establishing a cutting on its own roots is shortened. This is often very important, and may spell the difference between success and failure. Common materials are beta-indolebutyric acid and alpha-indoleacetic acid. They are sold under various trade names, and the directions supplied by the manufacturers are generally reliable.

There are three methods in common use, namely; (1) the dust method, (2) the solution method, (3) the concentrate-dip method. The cuttings are treated just prior to insertion in the propagation frame.

In the dust method, the bases of the cuttings are dipped into the dust mixture and snapped lightly to remove excess material. In the solution method the cuttings are stood upright in about an inch of a dilute solution of the chemical for 12 to 24 hours. In the concentrate-dip method the bases of the cuttings are briefly dipped in a concentrated solution just prior to planting.

The Juvenile Condition in Plants in Relation to Rooting

Plants undergo a series of growth phases during their development from seedling to senility. These phases bear a relation to ability to regenerate roots. Although development is progressive from seedling to senility, four rather definite stages or phases have been identified; namely, juvenile, transitional, mature, and senile. These differences include morphological deviations, such as leaf thickness, width of vein mesh, length of vein, spinyness, branch characters, bud break, leaf surface, leaf serrations, and leaf size. The juvenile phases resemble the wild form.

There are also differences in anatomical structure. The juvenile forms are more woody than mature forms, and the cells have a higher content of cellulose and lignin. Mature tissues contain more starch, sugars, and mineral nutrients.

Most important, however, from the standpoint of propagation, is the fact that plants in the juvenile phase have a much higher capacity for rooting than do plants in the mature phase. Thus it has been shown that cuttings taken from one-year-old seedlings of orchard trees may root readily, whereas the rooting ability of cuttings taken two or three years later falls off appreciably. It has already been pointed out that most cultivated varieties of fruit fail to root well from hardwood cuttings; yet at the same time, seedlings raised from seed taken from these same varieties will frequently root readily.

This generalization does not apply to all plants. Some plant material, such as ivy (*Hedera*), roots well from either young or older wood. At the same time, there is still a difference shown between the two ages. The juvenile phase produces vines from cuttings, whereas the mature phase produces small trees and shrubs.

The same situation as regards rooting applies to rootstock material. Some, of course, such as the Angers quince and the Malling and Malling-Merton apple stocks, propagate freely regardless of phase. But the *Malus robusta* 5 from Canada offers an interesting contrast. Thus, Blair, Mac-Arthur, and Nelson, of Ottawa, Canada, have reported that the young seedling from which this plant originated possessed rough, thorny twigs characteristic of juvenile growth. Plants rooted very well when layered. Stool beds were developed from this material, which were cut back to the ground each year in the normal method of propagation. The shoots retained their juvenile characters and also their ability to root freely.

However, trees were developed from *M. robusta* 5, and as these grew older they lacked the spiny and twiggy character. Moreover, when buds were taken from this mature growth and propagated in the nursery, the resulting trees produced smooth, spineless growth.

Since it was more pleasant to work with the smooth, spineless forms, a stool bed was developed from these trees. This was about eighteen years from germination of the seed from which *M. robusta* 5 sprang. The mother plants were much less vigorous than the juvenile form, less productive, and less well rooted. It was concluded that the advantages from smoothness and lack of spines were overbalanced by loss of good propagating characters.

Still further, there was a marked difference in the rooting ability of both softwood cuttings and of leaf-bud cuttings from the two sources. Softwood cuttings from the original stool bed, which exhibited juvenile characters, rooted 90 to 96 per cent, whereas softwood cuttings from stool beds from the seventeen-year-old trees rooted only 4 to 12 per cent.

Similarly, leaf-bud cuttings from the juvenile material rooted 50 to 90 per cent, whereas those from mature trees produced no roots whatsoever.

Apparently the severe annual pruning of the stool bed retarded physiological aging of the plants, and caused them to retain characteristics for rooting associated with juvenility. As has been noted, shoots from adventitious buds and etiolated shoots both tend to behave like juvenile material.

PROPAGATION BY HARDWOOD CUTTINGS

Hardwood cuttings are made from mature, dormant shoots. Of the three forms of cuttings mentioned (hardwood, root, and softwood), hardwood cuttings are the most readily procured and the most easily handled. The area occupied is less than for layering, and the attention required is less than for layering and for either softwood or root cuttings. Unfortunately, convenience and economy are offset by the fact that only a few of the desired rootstocks will produce roots readily from hardwood cuttings. However, those that will propagate by this method, such as the quince and some plums, produce excellent rooted material very economically.

Next to the inherent nature of the plant material itself, the most important factors are friable soil, adequate moisture, an equable climate, and a relatively long growing season. In the United States, these conditions are generally better provided in the South and in the Pacific Northwest than in the North and Northeast.

Sources of Plant Material

Dormant, moderately vigorous shoots of the current season's growth (one-year-old wood), supply the material for hardwood cuttings. The source is most important. If reliance is to be placed on this form of propagation, it is best to establish a row or a plantation especially for a supply of wood. Such a planting may be handled much like a stool block.

Each fall the aboveground shoots are removed just before severe freezing weather, and the mother plants are mounded with soil for winter protection. In the spring the mound is removed as early as possible, and the new shoots are permitted to develop under regular cultivation. Unless new shoots are kept arising year after year in this manner, the wood is likely to become old and will root less readily. The difference between a mother plantation maintained for hardwood cuttings and a plantation maintained as a stool block is in the detail of handling. The stool block requires additional special attention, which will be discussed fully under that heading.

Shoots for hardwood cuttings are also obtainable from a hedge of the desired material, although, as has been indicated, unless the plants are maintained in a good state of vigor the rooting percentage is low. Another source of material is the tops of budded lining-out stock, that is, the rooted plants that have been planted in the nursery row and budded to the desired scion variety. The tops will be cut off back to the bud the following spring, anyway. The best tops are those that have made a vigorous growth in the nursery row and have sent out vigorous lateral shoots. These laterals may be torn from the main stem and used as cuttings, and the main stem discarded. Tops that have not developed lateral shoots may also be used, but are less likely to root.

The tops should be cut at least 6 inches above the bud, and should be collected as late in the season as possible, but before severe freezing weather has set in. If the tops are removed too early, the maturity of the lining-out stock may be affected and winter injury may occur to the stock. It is better to err on the side of injury to the tops by the onset of winter than to hazard injury to the lining-out stock and the next season's stand of one-year-old trees. In fact, in locations where severe winters are common, the hazards to next season's one-year-old trees are too great to risk. Only in regions of moderately mild winters should this system be used. It is much better to have a hedge, a few stock plants, or a special planting, such as described above.

By far the best source of wood is from stool beds that have been established for layering. When the bed has been harvested for rooted shoots in the fall of the year, there will always be some shoots that are either sparsely rooted or not rooted at all. The bases of these shoots have probably been covered somewhat with soil so that the formation of root primordia has been induced. Cuttings of this type will root almost 100 per cent. In fact, it is in this way that the hardwood cutting method is most useful for some of the Malling apple rootstocks, namely, as a supplement to layering. With proper attention to details, all the unrooted shoots from such a stool bed can be made to root, so that in the final analysis a rooted plant is secured from practically every shoot arising from a stool bed.

Collection and Treatment

The material from which the hardwood cuttings are to be made should be collected as soon as growth is mature and well ripened, and before severe freezing temperatures have occurred. For example, the rooting of quince cuttings made from wood collected in November of 1934 was 82.94 per cent; whereas collections of the same material made in March of 1935, following a severe winter cold spell, rooted only 2.74 per cent. Early November is likely to be a favorable date for collection.

The wood is placed indoors in a nursery cellar or under similar conditions where temperatures are in the neighborhood of 40 degrees F., and where the wood will not dry out. The cuttings are made into sections 7 to 8 inches in length, the unripened stem tips being discarded. The basal cut is made square across, just below a node or bud. The top cut is made at an angle so as to identify the top of the cutting in planting.

Vigorous shoots will have sufficient length to be divided into three portions; namely, a 6- to 8-inch top portion, a similar length basal portion, and one or more median portions. Tips are poor material; they are often too small in diameter and are soft and best discarded. Basal sections are the best, especially if they are heel cuttings and are torn from the mother plant so that they have a piece of older wood attached. They may root from 90 to 100 per cent. Median sections are intermediate in rooting and may give 50 to 60 per cent rooted shoots.

The cuttings are graded and grouped according to size and portion used, and are bundled and buried upright deeply in damp sand, sawdust, or peat moss, so that the tops are covered with 2 or 3 inches of sand. The storage should be free of frost, the best temperature being 38 to 40 degrees F. Here they will form callous tissue over the base and be ready for spring planting. Many cuttings will initiate roots by spring. In fact, if the storage period is thought of as a period of rooting, and if conditions are provided favorable to rooting, a good percentage of well-rooted cuttings will be secured by planting time. Even the apple EM II, which does not root readily from hardwood cuttings, has given 100 per cent rooting of basal cuttings when packed in damp sawdust at 38 to 40° F. for three months.

Planting and Subsequent Treatment

In the spring, as early as possible, the cuttings are planted 2 or 3 inches apart in rows 18 inches apart, and placed so that the top will be about 2 inches above the soil level. Time of planting is most important. In mild sections, as in England or on the Pacific Coast of America, fall planting or mild winter planting is ideal, the planting being done as soon as the cuttings are made. While this cannot and should not be done in sections where the ground is frozen over winter, it does nevertheless emphasize the point that very early planting is essential for success with hardwood cuttings.

In commercial nurseries, a power-propelled trencher is used to open a vertical trench into which the cuttings are placed by hand and firmed by the heels of the planter. A weighted packer with two close-set wheels further firms the cuttings into position, and finally the soil is drawn up about them by a straddle cultivator. Tight packing of the soil against the cuttings so as to place them in intimate contact with soil moisture is very important.

If planting is done entirely by hand, a good practice is to make a trench with one vertical side, against which the cuttings are placed. A little sharp sand or peat moss may then be thrown against the base of the cuttings and the soil drawn up over them. Excellent results have been secured by the use of peat moss in this way on soil that might otherwise have been too heavy for cuttings.

Fruit tree rootstocks that propagate well by hardwood cuttings are Angers quince and Marianna, Pershore, and Common Mussel plums. English workers have reported as high as 95 per cent rooting of quince

EM C, as compared with 60 to 80 per cent for other quince clones. In New York State, Quince C has rooted 80 per cent as compared with 54 per cent for the common Angers.

PROPAGATION BY ROOT CUTTING

Propagation by root cuttings consists in planting small pieces of dormant roots upright in the soil; these then develop adventitious shoots from the upper end and roots from the lower end, to give a complete rooted plant (Fig. 64). The method is particularly suited to some of the Malling apple rootstocks, where there is a shortage of plant material and where a limited quantity is needed. Handling is not unlike that for hardwood cuttings, and the percentage of usable rooted shoots is high.

The chief limitation is the availability of root material suitable for making root cuttings, but this is not an insoluble problem. Roots may be taken from two-year-old nursery trees that have been fall-dug and placed in storage. Careful removal of a few roots, or shortening back of a few long roots, will not injure the tree for sale. Also, trees may be kept especially for this purpose, being dug each fall, trimmed both in root and top, and planted again in the spring.

Propagation by Root Cuttings

Roots should be selected that are about the size of a lead pencil—¼ to ⅜ inch in diameter—and cut into pieces 3 inches in length (Fig. 64). Sections as short as 1 inch may be used, and even material slightly under ¼ inch in diameter is useful. Though the resulting plants may be less vigorous the first season than where larger material is used, larger sections do not handle well and do not root well. The top of the cutting is cut straight across, and the base is cut at a slanting angle so as to ensure planting right side up.

Fig. 64. **Propagation of rootstock by root cutting: (Above) Section of root, which is placed upright in the soil. (Below) A shoot has arisen from the rootpiece, giving a rootstock that can be budded or grafted the next season.**

The cuttings should be bundled and buried until planting time in damp sand, sawdust, or peat moss in a nursery cellar or some similar location at a temperature of 36 to 38 degrees F., where they will not freeze.

Planting and Subsequent Care

In the spring they are planted as early as possible, as with hardwood cuttings, planting the rootpieces firmly 2 to 3 inches apart in rows about 18 inches apart. The procedure is essentially the same as with hardwood cuttings excepting that the rootpieces are planted with the square-cut upper ends just below the soil level rather than above. Also, the plant parts are smaller and more delicate, so that the operation is more intensive than with hardwood cuttings. The soil should have been carefully prepared in fall, and it should be well drained and ready for the plants early. Small beds or cold frames are very good. Supplemental irrigation is useful. A 2-inch top dressing of peat moss, sand, sawdust, or light soil that does not bake and crust over is almost a requirement so as to permit easy emergence of shoots, especially on heavy soils.

Once or twice during the growing season the plants should be examined, and any that have developed more than one shoot should have the number reduced to one. The plants should be dug in the fall, graded, stored over winter, and planted in the spring as lining-out stock for budding.

Some apple rootstocks, such as USDA 227, 312, 317, 323, and 329, are propagated readily and very successfully by this method. Rootstocks of the Northern Spy variety of apple are also reproduced in this way. Stands of 90 per cent are not uncommon. The Malling apple, plum, and quince rootstocks can be propagated by root cuttings also, but since the root systems of most of the Malling apple rootstocks are composed of small fibrous roots, it is not easy to secure sufficient material for quantity production.

An Emergency Measure

An example of how this method can be used to advantage may be of some interest. To meet an emergency situation, roots were taken from two-year-old apple trees on Malling dwarfing rootstocks and made into root cuttings in midwinter. They were planted the last of February in the greenhouse in bank sand, using root cuttings 2 inches in length and $\frac{1}{8}$ to $\frac{3}{16}$ inch in diameter. After 30 days they were transplanted to 2-inch pots. The house was kept cool (60 degrees F.) for a week or ten days, until the cuttings had become established, after which the temperature was raised to 70 degrees F. It was found that plants should not be forced too hard, that there is the possibility of burning on hot days unless shade is provided, and that losses from drying-out may occur unless watering is carefully attended to. Nearly 100 per cent rooting was secured this way from root cuttings of apple EM I, II, IV, V, and IX. Other types could probably be rooted similarly.

The potted plants were transplanted to the field the middle of May. The loss on transplanting was about 30 per cent. The plants were large enough by late August to be budded. The yearling trees produced were salable, but were small and lacking in vigor and not so satisfactory as stronger material from vigorous lining-out stock. Yet as an emergency measure, the operation was considered highly successful.

PROPAGATION BY SOFTWOOD CUTTINGS

Softwood cuttings are made from immature or growing shoots with leaves attached. They are more difficult to handle than either hardwood cuttings or root cuttings. They present the problem of preventing water loss until the severed plant has become established on roots of its own. This means that the cuttings must be rooted under conditions of high humidity, often with accompanying high summer temperatures, and that glass cases or shade or other similar equipment must be provided. Further, the plants require frequent attention to prevent injury from fungus attack, drying out, burning, and other hazards. The introduction of automatic mist propagation has greatly increased the usefulness of the method. Plants that are difficult to propagate by other means will usually respond to this method.

Wood is selected from vigorously growing plants, as from lining-out stock or a stool block. The time of collecting is most important. The wood must be just mature enough and stiff enough so that it offers resistance to the thumb at the tip when the shoot is held in the fist and the tip bent by pressure from the thumb. This condition is usually reached in late June and early July. A few weeks later, and the cuttings will root poorly if at all.

The tip portion of the shoot is likely to be too soft to use, and should be discarded. The rest may be made into cuttings 3 to 5 inches in length, making the lower cut just below a bud or leaf, and the tip cut a short distance above a bud or leaf. Two or three buds are preferable, although one is sufficient if the supply of material is low. For convenience in handling, the lower leaves may be removed. It has been found, however, that cuttings will root better if the foliage is reduced as little as necessary.

The cuttings may be kept from wilting by placing them under moist paper or in a pan of clear water as they are made. They should be planted immediately in a covered frame in sand or in a mixture of 3 parts sand and 1 part peat moss, 1 inch apart in rows 2 to 4 inches apart. The plants should be watered at once, and two or three times a day for a week to keep the leaves fresh. They should be shaded with newspaper or cloth. The principle involved is to provide warm, shaded, humid conditions for a week or ten days, and then gradually to supply a little more air so as to reduce mildew and fungus troubles and to give a little more sunlight as the plants develop. When they are well rooted—four to six weeks—they should be transferred to pots or flats containing garden soil.

An improvement in the propagation technique has been made by misting the cuttings automatically and intermittently with water. Fungus troubles are markedly reduced, and rooting is excellent. Since nutrients are leached from the leaves by misting, it is customary to supply a small amount of nutrients in the mist. While one time thought of as an emergency measure, to be used with a new plant material or a shortage of supply, mist propagation has reached commercial proportions. It has been used successfully in Italy for the olive, in Sweden for apple rootstock A2, and in the United States for several clonal rootstock material.

All of the Malling rootstocks can be propagated in this way, although the same types that propagate most easily by mound layering and from hardwood cuttings also propagate most easily from softwood cuttings. Even cherries and peaches, which are notoriously difficult to propagate vegetatively, will propagate readily from softwood cuttings. The plants must either be dug and stored in a nursery cellar over winter, or given winter protection if left outdoors.

Plants may also be propagated from softwood cuttings in March in the greenhouse from potted mother plants brought inside in January, and forced. The rooted plants should be transplanted to the field as soon as possible so as to get them established before the arrival of summer heat. As an emergency measure, such plants may be budded the same season when they have attained sufficient size.

PROPAGATION BY LEAF–BUD CUTTINGS

This method is useful when there is a shortage of propagating stock, where a limited number of plants is desired, or where other methods of propagation have not proved successful. It has been used commercially for propagating raspberries, but it has not yet been adapted to dwarfing rootstocks, although it could be.

Leaf-bud cuttings are made in early summer from current-season shoots, taken when the most of the leaves are fully expanded. A cut is made under a bud much as when a shield bud is made for budding, except that the shield is longer (1 to 1½ inches) and deeper and includes the attached leaf. Great care must be taken that the material does not dry out or wilt during the operation.

The cuttings are inserted in bank sand or a mixture of sand and peat so that the shield and the bud are covered and yet the leaf is fully exposed to light. The cuttings must be protected from direct sun by shading, and close attention should be given to watering and control of humidity for the first few days. A standard propagation frame is desirable.

Rooting is rapid, and the percentage rooting is high. The mist method is convenient, and an improvement over manual control.

Mahaleb cherry has been propagated very efficiently this way, and the A2 apple rootstock from Sweden has responded well. In regions not adapted to stooling, the leaf-bud cutting method offers real promise.

PROPAGATION BY LAYERING

Layering consists in establishing a plantation of the desired rootstocks and inducing new shoots to develop and root each season while still attached to the mother plants. The rooted shoots are cut from the mother plants each fall, and a new crop is induced to arise each successive year. The rooted shoots are used for lining-out stock on which to bud the desired scion varieties. The best grades of rootstocks are produced by this method, and the performances of both lining-out stock and of nursery and orchard trees worked upon them have been outstanding. The method has long proved successful both in Europe and in America.

There are two principal variations of layering used for the commercial production of dwarfing rootstocks; namely, (*a*) trench layering, or etiolation method, (*b*) mound layering, or stooling method.

PROPAGATION BY MOUND LAYERING OR STOOLING

In mound layering (Fig. 65), the mother plant is maintained as a closely cropped crown from which new shoots arise each year, as with gooseberry and currant bushes. The operation consists in cutting back a well-established and permanent "mother plant" or stool during the dormant season to about soil level, so that in spring a number of shoots arise from the crown of the mother plant. By mounding with soil, these shoots are induced to form roots along 2 to 4 inches of their growth at the base. They are removed at the completion of the growing season and the entire process is repeated the next year.

Mound layering is the method most in use for quantity production of clonal apple and quince rootstocks, but is not so well suited to cherries

Fig. 65. Steps in propagation of clonal rootstocks by mound layering: (Left to right) Rootstock planted out in spring of first year. Rootstock becomes established during first growing season. Top of rootstock is cut off in early spring of second year. Shoots arise from the mother plant and are mounded gently with earth during the second growing season in order to promote rooting. Shoots have been further mounded during the growing season, and by late fall have developed roots along the mounded portions. Rooted shoot or cutting removed from mother plant at end of the second growing season. The operation is repeated each year.

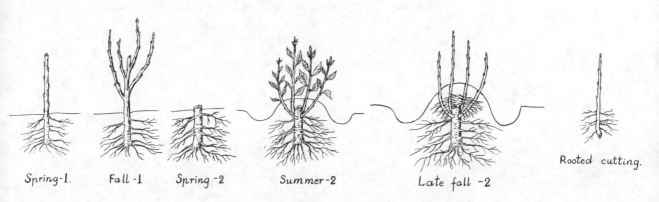

Spring-1.   Fall -1   Spring -2   Summer-2   Late fall -2   Rooted cutting.

and plums. It is especially well adapted to the less vigorous growing types as apple EM IX, and to those having a large number of growing points, as EM III, IV, V, VI, and VIII. It has also been used successfully for EM I, VII, X, XII, XIII, XVI, and MM 101, 104, 106, 109, and 111. There is no reason to believe that others of these types will not behave similarly; but these are the ones that have been proved successful under extensive field operations in the United States.

Establishing the Stool Block

A level site, in a sheltered location, well drained, should be selected for the location of the stool block. Eastern and northern exposures are good, since they reduce heaving, and injury from late afternoon sun. Rows running in the direction of the slope should be avoided, since washing and erosion are severe under such conditions. An area is preferred that will provide long, level rows adapted to machine operations.

A friable loam soil of good fertility, which will not dry out, is important. A clay loam has proved very successful. Lighter soils are easier to work but have not given so high a proportion of well-rooted shoots as have the heavier soils, perhaps because of lower water-holding capacity.

The incorporation of peat moss in the stool block at the rate of about 50 bales per acre has improved both the physical condition of the soil and the percentage of rooting. Cultivation and mounding operations are facilitated, and the emergence of shoots through the soil is encouraged. Sphagnum moss is especially good, but anything that will add humus to the soil and improve both its friability and its water-holding capacity is valuable.

In applying sphagnum peat it may be spread from a wagon or truck body after being thoroughly saturated with water. If applied dry it does not work easily into the soil, and blows badly. In the mounding and cultivating operations, the peat moss becomes incorporated in the soil of the mounds. The liberal use of manure or of manure composted with peat moss each year, applied between the rows, has maintained stool blocks in excellent vigor.

It is essential that the soil be well drained. Areas that have shown even slight indications of poor drainage have given low yields of plants. It must be remembered that operations in the block are carried on late in fall and very early in spring. A soil that is not well drained may puddle badly at such times, and be hurt in structure. Under such conditions all operations become difficult. In addition, new shoots arise early in spring, and rooting may continue late in the fall if the soil is properly aerated and not waterlogged.

The best stock to plant in the stool bed is a well-rooted No. 1 grade ($\frac{3}{16}$ inch in diameter at the collar). A No. 2 grade ($\frac{2}{16}$ to $\frac{3}{16}$ inch) is also satisfactory provided it is well rooted; in fact, some of the best stands occur with the smaller but well-rooted grades. The Extra grade (over $\frac{1}{4}$ inch) can also be used if it is well rooted. The choice, however, lies in the degree of rooting of the stock rather than in the size. A well-rooted No. 2 grade is preferable to a No. 2 grade that is only medium-rooted. Poorly rooted Extra grades are to be avoided; they are difficult to handle and do not transplant well. The plants may be set behind a spade or with a nursery trencher and packer, as is commonly done in a mechanized nursery.

The plants may be set upright 12 inches apart in rows 7 to 8 feet apart. And it is a good plan to set the original plants in a shallow trench 2 inches below the ground level so as to help keep the crown low in the soil. A customary planting distance in a fruit-tree nursery is 42 inches, and the planter is likely to think in these terms. However, this distance is not sufficient for a stool block. As the block becomes older and the mother plants tend to become somewhat higher in the soil, because of the mounding operation, it is increasingly difficult to draw sufficient soil from between the rows to mound the new shoots properly. Distances of 7 or 8 feet are much better than 42 inches, and it may be that even 10 feet will prove better adapted to American methods. In actual operation there seems never quite enough space between rows. At 1 foot $\times$ 10 feet, the number of plants per acre is 4,356.

Cultivation and Mounding

During the first growing season, the block should be cultivated and cared for so as to encourage vigorous growth and ensure a well-established mother plantation from which in subsequent years a good yield of rooted shoots may be cut. In preparation for winter the soil may be thrown up against the plants for winter protection by means of a shallow plow or disc cultivator (Fig. 66).

In early spring of the second year, before growth has begun, the soil

Fig. 66. Well-mounded stool block of Robusta 5. Sprinkler irrigation at left.

should be pulled away from the plants with a cultivator or shallow plow, and the tops cut back to about an inch above the crown. Cutting at this height helps, again, to maintain the plants low in the soil so that the mounding operation is more easily accomplished.

From the exposed crowns new shoots develop. When they have reached a height of 4 to 6 inches, soil is mounted up to and between them to half their height. This may be done with a shallow plow or straddle cultivator. When the shoots have reached 8 inches, more soil is again added, and the process repeated during the season once or twice more until the shoots are covered to a depth of 8 to 10 inches or more. At each mounding it is essential to sift the soil in and about the new shoots so as to exclude light and to spread them apart somewhat. This favors rooting. This operation is best done by hand, using a long-handled shovel and pouring loose soil in and amongst the erect new shoots of each stool.

Most of the work can be done with a straddle cultivator, a spring-tooth harrow, and a shallow plow. A 24-inch disc, straddle potato hiller can be adapted to mounding and has proved very effective. And a light cultivation between the rows to loosen the soil is worthwhile immediately prior to all mounding operations.

Stool blocks may be maintained in a profitable condition for a number of years. Quince blocks thirty or forty years old are not uncommon, and there seems no apparent reason at present why blocks of other materials may not be maintained successfully. They do require, however, the liberal use of fertilizer and of materials that will keep the soil friable and in good physical condition.

Barnyard manure is the ideal material since it supplies all essential nutrients as well as organic matter. It should be composted for a year. Peat moss incorporated in the compost pile, in equal parts of manure and peat moss, is especially good. A light dressing may be made each spring and worked into the soil during cultivation.

Disease and Insect Control

Disease and insect troubles do not seem to be limiting factors in propagation beds where plants are maintained in a vigorous growing condition. Various leaf spots, especially on quince and cherry, and leaf hopper and aphids have been the most troublesome in western New York and in Michigan. Insecticides and fungicides are available that give excellent control. Proper timing is, of course, essential; and the type and amount of materials used will depend on the pest and local conditions. For example, leaf hoppers and aphids have been controlled in Michigan with timely sprays of parathion and nicotine sulfate. Fermate has been effective against apple scab and most leaf spots. For moderate attacks of woolly aphid, chlordane and aldrin have proved effective when applied to the soil at the base of the plants. Annual applications over a few years have eliminated this pest. Apple EM I, IV, VII, and IX have seemed most susceptible to woolly aphid attack, whereas EM II, XII, XIII, and XVI have shown some resistance.

Crown gall has been troublesome at times on apple EM VII and IX, but has been controlled commercially by persistent roguing of infested plants.

Harvesting and Grading

In the fall of the year the mounds are pulled down by plowing away with a light plow, and are further leveled and cleaned out with a hoe so that the rooted shoots may be removed. This grubbing operation should not be carried out until the rooted shoots are matured, as shown by the yellowish casts to the leaves and their disposition to fall. Some clones mature earlier than others; further, considerable new root formation may occur quite late in the season under the cool moist conditions provided at that time of the year.

Quinces mature early, and may be harvested first; in fact, they should always be harvested before severe freezing weather because quinces are none too hardy and may in some seasons be injured by early cold. The EM I apple rootstock also matures early, followed by EM IV, VII, and XIII. The EM XVI and especially EM XII mature late, and may even of necessity be left unharvested until the following spring.

The rooted shoots are cut from the mother plant with a pruning shears, or may be torn off (Fig. 67). By some it is preferred not to remove any unrooted or poorly rooted shoots, but it is better practice to remove all the shoots excepting perhaps small thin, scraggly ones, which may be cut back to an inch or two in length. Any unrooted or poorly rooted material may be handled as hardwood cuttings, as described in an earlier section. They can be depended upon to have formed root primordia even though roots may not have been developed, and they will root readily when treated in this way.

As soon as the rooted shoots have been harvested, the soil should again be drawn up over the mother plant either by a plow or with a straddle cultivator. The exposed mother plant is very tender to cold, and serious damage to the stool block may result from even a few hours' exposure to freezing temperatures.

It goes without saying that shoots should not be removed when temperatures are at or below the freezing point (32 degrees F.) and that after removal from the mother plant the newly rooted shoots should be protected from drying out and from exposure to cold. Young roots are easily damaged by even a very few degrees of cold for even a very short time.

Fig. 67. Harvesting rooted shoots from mother plant similar to that shown in Fig. 65. The mounded soil has been pulled away, exposing the rooted sections of the shoots. These are torn or cut from the mother plant, and constitute the rooted shoots used as rootstocks upon which to bud and graft.

The harvested shoots are graded according to both size and degree of rooting. They are divided into sizes on the basis of diameter at the crown as for seedlings, namely Extra—$\frac{1}{4}$ inch and above; No. 1—$\frac{3}{16}$ to $\frac{1}{4}$ inch; No. 2—$\frac{2}{16}$ to $\frac{3}{16}$; No. 3—below $\frac{2}{16}$ inch. Each size grade is further divided into "well rooted," "medium rooted," and "poorly rooted" (Fig. 68).

For material that is to be lined out for budding, the size is of less importance than the degree of rooting. That is, a "No. 2 well-rooted" grade is superior to an "Extra, medium-rooted" or an "Extra, poorly rooted" grade. Even a No. 3 grade that is well rooted has done well as lining-out stock. Unrooted and poorly rooted grades, excepting for large sizes, should be planted out for a second season's growth and treated as hardwood cuttings.

Yields of Plants

The yield of rooted shoots per plant varies with the material. EM I, IV, VII, IX, and XIII apple rootstocks will produce more shoots than EM II and XII. MM 104, 106, 109, and 111 have rooted exceptionally well; namely, 16, 10, 20, and 10 shoots per plant respectively, and averaging 85 per cent rooting. Quinces will yield high. At the first harvest, the second year after planting, from two to six rooted shoots may be expected per stool. The third year after planting, the number may reach 10 to 15 and may increase to 15 to 24 in succeeding years. Records kept for eight years at the New York State Agricultural Experiment Station at Geneva, New York, show an increase in yield each year up to about five years of age, and a uniform yield thereafter.

Not all the shoots will be well rooted. The number depends upon the favorableness of the season as well as the nature of each particular clone. For example, from a stool block of EM IX in western New York the percentage rooting and the percentage salable rooted shoots over a six-year period was as follows:

Rooting (%)	81	100	100	100	91	88
Salable rooted shoots (%)	68	70	30	84	62	53

And some idea of the differences in rooting ability of clones is shown by the following record for one season taken from stool blocks growing under identical conditions:

Material	Rooting (%)	Salable rooted shoots (%)
EM I	78	43
EM II	50	11
EM IX	88	53
EM XII	42	17
EM XIII	72	43
EM XVI	52	24

Records from England give the approximate number of shoots per 100 feet of stools, as follows:

EM I   800 shoots, 650 rooted, 450 first grade
EM II   700 shoots, 560 rooted, 360 first grade
EM IX   450 shoots, 360 rooted, 200 first grade
EM XIII   1,200 shoots, 1,150 rooted, 900 first grade
EM XVI   750 shoots, 550 rooted, 350 first grade
Quince A   3,000 shoots, 2,800 rooted, 1,200 to 1,500 first grade

Average crop yields per acre per year for established stool blocks in England have been reported as follows:

Material	Fit to Bud	For Bedding	Total
EM I	30,000	20,000	50,000
Em II	35,000	25,000	60,000
EM IX	30,000	15,000	45,000
EM XVI	35,000	20,000	55,000
Quince A	70,000	30,000	100,000
Quince C	50,000	50,000	100,000
Brompton plum	23,000	15,000	38,000
Common Mussel	20,000	15,000	35,000
Myrobalan B	45,000	10,000	55,000
Common plum	35,000	15,000	50,000

Records taken in western New York give the following approximate number of shoots from 100 feet of stools, which are comparable:

EM I   550 shots, 475 rooted, 350 salable
EM II   400 shoots, 225 rooted, 200 salable
EM IX   550 shoots, 450 rooted, 300 salable
EM XII   300 shoots, 200 rooted, 150 salable
EM XIII   600 shoots, 500 rooted, 400 salable
EM XVI   500 shoots, 300 rooted, 200 salable
EM Quince A   1,800 shoots, 1,600 rooted, 1,400 salable

Fig. 68. **Typical rooted shoot harvested from stool block: (Left) well rooted; (Right) medium rooted.**

Brase and Way give the following production figures for five Malling apple rootstocks in stool blocks over twenty years of age, planted 16 inches by 10 feet, or 3,267 mother plants per acre:

Material	Shoots per Mother Plant	Plantable Rooted Shoots per Mother Plant	Calculated Plantable Rooted Shoots per Acre
EM IX	43	21	68,230
EM VII	26	14	45,070
EM II	37	7	21,750
EM I	29	15	48,460
EM XIII	33	15	47,500

The Importance of Climate

Propagation by mound layering is most successful in areas of equable climate, with neither extremes of winter cold nor of summer heat and drought, and with either a minimum of 2 or 3 inches of rainfall during the summer months or with standby irrigation available. Snow coverage is good protection against deep penetration of severe low temperatures. In fact, there are areas near Georgian Bay in Canada, in western New York south of Buffalo, and western Michigan near Traverse City, where the soil frequently does not freeze during the winter because of snow coverage. In spite of relatively low winter temperatures, such areas will produce good layered shoots. In Minnesota, where winters are cold and where stooling would not be considered likely to succeed, a mulch of straw in early fall protected a stool block effectively from the winter of 1958–1959. Stool blocks not protected in this way were a total loss.

In addition, the growing season must be sufficiently long to permit the shoots to root, as well as to provide an open period of weather adapted to harvesting the rooted shoots and carrying on the operations necessary to put the stool block into a good condition for winter. It is a common and disappointing experience to examine a stool block in late September or early October and find very few roots formed on the shoots arising from the mother plant. It is then just as common but a more satisfactory experience to examine the stool block two to four weeks later and find excellent rooting.

It is also essential that the season be sufficiently long and open following rooting, so that the rooted plants may be harvested. Anyone who has worked with a stool block of several acres in a climate where winter may suddenly descend and close the season will thoroughly appreciate the problem. The operator wishes to delay harvest as long as possible so that shoots may be well rooted. But unless there is time—several days or weeks—to complete the harvest operation before the onset of winter, he may be forced to sacrifice rooting. If winter suddenly descends, freezes the soil, and catches the stool block unharvested, the harvest operation must be postponed until the ground opens in the spring. Not only is the stock subject to winter injury; there is no opportunity to sort and grade the rooted shoots to prepare them for sale or for planting before spring. And there is sufficient to do in the nursery in the spring without having to add a difficult harvesting and grading operation in the stool block.

Because of such climate factors as these, the trend is to choose a climate of open winters. This is the situation in England and on a good part of the Continent of Europe. It then becomes possible to wait until plants are well rooted, even though it be midwinter, and to harvest at leisure.

The place in America most similar to these conditions is along the Pacific Coast west of the Cascade Mountains in northwestern United States and southwestern Canada—in Oregon, Washington, and British Columbia. Here exist all the climatic factors favorable to propagation by mound layering. It would seem that these are the areas in which large commercial enterprises must eventually be developed to meet the demand for clonal dwarfing rootstocks.

PROPAGATION BY TRENCH LAYERING
(ETIOLATION METHOD)

Propagation by trench layering differs from propagation by mounding in that instead of the new shoots arising from the crown of the mother plant, they come from arms of the mother plant that have been laid in the row in a horizontal position. In addition the plants are covered with an inch or so of light friable soil before bud break in early spring, so as to prevent light from reaching the young shoots for a period (Fig. 69).

The method is especially useful for materials that do not root readily from cuttings, stools, root cuttings, and hardwood cuttings, as some plum stocks, most cherries, and some apples. Some varieties of apples, as Northern Spy, may be rooted successfully this way.

To establish a propagation bed, one-year-old rooted plants are set 2 or 3 feet apart in rows 6 to 7 or even 10 feet apart. The plants are set at about 45 degrees with the horizontal, in the direction of the row, so as to facilitate layering the following season. The planting is cultivated and tended during the first growing season as for stooling.

The next spring, before growth starts, instead of being cut back to the crown as in stooling, the vigorous shoots are headed back lightly, and weak ones are cut back severely. A shallow trench or drill 2 inches deep is made along the row, wide enough to accommodate the entire plant when bent into position. The whole plant is then laid horizontal in the shallow trench and pegged in place. Wooden pegs and wire hooks made from a 6-inch length of 9-gauge galvanized wire are good for this purpose. The plant will thus be an inch or two below the general soil level.

Just before the buds break, an inch of fine soil is sifted lightly over the plant so that as the buds start out the new shoots will be forced to make their way through this coverage of soil. In this process, the base of the young shoots are kept from the light, remain whitish and soft, and in consequence readily form roots, as discussed at the beginning of this chapter. If the coverage is too heavy, the young shoots may be smothered. The incorporation of peat moss or sand in the soil used for this early-season coverage is most important if the plantation is on heavy soils. A second coverage is made before the young leaves have uncurled and expanded.

Subsequently, as in stooling, fresh soil is drawn up about the new

Fig. 69. The trench-layer or etiolation method of layering clonal rootstocks: (Left) Parent plant set at an angle to facilitate layering horizontal the next year for covering with earth. (Center) After one year of growth, the plant is pegged down. (Right) Steps in mounding shoots from the pegged-down mother plant so as to induce rooting at the base. (After *Fruit Tree Raising*, Min. Agric. and Fisheries *Bulletin 135*, London, 1949)

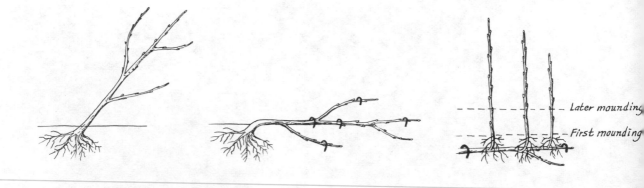

shoots when they are 4 to 6 inches tall, and the process is repeated until mounding is to a height of 8 to 10 inches by midsummer. A sifting of soil by hand, as with a shovel, in and about the shoots helps to spread the individual shoots apart, keep out light, and encourage rooting.

Removal of the rooted shoots and grading are in the same manner described under the discussion of stooling. Shoots that have not rooted, however, may be left attached to the parent plant and cut back to three or four buds and pegged down the following spring.

Yields of Plants

Records from England show very good percentage rooting by this method, as follows:

Material	Number of Shoots	Number Rooted	Number First Grade
Northern Spy	800 to 1,000	600 to 700	500 to 600
Apple A, C, and H	2,000	800	650
Common Mussel plum	500	400	350
Common plum	600	550	450
Myrobalan B	600	550	450
Black Damas C	700	650	400
Pershore plum	200 to 400	150 to 200	125 to 200
Brompton plum	700	500	450
St. Julien A plum	500	400	300
Pear B $\frac{1}{6}$	700	400	300
Pear B $\frac{2}{3}$	600	350	225
Pear B $\frac{2}{6}$	600	350	225
Mazzard F $\frac{3}{1}$	500	450	350
Mazzard F $\frac{5}{2}$	500	400	300
Mazzard F $\frac{9}{1}$	500	400	300
Mazzard F $\frac{7}{1}$	900	50	50
Mazzard F $\frac{7}{2}$	900	50	50

While trench layering has been successfully applied to apple clones in America, no success comparable to the above results with plum, pear, and cherry have yet been reported. It should be said, however, that most of the results that have been reported in the United States have come from the Northeast, where climate may not be well suited to the method.

There still remains the possibility that in regions more suited to vegetative propagation, such as the Pacific Northwest, the method might be equally successful.

It is noticeable also that trench layering is usually abandoned by nurserymen when sufficient planting stock is available to permit close planting and propagation by stooling. Extra labor is required for pinning the plants down. Perhaps more decisive, horizontal shoots or arms that develop from the mother plants become large, extensive, and lie close to the surface of the ground. They are thus poorly protected against winter cold, and are often caught by cultivating tools and ripped out, making great gaps in the block and causing considerable damage.

PROPAGATION FROM SUCKERS

Propagation by means of suckers is really a special form of layering. Suckers are shoots that arise from adventitious buds of roots. They are usually vigorous in growth, and spring up at the base of a plant and here and there from the underground root system. The method is not of large-scale commercial importance, but may be of limited usefulness when a small amount of material is desired, as by a home gardener or an orchardist interested in propagation of a limited number of special plants. It has been used in the propagation of the Stockton Morello cherry rootstock, which suckers freely from the root system of established orchard trees. The plants are dug either in the fall or the spring of the year, being separated from the mother plant by cutting or tearing. If sufficiently well rooted they may be lined out in the nursery for budding. If poorly rooted they should be grown for another year in the nursery row before lining out.

It has been reported that some cherry rootstocks that do not readily propagate by other means may be induced to produce suckers by wounding the root system superficially by means of a disc cultivator drawn through the orchard. Care must be exercised to see that the sucker is really a sucker from a part of the main root system, and not a seedling that has sprung from a germinating seed. Such a seedling would obviously not be true to the type of the clonal rootstock.

PROPAGATION BY THE NURSE–ROOT METHOD

Another method of increasing rootstocks, especially those difficult to raise by hardwood cuttings, is by the use of nurse roots. The method consists in grafting a dormant scion of the desired rootstock onto a short piece of some available root. The grafted combination is set deep to induce scion rooting, and the piece of root serves to sustain the scion until this is accomplished. Whip grafts are made during the dormant season, using seedling apple roots and scions of the desired rootstock material. The seedling roots are secured from commercial growers of straight-root apple seedlings raised especially for grafting. The top of the seedling is cut off and the root cut into sections 2 to 3 inches in length. Vigorous one-year-old rootstock scionwood is cut into sections about 6 inches in length. The rootpieces and the scions are matched for size, united with a whip graft, and securely tied with $\frac{1}{4}$-inch grafting tape or some other tying material, and stored until planting time in damp peat moss or sand.

The grafts should be planted as early in the spring as possible, in much the same way as hardwood cuttings, only deeper. That is, the top of the scion should be just at the surface of the soil, the object being to set the scion deep so that it has a good opportunity to form roots. As soon as new growth has developed to 3 to 5 inches, the cuttings should be hilled up; and a second hilling should be done if necessary to cover the basal 4 or 5 inches of new growth. Although roots will develop from the older wood of the scion, the new growth will root better.

At the end of the season, the grafts are dug, the seedling root is cut off, and the rooted scions are graded and stored for spring planting as lining-out stock. This method has been tried commercially with a fair degree of success with apple EM I, IX, and XIII. It is not unlikely that other types may respond similarly.

An improvement in this method is to tie a thin copper wire, such as is used for nursery labels, tightly around the graft at the union. As the graft grows, the wire tends to girdle the rootpiece and promote scion rooting. In tests with this method, the percentage rooting with Malling apple rootstocks was as follows. EM I, 65; EM II, 63; EM VII, 76; and EM IX, 41. As might be expected from the inherent ability of these types to produce roots, the root system of EM I was heavy the first year, EM II was poor, EM VII was sparse but sufficient, and EM IX was somewhat sparse. However, the second year showed all plants very well rooted. In about half the cases, the wire had cut the nurse root off. With the others it was necessary to snip the nurse root off.

15

DWARFED FRUIT TREES
IN THE NURSERY

THE production of dwarfed fruit trees in the nursery is not greatly different than for standard fruit trees (Figs. 70 and 71). Nevertheless, because there are a few differences that should be emphasized, and because there is a growing group of fruit growers and plantsmen who enjoy propagating their own dwarfed fruit trees, the general behavior of fruit trees in the nursery is briefly discussed in the paragraphs that follow.

General management in the nursery is similar to that with any clean cultivated crop. First consideration should be given to the selection of a well-drained soil of good fertility that has not been used for nursery stock within at least four or five years. Good soil structure is most important. It is not feasible by commercial fertilizers and artificial means to correct a poor soil or a poor location after the plants are established. It must be remembered that with nursery stock it is the vegetative plant that is of chief concern. The object is to keep the plants growing vigorously, supplying sufficient moisture and nutrients, and keeping the foliage free from insect and disease troubles.

Sizes and Grades of Lining-Out Stock

Dwarfing rootstocks may be procured from nurserymen who specialize in their production, or they may be propagated by the individual himself in the manner described in preceding chapters. Since the plants

257

are planted in rows, the operation is called "lining out," and the plants are called "lining-out stock."

Generally speaking, the No. 1 grade ($\frac{3}{16}$ to $\frac{1}{4}$ inch) is the most desirable. Stands of 80 to 90 per cent should be expected. The Extra grade ($\frac{1}{4}$ inch and above) gives a poorer stand because it often does not transplant well. The No. 2 grade ($\frac{2}{16}$ to $\frac{3}{16}$ inch) and the No. 3 grade (below $\frac{2}{16}$ inch) may give excellent stands as stock, but some of the plants, especially of the No. 3 grade, may not reach sufficient size to be budded. The preference for the medium and smaller grades (No. 1 and No. 2) applies especially to cherry and plum rootstocks. They grow rapidly, and larger sizes are likely to overgrow.

Relatively more important than proper rootstock size in securing a good stand is the degree of rooting. For example, in a test planting with the apple EM XIII, the stand for the "well-rooted" grade was 82 per cent, while for the "medium-rooted" and "poorly rooted" grades it was 72 and 52 per cent respectively.

As a general rule, clones that root most readily in the stool bed give the highest stands in the nursery as lining-out stock. Thus, in a planting of EM I, IV, VII, IX, XIII, and XIV, which root readily, the percentage stands ranged between 94 and 98 per cent. On the other hand, plantings of EM II and XII, which root less readily, the percentage stand was 72 and 61 per cent, respectively. The ability to regenerate roots that is so characteristic of many of the clonal dwarfing apple and quince rootstocks makes them prime favorites in nursery operations. This is especially so in adverse seasons, as when spring moisture is low. Under such conditions, many of the clonal rootstocks will give a sufficiently greater stand than seedlings, as lining-out stocks in the nursery, to warrant their use for this reason alone.

BUDDING AND GRAFTING COMPARED

Four principal methods are used in the propagation of dwarfed fruit trees in the nursery; namely, (1) grafting onto a root or stem of a dwarf-

Fig. 70. (Upper left) A fruit tree is composed of two individuals growing together as one. The arrow points to the union between the Montmorency cherry scion and the Mahaleb rootstock (dwarfing).

ing rootstock, (2) budding onto a root or stem of a dwarfing rootstock, (3) double-working a congenial body stock between an otherwise uncongenial scion/dwarfing rootstock combination, and (4) double-working an intermediate stempiece of dwarfing material between the scion variety and the rootstock.

Fig. 71. Steps in the production of a fruit tree: (Left to right) First year—Seedling rootstock (or layered rootstock) produced from seed (or layers) and sold to fruit-tree nurserymen as lining-out stock (1961). Second year—Seedling rootstock planted (lined out) in spring in nursery row and budded just above the crown with a bud of the desired scion variety (1962). Third year—Top of seedling cut off down to bud in spring, which forces bud to develop into vigorous one-year tree (1963). Fourth year—One-year tree trimmed and grown a second year to develop a two-year-old tree (1964) for planting in the orchard the next (fifth) year (1965).

Although satisfactory dwarfed trees are produced by all four methods, each has a special value and use. Budding onto dwarfing rootstocks is the most widely employed method for large-scale operation. Difficult stock-scion combinations are more successful by budding than by grafting. On the other hand, grafting has the advantage that a salable one-year-old tree can be produced in one season under good growing conditions, whereas budding requires two seasons. In general, buds grow more vigorously than grafts, and difficult combinations are more successful by budding than grafting. Further, grafts do not result in sufficiently large trees in regions of short growing seasons and relatively cool summers, such as prevail in northern Michigan and New England; whereas in warm regions with long growing seasons buds are inclined to develop into overly large trees that are difficult to handle. For these reasons, grafting is more common in the southern states, and budding is more common in the northern states.

ROOT GRAFTING

Root grafting involves the grafting of a scion directly onto a piece or root. The method is not generally adapted to producing dwarfed trees, for the reason that when the graft is planted the union is at or below the ground line, and the scion is likely to root. Established on its own roots, the tree loses the dwarfing influence of the dwarfing rootstock. On the other hand, this may be exactly what is sometimes desired, as with dwarf pears. The young tree is somewhat dwarfed in its first few years, comes into bearing early, and then becomes established on its own roots with attendant longer life and larger size. Root grafting is employed, also, in some double-working operations with dwarfing rootstocks, as will be discussed later.

Root grafting is performed indoors in the winter, using dormant wood (Fig. 72). Scions of the desired variety are cut into sections 4 to 6 inches in length and grafted onto the root portion of the rootstock, the top portion of the rootstock being discarded. One entire rootstock may be used for one scion, in which case the graft is termed a "whole-root graft"; or the rootstock may be sufficiently long to be cut into two or sometimes three sections, with each one used for the production of a graft. In this case the graft is called a "piece-root graft." Clonal root-stocks propagated from stools or layers are rooted only at the base, and generally provide material for only one graft.

The type of graft commonly used is the whip or tongue graft. It is easily made, and places the cambiums in favorable position for good union. The upper end of the rootstock is cut off with an upward slanting cut about 1 inch or 1½ inches long. Beginning ½ inch down the slanting surface of the rootstock, a tongue or cleft is cut downward parallel with the length of the rootstock. The base of the scion is prepared in a similar manner but with a downward slanting cut and an upward tongue, or cleft. The stock and scion are dovetailed together, so that the cambial regions of each, just beneath the bark, will be in contact. Where root-stock and scion are of the same diameter, no difficulty will be experienced in this regard. Where rootstock and scion are of unequal diameter, pains must be taken to match the two along one side. Careful matching of rootstock and scion will reduce loss from callous knot formation that renders trees unsalable.

The union should be wound with common twine, grafting tape, waxed string or waxed cloth strips. The grafts are packed away in damp moss or sand in a cool cellar until spring, when they are planted in nursery rows, leaving only the top bud aboveground. The grafts must be set firmly and in intimate contact with the soil. Early-season planting is very important. A good practice is to make a trench with one straight side against which the graft is set. A little peat moss distributed in the trench helps the plants to become established, and a light dressing on the surface of the soil is also beneficial.

STEM GRAFTING

The procedure with dormant stem grafting is exactly the same as for root grafting, except that the entire rooted shoot, rather than a piece of root, is selected for the root stock. The scion is set into the stem of

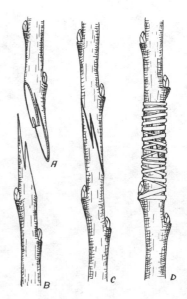

Fig. 72. **Whip grafting. A, prepared scion; B, prepared stock; C, matched graft; D, graft tied with string to be thoroughly covered with wax before completed.**

the rootstock 6 to 8 inches above the crown. Tying and handling are as described for root grafts in the preceding paragraphs.

The big difference is in the depth of planting. That is, the graft should be set so that the crown is an inch or two below the surface of the soil. This places the union 5 or 6 inches above the soil so that it will not scion-root.

Stem grafting is also employed with established rootstocks where a long body stock is desired. The plant to be top-worked is grown for one season in the nursery. It is then whip-grafted in place in early spring with the desired scion.

It is important that the scion wood be completely dormant. The stock may have begun growth; in fact, some of the best results have been secured when the grafting operation is done just as growth starts in the stock, but the scion must be dormant. This method is used successfully for dwarfing cherries by grafting onto mahaleb seedlings at a height of 20 to 24 inches. It has also been used with double-worked trees involving a short stempiece of Clark Dwarf worked onto a hardy body stock.

BUDDING

Budding is a form of grafting in which a bud rather than a scion is inserted upon the rootstock. The operation is performed during the summer on actively growing rootstocks. Since only a single bud is used, the method is economical of wood. Stone fruits (apricots, cherries, peaches, and plums) are more easily propagated by budding than by grafting, since the wood does not split well when grafted, whereas the bark slips readily for budding. Further, wound gums and resinous materials are likely to form in grafting, interfering with good and rapid union of stock and scion.

The rootstocks are procured during the winter season and trimmed in both top and root to an overall length of 8 to 12 inches. Trimming is done largely to expedite planting, since, in commercial nursery operations the rootstocks are planted by insertion in a narrow, vertical trench, opened by a special plowlike device called a trencher. The stock is set 8 inches apart in rows 42 inches apart, and cultivated as is any other nursery crop.

Time of Budding

Budding is done in the summer after the scion buds of the current season have matured and before the bark of the rootstock has tightened. Following are given the proper times to bud various rootstocks in western New York and southern Michigan. The dates may be earlier in a dry season or on light soils and in southern regions:

Pear	July 10 to July 20
Apple	July 15 to August 10
Apple (EM IX)	August 10 to September 1
Plum (St. Julien)	July 15 to August 1
Plum (Myrobalan)	August 15 to September 1
Cherry (Mazzard)	July 20 to August 1
Cherry (Mahaleb)	August 15 to September 1
Cherry (Sand cherry)	August 1 to August 10
Quince	July 25 to August 15
Peach	August 20 to September 10

It should be noted that special mention is made of the apple EM IX rootstock. If this stock is budded too early, the buds may start the same season they are set, instead of remaining dormant until the opening of the next season. The resultant soft growth is subject to injury from winter cold. Serious losses have been sustained from this cause.

On the other hand, the sand cherries tend to mature early, and unless budding is completed before the bark of the rootstock tightens, the stand of buds will be poor. Similarly, leaf-spot diseases may cause premature defoliation and tightening of the bark, especially on quinces and cherries. These troubles can be controlled by proper spraying; but if they are not, close attention should be given to completing the budding operation before defoliation has become severe.

Height of Budding

In the propagation of standard trees, the bud is placed about 2 inches above the collar of the rootstock. For dwarfed apple trees, however, the trend has been to place the buds an inch or two higher (3 to 4 inches above the collar) so as to reduce the chances of scion-rooting when the trees are planted in the orchard. There is much to recommend the higher budding. In fact, the bud may be placed 6 to 7 inches above the collar or at about 4 to 5 inches higher than used for standard trees. Then the trees can be set several inches deeper in the orchard, yet the point of union will be sufficiently above the soil level to avoid scion-rooting. Some nurserymen regularly bud dwarfed apple trees as high as 8 inches, so as to provide a stempiece of rootstock that is 12 or more inches in length. This permits deep planting in the orchard and has been shown to give better anchorage for the first few years for trees that are inclined to lean. New roots arise from the stempiece to provide better bracing.

However, one must always bear in mind the consequences of placing the union so high above the ground line that it will lack the protection of radiation from the soil or protection from snow, mulch, or weed growth. The union is a region of immaturity. In favored horticultural regions this may not be an important consideration, but in areas where early fall freezes or severe winter cold are possibilities, this point must be borne in mind. Further, some rootstock materials, such as the quince, are less hardy than the scion varieties budded upon them. To bud quince rootstocks high in this way, and to place the union high above the soil in the orchard, are to run the risk of severe injury. Further, it should be pointed out that unless the rootstocks are vigorous and the bud growth is strong, the bud often starts out at an awkward angle, and results in an inferior tree.

On the other hand, high budding can be used to advantage with varieties of apple that are susceptible to collar rot (*Phytophthera cactorum*). The Coxe's Orange Pippin is, for example, subject to this disease; whereas the following rootstocks are resistant: EM IX, VII, II, and XXV, and MM 104, 106, and 111. Coxe's Orange has been successfully budded on these rootstocks at a height of 12 inches above the ground line. The trees have been satisfactory in the nursery and have performed well in the orchard.

The Budding Operation

Shoots are cut from the vigorous growth of the scion variety (Fig. 73). The leaves are trimmed away, leaving about ¼ inch of the stem (pedicel) as a handle to the bud. These are called bud sticks. They are wrapped in damp burlap; once dried, they are worthless. The buds at the tip and the base of the year's growth are discarded, and only plump, hard buds near the middle are used.

At the point where the bud is to be inserted, a T-shaped incision is made. The transverse cut is made with a rocking motion of the knife, and the vertical one by drawing the knife upward lightly from a point about an inch below the first cut. Before removing the knife a slight twist of the blade will loosen the edges of the bark.

The bud is cut from the bud stick with an upward-drawing motion of the knife from below the bud. The entire thickness of the bark is cut at the point of the bud so that it will not crumple when inserted into the stock.

No wood is cut away with the bud unless just under the "eye"; on the other hand, the bud must not be cut so thin that the soft growing tissue between the bark and the wood is injured. Grasp the shield firmly between the thumb and forefinger, and carefully lift it from the wood, without tearing the bud. With the leaf stem as a handle, the bud is inserted into the T-shaped incision and pushed down until its "heel" is flush with the transverse cut.

No waxing is necessary, but the bud must be tied. For this purpose raffia and rubber budding strips are used. Raffia is cut into lengths of 18 or 20 inches and moistened to make it pliable. Rubber strips are sold already cut into 2-inch lengths, ⅜ of an inch in width. The strand is first brought firmly across the upper end of the bud to keep it from working out. Beginning at the bottom of the slit, the material is wound smoothly upward, covering everything but the "eye," and fastened. It is essential that this winding be tight, for it must hold the bud immovably in place until it has united with the stock. In from two to four weeks, depending on the growth of the stock, the tie should be severed so as to prevent girdling. The bud unites with the rootstock and remains dormant until the following spring.

In the case of the peach, budding is done on the seedling rootstocks in the same year that the seedlings develop from the seed. This can be done because of the relatively faster growth of peach seedlings. The

Fig. 73. **Budding. A, bud stick, with leaves cut off and buds cut; B, single bud cut from bud stick; C, stock prepared for bud with T-shaped incision; D, bud pushed down into incision; E, bud tightly held by tying with string.**

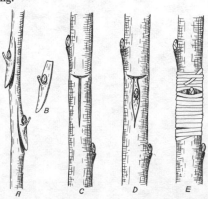

plum and cherry can also be grown in this way, and the apple and pear can be budded the first season from seed as an emergency method, particularly if the seedlings are started in the greenhouse.

Care the Following Season

In the spring before growth has started, the top of the seedling is cut back to a point just above the dormant bud. This forces the bud into rapid growth. Care must be taken to keep any suckers and buds from developing. All should be rubbed off as promptly as they appear so that all the strength of the plant goes into the development of the single inserted bud. The "budling" by fall has become a strong one-year-old, or "yearling," tree.

Growth of Budlings on Various Dwarfing Rootstocks

It is of interest to note that the dwarfing effect of a rootstock is of much less magnitude in the nursery than in the orchard. In fact, a striking feature of dwarfing rootstocks in the nursery is the similarity in height and diameter of yearling trees worked on different dwarfing rootstocks that in the orchard give trees of quite unlike sizes. This is well illustrated by measurements of height of yearling Cortland apple trees worked on nine different dwarfing rootstocks, arranged in order from the rootstock that produces the most dwarfing tree in the orchard to the rootstock that dwarfs scarcely at all, as follows:

Cortland/EM XII	106 centimeters	Cortland/EM V	114 centimeters
Cortland/EM XVI	121 centimeters	Cortland/EM III	110 centimeters
Cortland/EM XIII	116 centimeters	Cortland/EM IV	116 centimeters
Cortland/EM I	112 centimeters	Cortland/EM VII	108 centimeters
Cortland/EM II	121 centimeters		

DOUBLE-WORKING

Double-working, as the name implies, involves more than a single union between two consorting parts. It frequently involves two unions, with three different parts; and it may involve more. The method is often used to bring together a scion and a rootstock that are otherwise not congenial, as with some varieties of pear on quince. By interposing a congenial stempiece between the two, a successful combination is made. It is also used to develop a hardy or disease-resistant body stock, as well as to produce a dwarfed tree by placing an intermediate stempiece of dwarfing material between a rootstock and the scion, as with the Clark Dwarf apple.

Differences in results are reported. Some propagators claim that trees produced in this way are as dwarf as those produced directly on dwarfing rootstocks. On the other hand, experiences in western New York have shown trees to be less dwarfed, often substantially less dwarfed, than when worked on a dwarfing rootstock direct. Perhaps by employing a longer stempiece of dwarfing stock, a greater degree of dwarfing may be obtained.

In addition, trees produced in this way are often lacking in uniformity. It is difficult enough to make one good graft union, let alone two or even three. Further, the intermediate stempieces have not always been uniform in length, and this may introduce a factor of variability. With express care given to matching the wood used in grafting, with more exactness as to the length of the interstock, and with closer attention to details in the nursery, much greater uniformity and better performance may be expected.

The operation may be done either by budding or grafting or by a combination of both in a variety of ways. The choice lies with individual preferences and with the nature and amount of plant material available.

Double-Working Pears on the Quince

This procedure is necessary only when the desired pear variety is not congenial with the quince rootstock used for dwarfing. An intermediate pear stempiece compatible with the quince is first worked upon the quince. The stempiece in turn is then worked with the scion variety.

The most popular method of double-working is to line out the quince rootstocks in the nursery row and to permit them to become well established during one growing season.

During the late winter, prior to spring growth, a graft is made composed of a 3-bud scion of the chosen variety and a 5-inch length of the congenial pear that is to serve as the intermediate stempiece. The combination is made in the usual manner previously described, sealed and stored. It must remain dormant.

In early spring, just as growth starts, the rootstock is cut off several inches above the ground and the whip grafted in place in the nursery row. The scion will commonly produce a tree the same height in one season as is produced by a single-worked tree.

Another procedure is to bud the quince rootstocks in the nursery row with buds from the pear that is to serve as the intermediate stempiece. The next spring the quince top is cut off just above the pear bud, forcing the intermediate into growth. In turn, the pear intermediate is budded to the chosen pear variety and similarly forced into growth the following spring.

It will be noted that the double-budding method requires three years in the nursery to produce a one-year-old tree, whereas by the grafting method, only two years are required. Both methods, however, have their advocates and are successful.

Double-Working Dwarfed Apple Trees

A common method of developing dwarfed apple trees by the use of an intermediate stempiece of dwarfing apple is to double-bud. That is, the dwarfing intermediate is budded onto a rootstock, and the chosen scion variety is in turn budded upon the intermediate.

The procedure more in detail is to line out the desired rootstock material in the nursery. This may be a seedling rootstock, as French Crab or Delicious seedlings, or it may be a clonal rootstock, as EM XVI. Budding is done as for standard trees, the bud of the dwarfing intermediate being placed about 2 inches above the ground line. It is not necessary to place the bud higher, because in this instance it does not matter if the dwarf intermediate tends to scion root.

The next season, the top of the rootstock is cut off just above the bud, forcing the development of a shoot from the dwarfing intermediate material. This is in turn budded during the summer with the chosen scion variety at a height on the intermediate which will provide a 2- to 4-inch length of stempiece in the finished tree.

In the third season the top of the intermediate is cut off so as to force the scion variety to develop as a strong budling. The tree is thus composed of a scion variety and a rootstock, between which is interposed, just above the crown, a 2- to 4-inch section of dwarfing stock. The results have been quite satisfactory.

It will be observed that three growing seasons are required to produce a one-year-old dwarfed tree by this double-budding method. A method of producing a similar tree in one season less is by a combination of budding and grafting.

In this method, a dormant graft is made in winter, composed of a 5-inch piece of dwarfing interstock, as Clark Dwarf, whip-grafted onto either a seedling or a clonal rootstock. The grafts are tied, waxed, and stored, and planted out in the spring with the union just above the ground line.

During the summer, the top of the graft is budded with the chosen scion variety at a height of 2 to 4 inches above the union. The bud is forced into growth the following season by cutting of the dwarfing stempiece just above the bud.

Both methods have given good results. The double-budding technique is more conservative of wood and is better adapted to regions of short growing season, whereas the grafting-budding technique is adapted to regions of longer growing seasons. The double-budding method produces a strong one-year-old tree with a three-year-old root system, whereas the grafting-budding method produces a similar tree with a two-year-old root. As has been said, the choice of method rests largely with the preferences of the individual.

THE "CLARK DWARF" APPLE TREE

The so-called "Clark Dwarf" apple tree is essentially a dwarfed apple tree produced in somewhat similar manner to that described in the preceding paragraphs. It is a combination in which a stempiece of dwarfing apple is interposed between a vigorous scion and a vigorous root system. However, a hardy body stock is also included in the combination, so that instead of a combination of three individuals and two graft unions, it is composed of four individuals and three graft unions.

Historically, this is of some interest because it represents the first attempt to produce dwarfed fruit trees extensively by the use of dwarfing interstocks. The dwarfing interstock material used is called "Clark Dwarf." As a consequence, the trees so produced have been commonly called "Clark Dwarf" trees, and the method used in their propagation has even been called "Clark Dwarf Method." Since its introduction, "Clark Dwarf" has been found indistinguishable from EM VIII, as discussed in a previous chapter.

The first operation in this method of producing a dwarfed apple tree is the grafting of a dormant scion of hardy body stock (Virginia Crab) onto a piece of dormant seedling apple root in midwinter. The graft is planted the next spring with the union below the ground level so as to induce scion-rooting in the Virgina Crab. The graft is grown for one season in the nursery, and trained to a single stem.

During the second winter a second graft is made, involving a 4-inch stempiece of Clark Dwarf and a scion of the variety chosen for the tree top. The grafts are tied, waxed, and stored as with other grafts. In early spring the grafts are in turn grafted onto the Virginia Crab seedling growing in the nursery row. This is done at a height of 20 to 24 inches on the Virginia Crab stem so as to introduce the hardy body stock. Sometimes all four materials are grafted together in midwinter in one operation, and treated similarly to other grafts.

The result, then, is a four-story tree, namely, the scion variety/Clark Dwarf/Virginia Crab/seedling root. The Virginia Crab is supposed to become scion-rooted, thus developing a strong root system with good anchorage for the tree, but scion-rooting is not uniform, and thereby introduces variables. The Virginia Crab also serves as a hardy body stock. The idea was thus to develop in a single method a strongly anchored, dwarfed tree with a hardy trunk or body. There have been some modifications in the method and the materials used. Thus, Hibernal has sometimes been substituted for Virginia Crab for the bodystock, but the method is essentially as described.

MISCELLANEOUS DOUBLE–WORKING TECHNIQUES

Among miscellaneous double-working techniques is an interesting form known as double-shield budding (Fig. 74). It consists of placing

one bud shield upon another so that they unite with the rootstock as well as with each other, and so form a three-tier tree in one budding operation. The intermediate tier may be a dwarfing rootstock, so that a dwarfed tree is formed by the introduction of the stempiece of this material. Or it may be a portion of compatible pear to be placed between a quince rootstock and another pear variety that is uncongenial directly on the quince.

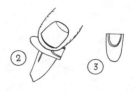

The operation is essentially the same as for shield budding as previously described. However, the T-shaped incision must be sufficiently large to accommodate the two buds, one upon the other. A small budless shield is cut from the plant material that is to supply the intermediate. It is prepared in such a way that a portion of the bark of the shield is removed from the top side at the apex, thus exposing the cambium on both the inner and the outer sides of the bud.

The bud is slipped well down into the bottom of the T-shaped incision. Then, a bud shield of the chosen scion variety is cut from the budstick and inserted in the T-shaped incision in such a manner that it rests not upon the stock but upon the outer exposed cut section of the budless shield.

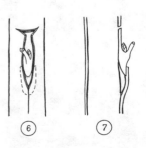

The intermediate bud unites with both the stock and the scion bud shield. From here on, the operation is conducted exactly as for regular budding. The next spring, the top of the rootstock above the bud is removed, and the scion bud forced into active growth. With a good season and good technique, a double-worked tree will be formed that is approximately the same height as a single-budded tree.

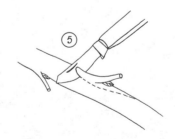

Fig. 74. Double-shield budding, a rapid method of double-working: 1. Cutting shield of bark and wood from a shoot of the variety that is to serve as the intermediate stempiece. 2. Shield of bark and wood from the variety that is to serve as the intermediate stempiece. 3. Same shield showing beveling of upper surface so as to provide region with which shield bud of scion variety is to unite. 4. Shield inserted in the T-shaped incision in the rootstock. 5. Shield bud being cut from shoot of scion variety (differing from previous shield by the inclusion of a bud of the scion variety). 6. Shield bud inserted in T-shaped incision in rootstock so as to rest on shield of interstock. 7. Shield of interstock and bud shield of scion variety in place in T-shaped incision on rootstock. (Adapted from *Fruit Tree Raising*, Ministry of Agric. Fish. and Food *Bulletin 135*, London. After Garner)

A modification of this technique was proposed by Nicolin of Germany about 1953. The method has assumed the name "Nicolieren," after its originator. A budless shield is taken from a stem of the interstock material as described in the paragraphs above. However, care is taken to remove a thin strip of bark from the outer face of the shield. This exposes an oval ring of cambium on the outer face of the shield.

The budless shield is then slipped into the T-shaped incision. Then the shield bud of the chosen scion variety is similarly slipped into the incision but on top of the prepared budless shield. Here again the budless shield unites with the stock as well as with the scion bud. The next year, a tree forms from the double-budding combination that consists of the rootstock, a short intermediate stempiece, and the scion variety. Growth may be as good as with a single-budded tree. Needless to say, these techniques are successful only in the hands of expert technicians.

PRUNING AND TRAINING TREES IN THE NURSERY

No special pruning or training is necessary during the development of the budling in the nursery, excepting that with grafts, only one bud should be permitted to develop, so as to form a straight, single stem. Occasionally, because of injury to the terminal bud by some insect or some mechanical misfortune, lateral shoots may develop. They should be removed, and the central bud, or a replacement of the central bud, should be encouraged.

Sometimes the shoot from the bud grows at an awkward angle. This is especially so with the apple EM IX rootstock. The resulting trees will be greatly improved if such shoots are tied upright to a light bamboo shoot or other support.

When the trees have matured in the fall of the year, as indicated by yellowing of the leaves and a tendency to drop, they may be dug for fall planting or for storing over the winter for spring planting. Or they may be left for spring digging and planting. No special pruning is required. Any subsequent training or pruning is done by the planter to suit his own ideas of shape and form. Many planters desire trees of this kind. They are called yearlings, or one-year-old trees.

Heading the Tree

When the trees are grown a second year, for sale or planting as two-year-old trees (Fig. 75), it is necessary to do some training in the nursery. Commonly, trees are cut back in early spring to a height of about 30 to 33 inches. As the buds start growth, the top six buds are left, and those below these six are rubbed off by a downward sweep of the hand.

This height of heading is satisfactory for semistandard and standard trees, but is too high for semidwarf and dwarf trees. For example, semidwarf apple trees (EM VII) should be headed at 26 to 28 inches, and

Fig. 75. Good stand of Montmorency sour cherry trees in the nursery row, budded on Mahaleb (dwarfing) rootstocks. Dwarfed fruit trees frequently make as vigorous growth in the nursery as do standard trees on vigorous rootstocks.

dwarf trees (EM IX) should be headed at 20 to 22 inches. It is difficult to school one's self to head trees this low, especially if they have made a strong growth as budlings. But much of the trouble experienced with dwarfed trees can be traced to too high heading, so that the tree becomes leggy and topheavy. It is better to err on the side of heading the trees too low. Some excellent dwarf trees (EM IX) have been developed as bushes without staking, by heading them at 16 to 18 inches.

Special Training

Nursery trees will commonly develop three to five lateral branches during this first year and will require no further care other than to rub off any feathered growth that may persist from the trunk, and any suckers that may come from the rootstock. However, more attention could properly be given to training in the nursery, provided planters could be educated to receive and plant trees trained for them by the nurserymen. As has been indicated, most orchardists prefer to train the trees to their own standards. They are not so much concerned about the branching of the tree in the nursery so long as the tree is stocky, "fat," and well grown.

For less experienced planters and for those who desire only a few trees, the nurseryman could help materially in shaping the tree. For example, if six to eight buds are left on the yearling tree when it is cut back, the No. 1 bud may be left to develop as the leader. No. 2 should be rubbed off inasmuch as it will tend to grow upright at a sharp angle. No. 3 and No. 4 will be nearly opposite each other, and should be left to develop as laterals. No. 5 and No. 6 may be rubbed off to allow a space between the next set of laterals provided by No. 7 and No. 8. The laterals, moreover, should not be one directly over the other. This is, of course, idealistic; no tree grows exactly this way. But the conception is useful.

Furthermore, trained trees could well be transplanted in the nursery and grown and trained a third season and even a fourth season. Planted 4 feet apart in rows spaced 10 feet apart, they may be handled more as ornamental trees are handled. Such trees are relatively large and are more difficult to handle. Accordingly, this kind of operation is better suited to a nursery located close to the potential customer, as in a large metropolitan district. The cost is also greater, so that the customer is likely to be an individual who desires only a few trees—perhaps two to ten in number. The trees can then be transplanted in fall or in spring directly from the nursery with great satisfaction to the purchaser. A three-year-old Coxe's Orange/EM IX apple tree trained in the nursery to a main leader and four to five branches is a choice article when transplanted to the home grounds.

Dwarfed trees may be trained to special forms. The operation may be done in the nursery or it may be done by the planter for his own satisfaction and pleasure. This topic is treated in a separate chapter.

Part Five

ESTABLISHMENT AND MANAGEMENT OF THE ORCHARD

16

SITE, MICROCLIMATE, AND CLIMATE CONTROL

The Importance of Site

SITE is the most important single factor in commercial fruit production. Even though trees will grow, it is impossible to stay in business against competition with growers who have more favorable locations. There are, for example, peach orchards in Michigan that have missed only one crop in fifty years owing to damage from either winter cold or spring frost. And in Pennsylvania there is record of a peach orchard that did not skip a crop in fifty-three years. A location that misses a crop every three or four years cannot bear this sort of competition.

In finding a good site, it is often surprising how much local information can be secured if one will but take the time to ferret it out. Where a section has been in fruit for several generations, this is no serious problem; there is abundant experience—not only in a general way but also specifically as to variety and as to particular parcels of land. In relatively new sections, it is often assumed that no such information is to be had; but this is often an unwarranted assumption, as careful inquiry of older residents will soon indicate.

Preferred Sites and Locations

A preferred site is one that is higher than the surrounding land so as to provide good air drainage against both severe winter cold and late

279

spring frost. This has led to the observation that the best orchards are found in scenic regions. Low spots are a hazard. Protection from wind is an asset. Southern and easterly slopes are better than northern and western exposures.

As one goes south an appreciable distance, danger of late spring frost increases, but the danger of winter damage to tree and bud lessens. As one goes north, just the reverse is true. Since the peach is tender to winter cold, it is not surprising that it is found most at home in southern sections, or in northern areas where large bodies of water moderate the climate. Low temperatures of -10 to -15 degrees F. will injure peach buds, and -15 to -20 will injure the trees. On the other hand, sour cherries are relatively hardy to winter cold, and grow well in more northern areas. Further, since the sour cherry blooms relatively early, and since late spring frosts are less common in the North, the sour cherry is doubly favored there.

The apricot, the plum, and the sweet cherry also bloom early, so that protection from late spring frosts is most important for these fruits. On the other hand, some varieties of apple, such as Rome Beauty and Northern Spy, bloom quite late and often can succeed where late frosts preclude other varieties and other classes of fruits. These are the types of evaluations that must be carefully made before an individual plants a tree.

Soil

As in the case of site, an individual who already possesses an area of land, or is bound to it by market or other conditions, has little choice as to the best soil for his trees. He can, however, make the most of what he has. For example, peaches and Japanese plums do not do well on heavy soils; they prefer a light, warm soil. Cherries will tolerate heavier soil if it is well drained, but they are thoroughly intolerant of poorly drained soil. It is often observed that cherry trees succeed in a fence row. This is largely because a fence row is frequently ridged and provides better soil drainage. Pears and the European plums enjoy a relatively heavy soil. Apples will grow in a much wider range of soil type. With this general information, the planter can adjust his choice of fruits, at least, to the site and soil he possesses.

For extensive commercial operations, experience emphasizes deep and well-drained soil, into which tree roots can penetrate. A severely fluctuating water table is to be avoided. On the other hand, a high water table is not unacceptable provided the level is fairly constant. For example, citrus trees can be found growing on the delta of the Mississippi River where the water table is scarcely more than 12 or 18 inches below the soil surface, but the level is stable. Further, the author recalls a highly productive eighty-year-old apple orchard in western New York that was growing on relatively shallow soil (3 feet) covering a layer of shale; but the rock layer sloped gently toward an adjacent lake, and water drained along this rock stratum so as to provide a remarkably uniform supply of moisture by what might be called natural subirrigation.

Soil Moisture

One of the most important considerations is good water-holding capacity of the soil. With close planting of trees and surface feeding, as is often the case with dwarfed fruit trees, the need for water is high. Trees will not long stand "wet feet," and they will not stand a droughty soil. Fortunately, much can be done in this situation. A clay loam soil is in itself a good holder of moisture, and all soils can be markedly improved in this respect by a high organic matter content. If the soil is low in organic matter, it should be built up before planting. It is very difficult to improve the soil once the planting is made. Good organic content can be maintained, but not easily increased. Supplemental irrigation can be used to good advantage during droughty periods. Heavy wet soils can be improved by tiling.

pH of the Soil

Much has been made of the pH value of the soil. While a pH of 5.5 to 6.5 is commonly found in good orchard sections, this is less a factor than it has been assumed to be. Excellent orchards can be found growing at a pH of 7.5 to 8.0, and similarly good ones can be found at 4.0 and 4.5. It is not the pH of the soil that is so important as it is the availability of nutrients to the tree, and this may be markedly altered by practices that alter the pH of the soil, such as liming. But, as has been said, pH in itself (within reasonable limits of 4.0 to 8.0) is not the most important factor.

Soil Fertility

Fertility of the soil is a consideration, but not a very serious one. Nutrients can be supplied. But the native structure of the soil, its water-holding capacity, drainage, and water table are matters inherent in a general situation, and are likely to be that way for a relatively long time, with little opportunity for correction. It has been well said that a good soil and a weak rootstock are to be preferred to a poor soil and a strong rootstock. That is, the stock will not make up for the soil.

It has often been noted that fruit trees succeed where shade trees and forest trees do well. This should be expected, since the soil factors that favor the one generally favor the other. This is a good association to look for and observe when one is considering a planting of fruit trees.

SPECIAL CLIMATIC PROBLEMS WITH DWARFED FRUIT TREES

The dwarfed or size-controlled tree introduces some new considerations into the already present problems of location, site selection, spring frost, winter cold, wind, and other climatic factors. This is due to at least four considerations. First, because dwarfed trees are relatively low-growing and close to the ground as contrasted with high-headed and taller-growing standard trees, the problems of cold-air stratification, frost, and soil-radiation differences are accentuated. Second, because the trees are closely planted, the local climate or microclimate within the planting is also accentuated, involving air drainage, air temperatures, and insect and disease populations. Third, because dwarfed fruit trees are commonly not so deeply rooted and not so well anchored as standard trees, they are more susceptible to the effects of wind. Finally, dwarfed-tree plantings are not infrequently made on relatively poor orchard sites by surburban dwellers and small or part-time growers close to market. While they might prefer a better location, the fact is that they already possess the one they have, and must adjust to it and make the most of it.

Climate Versus Microclimate

Climate, of course, is a general term that deals with relatively large areas, such as the climate of Michigan or of California. Microclimate deals with very small areas, as a single fruit farm, particular areas of that farm, or even various positions on an individual tree, such as the top branches compared with lower branches. The temperature for a region may be officially 28 degrees F., but within areas of this region there are often considerable variations in temperature. These are due to differences in elevation, air drainage, radiation effects, proximity to bodies of water, and the like.

Local differences reach surprising proportions at times, especially under extreme conditions. The writer recalls observing the unusually low temperature of −31 degrees F. at his home in western New York during the severe winter of 1933–1934; yet less than a mile away, somewhat closer to a lake, the temperature was −16 degrees F. This difference of 15 degrees was sufficient to explain severe winter injury in the one place and only mild injury in the other. Orchardists who have been confronted with a spotty fruit set in the orchard and who have investigated, have similarly found temperature differences of 5 to 10 degrees at relatively close locations in the orchard both in the winter during periods when winter injury may be induced, and in spring when frost damage may be involved.

Careful studies in Pennsylvania have shown daily variations of 5 to 8 degrees on clear nights in the spring in different parts of an orchard. In fact, the differences were often as great as 11 degrees. One spot in this orchard was found to average 3.04 degrees lower than one immediately adjacent; and on occasions the differences were 8 degrees. When temperatures at the one site were 30 to 32 degrees F., the other site might register 24 to 25 degrees. This was enough to reduce the crop appreciably from frost damage. In another instance, one site was at 28 degrees while another was at 25 degrees; and the crop was entirely destroyed at the lower temperature. Again, another location in the orchard was 10 degrees lower in night temperature in spring than a spot only one-tenth of a mile away. The former site produced only two crops of fruit in twelve years.

Similarly, the direction and velocity of winds vary not only between general regions but also between locations on a given farm. Hail, thundershowers, depth of snow, and other climatological phenomena also differ from region to region and from farm to farm, both in character and in frequency.

An added consideration is the closeness of planting with small trees. With standard-sized trees set 40 feet × 40 feet (27 trees to the acre), the movement of air in the orchard is less likely to be impeded. On the other hand, with trees planted 12 feet × 12 feet, which is not uncommon (302 trees to the acre) with dwarfed trees, the free movement of air is restricted and there is likelihood of damage from spring frost on cool, clear nights. Triangular planting when oriented at right angles to the slope is more restrictive to air circulation than planting on the square, and is that much worse.

Cordons planted 2 × 6 (3,630 trees per acre) create almost a solid mass of foliage within the row. And one planting is recorded of cordons set 1 foot × 3 feet, or 14,520 trees per acre. A stagnant air pocket may be created. Such situations invite frost damage. All this is quite apart from light, water, and nutrition.

Accordingly, an understanding of a few of the general principles of radiation, airflow, and microclimate is useful in attempting to rationalize these experiences, and to assist in the selection of planting sites, the arrangement of the planting, and the closeness of the trees.

SPRING FROST

To begin with, a distinction must be made between cold temperatures due to a drop in temperature of a large air mass over a large area, such as accompanies the airflow of a cold front, and between local cold temperatures, over a limited area, due to specific local conditions, such as radiation on a clear still night. Very little can be done to counteract cold injury from the former, but considerable can be done to avoid serious injury from the latter.

Some Principles of Airflow and Radiation

Chilled air follows the laws of gravity, and, being denser than warm air, flows downward. One need only recall that when the oven door is opened there is a rush of hot air upward, whereas when the refrigerator door is opened the cold air flows out onto the floor. And so cool air flows downward from the hilltops, like water, and follows the valleys and the gullies.

Heat is constantly being radiated in all directions from one object to another. A warm object radiates toward a cooler one, and a cold object similarly radiates to a warmer one. This is why an open fire warms the face and hands of the individual facing it, while his back may be chilled by cooling radiations from behind. This is why an individual may feel cool in early spring when the ground is cool even though the air is warm. This also is why he may feel comfortably warm in late fall when the ground is still warm even though the air temperature is cool.

The earth receives radiations from the sun during the day, and in turn radiates heat upward toward outer space at night. Cool air develops at the top of a hill because of radiation. As it flows down the side of the hill, it picks up additive cool air from the sides of the slope, which all settles at the bottom. Its behavior is, again, much like that of water. The cold evening wind that blows down the steep fjords of Norway is another example. Radiation from the smooth rock surfaces is rapid, and the cold air accumulates and flows swiftly down the fjord.

Importance of Gullies and Air "Draws"

If there is a gully or a ravine large enough to accommodate the cold air mass, the flow continues down and spreads onto the land below, and the sides of the slope are spared frost damage. On the other hand, if there is no "draw" down which the cold air can flow, or if the air mass is too great to be contained by the "draw," then the sides of the slope may experience frost, especially the lower reaches of the slope.

Accordingly, trees should not be set in the natural draw that drains cold air from an area. Not only may the planting impede the drainage of air; the trees are subject to spring frost hazard from the cold air that drains down the slope. The steeper the slope, the more cold air the gully can handle. It will rise in the gully just high enough to carry away, much like water in a canyon. If the slope is very gentle, the spread of cold air is wider in the draw. A greater width either side should, accordingly, be left unplanted on gentle slopes than on steep slopes, so far as protection from spring frost is concerned.

Obstruction to Airflow

As cool air flows down a slope, it hugs the ground and slides along. A smooth surface facilitates its passage; and a rough or uneven surface slows it (Fig. 76). When an obstruction is met, such as a low-growing tree or bush, the cool air mass flows upward and over it and engulfs it to a height greater than the depth of the cool layer of air (Fig. 76). This is why a compact low-growing tree is sometimes frosted out to a greater height than a more open or high-headed adjacent tree.

Like water, too, cool air can be dammed up. An embankment at the bottom of a slope can create a pool of cold air that backs up the slope to the height of the embankment (Fig. 77). Fence rows with trees and brush can act similarly. If there is obstruction to airflow from any cause, frost is more likely to occur. Sometimes shelter belts or rows of tall-growing trees are planted to break the force of prevailing wind. They are useful in reducing wind whip of leaves and fruit, and they protect trees against wind breakage and blowing over. But they can also introduce an increased hazard from spring frost by obstructing the flow of cold air and impeding natural air drainage.

Deciduous trees are preferred to evergreen trees for windbreaks in regions of severe winter cold because their bare branches do not materially obstruct air movement. If evergreens are used, a good plan is to remove the lower branches. The force of wind is broken by the tops, yet cold air on quiet nights can flow easily beneath the branches. Otherwise, gaps should be cut in the shelter belt or hedgerow.

Similarly, if fruit trees are closely planted on the contour or in rows across a slope they may impede air drainage. It has been suggested as a general rule that trees should be planted twice as far apart as their height is expected to be at full bearing age. Thus, where trees reach a height of

Fig. 76. **A flow of cold air that meets an obstacle, such as a low-growing bush or tree, may engulf it and cause cold injury; whereas an object which does not impede the flow of air, such as an open or high-headed tree, may escape uninjured. If a slope is planted with trees of different sizes, the smaller ones should be at the top and the larger ones at the bottom so as to facilitate air drainage and reduce cold hazard. (After _Frost and the Fruit Grower_ by Raymond Bush, Cassell and Company, London, 1945)**

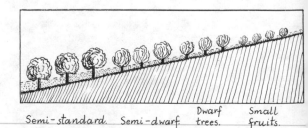

Small fruits.   Dwarf trees.   Semi-dwarf.   Semi-standard.

Semi-standard.   Semi-dwarf   Dwarf trees.   Small fruits.

10 feet, they should be planted 20 feet apart as a minimum. However, if there is sufficient land available, a distance of 2½ to 3 times the height would be better, so that trees that reach 10 feet would be planted 25 or even 30 feet apart. While these instructions are given primarily with reference to nutrition and available water for the tree, they also apply to air drainage.

A good procedure is to plant the trees closer in the rows than the distance between rows, such as 15 × 20, 20 × 25, or 20 × 30. If the site is frosty and the slope not very great, the rows should run parallel with the slope, and not across it. If the slope is relatively steep, interference with air drainage is not such an important factor, and is overbalanced by a possible increase in soil erosion and the difficulty of traversing the orchard.

With the hedgerow system of growing fruit trees as practiced in Italy, some attention must be given to how the rows are run. In this system, trees are set close together in the row so that they present a solid hedge or wall of foliage with no breaks between the trees in the row. If the slope is gentle, so that air drainage is naturally slow and yet soil erosion is not a problem, the rows should run parallel to the fall so as to favor air drainage. If the hedgerow planting is run across the slope, the movement of air is greatly interfered with.

If a slope is planted with trees of different sizes, the smaller ones should be at the top and the larger ones at the bottom, as shown in Fig. 76.

Protection by Clouds, Buildings, and Other Objects

Radiation effects, and protection from losses in heat through radiation, are well illustrated by the effects of clouds. On clear nights, with no cloud covering, the drop in temperature due to radiation may be sufficient to cause frost. If, however, there is a blanket of clouds, the radiation is reflected back to the earth, and frost is avoided. The point is illustrated by a cloud moving from left to right. As the cloud covers

Fig. 77. Like water, cool air can be dammed up to create a pool of cold injurious air that backs up the slope to the height of the obstruction. (After *Frost and the Fruit Grower* by Raymond Bush)

an area, the radiations below are trapped and reflected back, and the temperature rises. Behind the cloud, as it moves, the temperature falls, and ahead of it the temperature rises.

The overhanging branches of a tree intercept warm radiations from the earth, and are warmed. A ground covering of mulch or weeds decreases this warming effect and increases the incidence of spring frost because it serves as an insulation and reduces warming radiation upward to the tree. Clean cultivation provides greater spring frost protection. In fact, this may be such an important consideration that it may tend to swing orchard-management practices back to clean cultivation for dwarfed trees.

Similarly a wall or a building provides protection. It is not uncommon for a fruit tree close to a building, such as an early-blooming apricot, to escape spring frost and develop a crop while nearby fruit plantations are a total loss. Also, the side of the tree facing the building may bear fruit, whereas the side away from the building may be bare. The explanation here again is the warming radiation from the building toward the tree, coupled with the interception of warming radiations from the ground that are reflected back to the tree. An experience is recorded in Pennsylvania where a full crop of fruit was borne by trees in the second and third rows adjacent to a woodlot, whereas the remainder of the orchard carried little or no fruit.

This is one of the reasons, also, that fruit trees are grown so successfully against walls and in enclosed gardens (Fig. 78). An individual may wish to grow fruit on property that provides no natural elevation or air drainage and where frost is a hazard. The enclosed fruit garden, such as one finds in Europe, is well suited to this situation.

Radiation phenomena apply equally to individual fruits, leaves, branches, and twigs. Each is involved in receiving and giving off radiation. A fruit or blossom close to a large branch may be raised in temperature by radiation from the branch to just the amount necessary to avoid frost damage. And a leaf covering a fruit or blossom may give similar protection. With air temperatures of 28 degrees F., fully exposed fruit has been measured at 23 degrees F. This is the reason for survival of individual blossoms here and there on a tree that seemingly defies explanation. The russeting that appears on fruits is frequently on the most highly colored side or at the apex. Partial explanation lies in the fact that

Fig. 78. Grapes growing against protecting wall with frame for matting or canvas against cold, frost, or other inclemencies.

the well-colored areas are exposed and radiate outward so that their temperature is lowered; and young fruits of some varieties stand erect, thus also favoring radiation from their tips. The resultant lowered temperature and frost damage expresses itself in russeting at the calyx end of the fruit.

Orchard Heating

Orchard heating to dissipate spring frost is most effective on level areas and where air movement is at a minimum. On a clear, calm night, there is often only a relatively thin layer of cold air hugging the ground, with air of higher temperature up to a height of 300 to 800 feet. This condition is known as temperature inversion (Fig. 79).

If an orchard is heated under such conditions, the hot gases from the heaters mix with the cold air and rise slowly until they reach air of the same temperature. They rise no farther, and the warm air

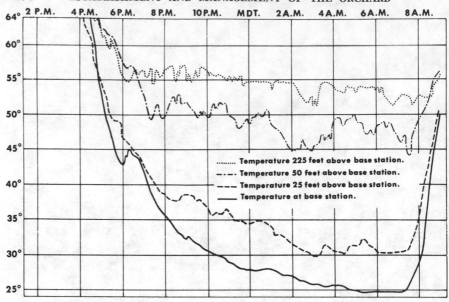

Fig. 79. Great differences in temperature may occur at different elevations on a clear still night. Graph shows records of temperatures from 4:00 P.M. to 9:00 A.M. at the base of a steep slope and at three heights above the base (25, 50, and 225 feet). Although the temperature at the base was low enough to cause considerable damage to fruit (24 degrees F.), the lowest temperature at 225-foot level was only 51 degrees F. At higher levels in the atmosphere the change in air temperatures between day and night is even less. *Courtesy of U.S. Weather Bureau.*

above the orchard then acts as a roof that stops the ascent of the heated air. Thus the heat from the burning heaters raises the temperature of the air within only a relatively few feet off the ground (often 30 to 40 feet) and protects the orchard from frost with fairly nominal expenditures of energy.

The degree of temperature inversion near the ground determines the depth of the layer of air that must be warmed.

The amount of temperature inversion is determined mainly by the amount of fall in temperature from afternoon to early morning. If the temperature in the afternoon is high, and falls to freezing on the following morning, the inversion is likely to be great, and orchard heating unusually effective. The most difficult nights for heating are those following cold afternoons when the inversion in temperature is slight. A large number of small fires is more efficient than a small number of large fires.

Closely planted orchards of small trees, set on fairly level ground and protected by shelter hedges or the general topography of the land, can be greatly helped by orchard heating. It is not likely that large-scale commercial orchardists in the United States will fall into a pattern of this kind as long as there are still frost-free sites available. But home fruit growers and part-time orchardists with small plantings near local markets might find themselves in such a position. If they do, heating is a reasonable suggestion for assistance in protection from frost damage.

Another system of orchard heating that has not been fully tried but offers promise is dependent upon a very different principle, namely, radiant energy from radiant heaters. Such heaters do not attempt to heat the air mass. Rather they send out infrared rays that heat whatever intercepts the rays. Similarly, branches and leaves may be warmed by the rays from infrared heaters although the surrounding air mass itself is relatively cool.

There are definite limitations to this method of heating, however, for branches "shielded" from the rays by other branches are not warmed. Nevertheless, such heaters might be expected to be useful with small-growing plants, such as dwarfed fruit trees.

Orchard Sprinkling

Overhead sprinkling with water is another practical method of reducing spring frost damage. The latent heat of fusion of water is such that considerable cold is absorbed by the water in the freezing process. Thus, it takes as much cold to freeze a pail of water that is at 32 degrees F. as it does to lower the temperature of the water from 144 degrees F. to the freezing point. This is considerable protection. In addition, the water vapor and the droplets themselves reduce radiation and thereby provide some warming influence. The temperature of sprinkled flowers has been held at 32 degrees F. where nonsprinkled flowers have reached 26 to 27 degrees F., with considerable attendant frost injury.

Sprinklers should not be started until the temperature of the plant tissues has dropped to the freezing point or even slightly below, and should not be run longer than necessary. The soil may become water-logged; also, heavy ice formation may cause breakage in the tree.

WIND AND WINDBREAKS

Wind may cause considerable damage in the dwarfed orchard. Some rootstocks, such as the EM IX, have brittle roots and are subject to leaning and blowing over, and the EM IV has a high top-to-root ratio and can be uprooted by strong winds. Frequently, high winds are accompanied by heavy rainfall that saturates the soil, weakens soil anchorage, and makes the trees susceptible to blowing over. The hurricane of 1944 that swept through New England caused severe damage because of just this combination of high soil moisture and high wind. Large long roots were slipped cleanly from the soil as though they were greased cables.

The situation is aggravated when the fruit trees are set close together in hedgerow fashion across the direction of the prevailing wind. Instances are known where an entire row of trees has been blown over at one time.

While not confined solely to dwarfed trees, whipping of the leaves and fruit by strong winds can also injure both the leaves and the fruit. And fruit may be caused to drop prematurely or may even be forcibly blown from the tree. Further, wind may reduce fruit set and increase drought injury, and make spraying more difficult for pest control.

Some idea of the force of wind can be gathered from a few illustrative figures. A wind of 10 miles per hour exerts a force of $\frac{1}{3}$ pound per square foot. At 30 miles per hour the force is increased to 2.5 pounds, and at 60 miles per hour it reaches 9.0 pounds per square foot.

When winds prevail mostly from one direction, considerable can be done to reduce hazards and damage to both tree and crop. A little familiarity with a prospective orchard site will show the frequency and velocity of high winds. The lee side of a hill often gives excellent protection, especially where the windward side of the hill is steep. Under such conditions a strong wind tends to rise and to clear an orchard planted on the lee side for a considerable distance.

Windbreaks are effective in reducing damage from wind. Relatively quiet "pools" may be created in the area just above the earth. Though comparatively small, these pools extend for considerable distances with strong winds, and a small reduction in current may carry over to another windbreak as much as one-fourth to one-third mile away. Planted at or near the crest of a hill, they give good protection for several hundred yards. Careful study has revealed how much protection can be expected from windbreaks of different heights, widths, and linear extension. Thus, if the wind does not shift widely, but strikes a planting squarely, the protected area will be triangular in outline (Fig. 80). The longer the line of trees, the deeper is the area that will be protected. If the prevailing wind is from the south and west, and the trees are planted north and

Fig. 80. The area protected by a windbreak is triangular in shape, and maximum protection is provided by long windbreaks—contrast A, B, and C. (From U.S.D.A. *Farmers' Bulletin 1405*)

Windbreak length: *Area protected*
   6 rods... 9 square rods.
  12 rods... 36 square rods.
  72 rods... 1,296 square rods, or 8.1 acres.
160 rods (one-half mile) .... 4,464 square rods, or 28 acres.

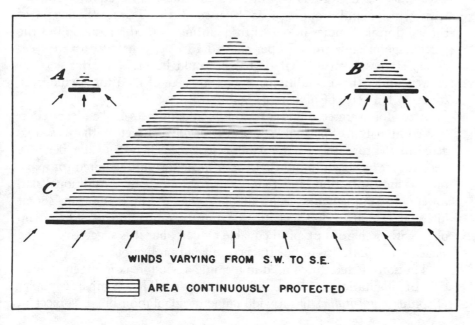

WINDS VARYING FROM S.W. TO S.E.

▤ AREA CONTINUOUSLY PROTECTED

south according to the compass, the windbreak should be along two sides in an "L" shape.

There is also a definite relation between the height of the windbreak and the depth of wind protection. Although some protection may extend up to 40 times the height of the windbreak, the most effective range is more like 15 to 20 times, and maximum protection is provided at 5 to 10 times the height.

Accurate wind-velocity studies have shown a reduction of wind by 75 per cent at 10 times the height of the windbreak, 50 per cent at 15 times, and 25 per cent at 27 times.

In other tests, ground winds have been reduced ten per cent at 30 times the height of the windbreak, so that trees 50 feet tall have given some protection from excessive winds to a distance of 1,500 feet. As a rough guide, it may be said that windbreaks placed a quarter of a mile apart will give considerable protection to an extensive area.

The extent and the width of windbreaks must be balanced against the loss of the land they occupy. Single or double rows of eucalyptus trees have been used successfully in California to protect orange groves. Trees may reach 60 to 80 feet in 10 to 20 years and have protected fruit from bruising and dropping to a distance of 5 to 7 times the height of the windbreak. Studies in one citrus planting showed a reduction in the percentage of culls from 50 per cent to 18.5 per cent over a three-year period. It has been estimated that it would be profitable for a citrus grower, under these conditions, to use one acre of land for a windbreak for every ten acres of orchard (Fig. 81).

A row of evergreens and a row of deciduous trees make a very effective combination. Spacing may be 10 feet by 10 feet, with trees staggered in the row. In extreme windswept areas where a shelter belt is a necessity, it has been found that five rows of trees are required for maximum wind protection, and that seven rows are better. Additional narrower and supplemental belts of one, two, or three rows are set at intervals of 200 to 300 feet. Such extensive protection would find no place with orchards in typical equable climate, but it is suggestive of the value and importance to be attached to wind protection.

The sort of trees to be used in forming a windbreak is largely a question of adaptation to local soil and climate. The Lombardy poplar (*Populus nigra italia*) has a wide range of adaptation but is subject in

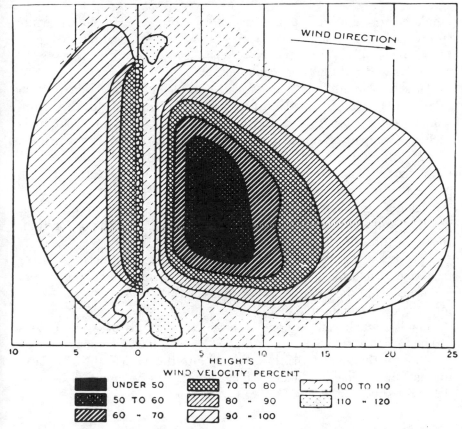

WIND DIRECTION

HEIGHTS

WIND VELOCITY PERCENT

UNDER 50	70 TO 80	100 TO 110
50 TO 60	80 - 90	110 - 120
60 - 70	90 - 100	

Fig. 81. Diagram of an actual field of protection provided by a windbreak 19 heights long, with an average density of 50 per cent but more open in the lower than in the upper half. The air pushes through the opening at the bottom and shoots upward on the leeward side. (From U.S.D.A. Farmers' Bulletin 1405)

America to several disfiguring diseases. For this reason *Populus simonii* and *Populus alba pyramidalis* (Bolleana poplar) are preferred.

The Chinese elm (*Ulmus parvifolia*) and the Asiatic elm (*Ulmus pumila*) succeed in the Midwest where hot, dry conditions prevail. Cottonwood (*Populus* spp.), honey locust (*Gleditsia triacanthos*), American elm (*Ulmus americana*), and box elder (*Acer negundo*) are also used successfully. Among evergreens the following are commonly employed for windbreak purposes: Norway spruce (*Picea abies*), white spruce (*Picea glauca*), Scotch pine (*Pinus sylvestris*), and red pine (*Pinus resinosa*). White and Norway spruces are good on heavy soils, and Scots and red pines on light soils.

Closely planted fruit trees provide considerable protection to each other. Tests have been made in England on the effect of tree density on wind-flow, tree growth, and fruit yield in an exposed commercial apple orchard. Comparison was made on trees on the EM II rootstock planted (1) 25½ feet square, and (2) 25½ feet square with a tree on EM VII in the center. Compared with bare land, the wind was decreased 21 to 28 per cent in the low-density planting, and 38 to 58 per cent in the closer or high-density planting. Growth of trees was greater on the closer planting, and yields were increased up to 67 per cent for the period from the seventh to the thirteenth year of age.

A common sight in some parts of Europe is the palmette hedge of fruit trees enclosing a planting and providing some wind protection. Delicious/Robusta 5 has been suggested for this purpose in some of the more exposed regions in America.

Finally, much can be done to lessen the hazards of wind by low-heading of the tree, by pruning the leader to a bud that points in the direction of the wind, by tilting the tree into the wind slightly at planting, and by staking and bracing. These points will be discussed more in detail on subsequent pages.

HAIL, MULCH, AND WINTER COLD

Hail

As has been already noted, hail tends to develop characteristically in certain localities and areas. Local inquiry among long-time residents may reveal these sites, so that they can be avoided. For example, in western New York where the author once resided, there is a swath several miles wide just north of the foot of one of the Finger Lakes where hail is characteristic, although infrequent in the general region. Summer thunderstorms can be observed regularly approaching from the southwest, and veering sharply to the north as they reach the lake, which extends north and south. The storm swings around the north end of the lake, dropping hail in the belt it traverses, and then continues eastward.

Little can be done to alter such situations. Yet in northern Italy, where hail is frequent in summer when crops are maturing, rockets are fired into hail clouds as they move down into the valley from off the Alps. With an organized program of community participation, information is relayed to growers as to the size of rocket to use to reach the required height of the cloud, and the time and number of rockets to fire. The program is effective.

Mulching, Snow Protection, and Winter Cold

Mulching of the soil may be thought of as a microclimatic influence. As has been mentioned elsewhere, a heavy mulch tends to interfere with the warming radiations from the soil to the tree and the surrounding air, and so may be responsible for greater spring frost damage than would be the case from clean cultivation.

On the other hand, mulching gives very good protection against root injury from deep soil penetration of severe winter cold. Records from Kansas reveal penetration of cold and freezing of the soil to a depth of 26 inches when the ground was bare. With a good snow cover, the depth was reduced to 12 inches, and with a straw mulch and accumulated snow covering it was only 6 or 7 inches. In Minnesota, fruit trees have been killed outright by prolonged subzero temperatures during which the soil has become frozen and the roots killed. Adjacent orchards with good mulch covering were not injured.

Snow, similarly, gives good protection against penetrating winter cold. If the fall of snow is early and remains or is replenished throughout the winter until late spring, it is not uncommon to find that the soil has not been frozen during the entire winter. There are such areas in northern Michigan close to Lake Michigan, along the east side of the lake. Similarly, such conditions prevail on the eastern side of the Lake Huron and in the Georgian Bay region of Ontario in Canada. EM IX apple roots have been reported to have experienced severe winter temperatures of −50 degrees F. with no damage when a good snow cover prevailed.

It is also well known that certain localities, as well as certain spots on a farm, accumulate greater snowfall and snow coverage. Closely planted trees tend to hold the snow, and in this respect provide a favorable microclimate for root protection. A windbreak serves effectively in similar manner, and trashy cultivation or a cover crop in the orchard tends to catch and hold the snow.

Low winter temperatures, as has been suggested, vary appreciably from one location to another in an orchard. The factors responsible are similar to those associated with the occurrence of frost and frost damage. That is, cold air, even at subzero temperatures, flows downward, collects in pools formed by obstructions, and may stratify on quiet nights.

The greatest protection is provided similarly by good altitude. The lowest temperatures are found at the lower levels in the orchard. Attention to the details outlined in preceding paragraphs for frost protection are operative here, too. In addition, soil temperatures and possible root damage become of concern, with rootstocks that are shallow-rooted.

17

SELECTING, SPACING, AND PLANTING THE TREES

SECURING THE TREES

TOO many orchards have been planted with trees bought with little appreciation of the nursery enterprise that grew them or the quality of the stock. When one considers the years of labor and expense that must go into a successful orchard enterprise, it is only good sense to give considerable attention to the trees upon which the enterprise is to rest. Price becomes a second consideration.

It is true, of course, that one of the retarding factors in the freer use of dwarfed fruit trees by orchardists has been the relatively high price of trees. This is especially so when one considers the large number of trees required per acre for close planting. The standard commercial planting of apple trees, for example, at 40 × 40 feet, requires 28 trees per acre, whereas a planting 15 × 20 feet requires 145 trees. There is no reason that dwarfed trees should cost greatly more than trees on standard seedling rootstocks. Once the supply of dwarfed trees is sufficient to satisfy the demand, prices may be expected to be comparable to those for other stock.

Yet there is no virtue in "bargain stock." By far the most costly mistakes have been made in buying nursery stock, sight unseen, from unidentified transients at bargain prices. The more reliable and better-known nursery organizations and dealers, who are in the business permanently, and particularly those with whom the planter has an understanding or acquaintance, are the ones that will ordinarily give the greatest satisfaction.

Types of Nursery Enterprises

Nursery enterprises may be grouped as wholesalers, catalogue houses, agency houses, and local peddlers or jobbers, each representing a different class and type of service. As with any sales operation, the satisfaction of the individual purchaser rests largely upon the selection of the type of service that best meets the purchaser's requirements.

The wholesale houses are among the largest growers of nursery stock. They may sell in large quantities to other nurserymen or to large planters. The catalogue house issues an attractive catalogue, operates largely by mail, and handles many small orders. It must receive a price relatively higher than the wholesale price for the extra service of handling small shipments. It will usually furnish a special, reduced, or "wholesale" price to large planters. The agency house sells through local salesmen who visit from house to house. It may offer "free" landscape services, planting advice, free delivery, and other services; in turn, it asks a higher price. The local dealer or jobber operates as an independent agency representative. He purchases stock from a producing company and peddles or jobs it himself. The stock he sells is only as good as his integrity and his reputation, which should be of prime concern to the purchaser.

There is still another type, which is often a combination of fruit grower and nurseryman. They are located in the major fruit-producing areas, and usually grow trees to order for local fruit growers. They can survive only because they satisfy their neighbors. They are among the most reliable and satisfactory nurserymen. This is especially so with dwarfed fruit trees, since these men understand some of the complexities of growing good dwarfed fruit trees, as well as appreciate their uses and requirements.

Trueness to Name

Excellent regulatory laws protect the planter from receiving stock infested with insects and most diseases. Virus diseases, however, pose a more difficult problem. They may be especially acute in dwarfing root-stocks that are clonally propagated. Once a virus is introduced into the stool block or other clonal propagating material, it is perpetuated and disseminated in the rooted plant. On the other hand, once a virus-free clonal line has been developed, it can be maintained in that condition. This is the next major step in inspection and certification of both root-stock and scion variety. Some supplies of both have already been indexed and certified for certain virus diseases, and progress is constantly being made. A prospective planter should make every effort to secure virus-free stock insofar as possible.

Concerning trueness to name of the scion variety, reliable nursery firms have adopted the practice of having their fruit trees inspected annually by experts for mixtures or misnaming. Varieties can be identi-fied in the nursery by plant characters. A planter can be fairly secure in the belief that he will receive nursery trees true to name as regards the scion variety.

On the other hand, there is less certainty regarding the authenticity of the rootstock. Although every precaution may be taken, mixtures and errors in labeling may still occur. Once the rootstock is in the ground, any possible mistake is difficult to detect. Research workers make it a practice to authenticate each rootstock by making a piece-root cutting from each tree, as described in Chapter 14, and certifying each plant according to leaf and shoot characters. Such a practice is out of the question for extensive plantings, but might be resorted to for a few trees for the satisfaction and interest of the planter.

Fortunately, some states, such as Michigan, have extended certifi-cation to dwarfing rootstocks as well as to scion varieties. The procedure is to have the lining-out stock examined in the nursery row before the rootstocks have been budded and while they are in leaf. At this time any misnaming or mixtures can be detected, and either corrected or rogued out. This is a most important matter. The author has seen some very serious cases of misnaming. A customer who orders a very dwarf tree, as apple EM IX, and receives a semistandard, as apple EM XIII, is not only disappointed but also may disseminate erroneous reports as to the performance of the rootstock that he in all honesty assumes he received.

A planter has every right to insist upon certification of the rootstock for trueness to name. If steps are not taken to enforce such a system of certification, the dissatisfaction arising from unhappy customers could easily discredit and cripple interest in dwarfed fruit trees.

It is now possible for nurserymen to secure from governmental regulating agencies a supply of plant material that is free from many viruses and other disorders and that is true to name and type. This should serve as the mother stock for propagation. The aim should be to provide a tree with a "birth certificate" even though the price might be somewhat higher. Mistakes that are planted in the orchard are there for a long time.

Grade of Stock

American nurserymen have adopted certain standards for nursery trees, based on the common sizes secured under average nursery conditions. All trees are expected to have reasonably straight bodies according to the growth habit of the scion variety, and they must have attained a certain caliper just above the bud. These standards are vigorously held to, and a prospective planter can be assured of receiving what he has ordered.

In general, the size of the stock is of less importance than the quality and the condition. Yet, everything else being equal, the No. 1 commercial grades may be expected to give the best results. Overly large trees may be offered, but they do not transplant well. Very small trees, on the other hand, may be suspected of some inherent weakness, and are to be questioned.

The most desirable tree is one that has a natural appearance of stockiness. It has good caliper and is best described as being "fat." Too many dwarfed trees that are weak, miserable creatures have been sold and planted. Such trees will never give a good performance; they should under no circumstances ever be planted. The author has seen too many dwarfed trees of this type, and cannot emphasize the point too strongly. It does not follow that just because dwarfed trees carry this designation, that they are small and weak as nursery trees. On the contrary, as has been noted in Chapter 15, dwarfed trees are remarkably similar in nursery growth, regardless of the degree of dwarfing.

Height of Budding

To avoid the possibility of scion-rooting, some nurserymen prefer to bud the dwarfing rootstock at a height of 7 to 8 inches above the ground line so that there is a rootstock stempiece of 15 to 16 inches. The root system of the tree can then be placed deeper in the soil, so that otherwise typically shallow-rooted stocks may be sufficiently deep to give good anchorage. This factor, quite aside from the possibility of scion-rooting, is of special concern in areas where strong prevailing winds are customary, as in the middle western and central western parts of the United States. Reports on the performance of high-budded trees have vindicated their judgment. On the other hand, where trees are grown on trellises or are regularly supplied with some system of support, there is no special merit to high budding. In fact, there are disadvantages in the case of rootstocks that are tender to cold or subject to crown troubles.

Age of Tree

The age of a tree is told by the seasons of growth of the scion, regardless of the age of the rootstock. Peach trees reach sufficient size in one year to be sold as one-year-old trees. Sweet cherry trees and plum trees also attain salable size in one season, but are more commonly grown a second year in the nursery and sold as two-year-old trees. Sour cherry, apple, and pear trees are frequently sold as two-year-old trees.

Where trees are trained to special forms in the nursery, the age at planting quite naturally differs considerably, depending upon the type and degree of training. Some espaliers may be four to eight years old. On the other hand, where the training is done by the planter, one-year-old and two-year-old trees are desired.

As has already been intimated in earlier paragraphs, there is much to be desired in the training of dwarfed trees in the nursery, especially to meet the needs of small planters. A one-year-old tree takes several years to come into bearing. If such a tree is left in the nursery and trained properly by the nurseryman until two or three years of age, fruiting is speeded up just that much. Such a development awaits demand from planters who desire this sort of tree. When raised in a local nursery not far from the orchard site where it is to be planted, they can be dug and planted the same day with excellent success and enormous satisfaction.

PLANTING DISTANCES, INTERPLANTING, AND NUMBER OF TREES PER ACRE

Spacing

The spacing or planting distances for dwarfed fruit trees is dependent upon a number of factors. In rich soils with good moisture supply, trees may be placed closer together than on soils of lower fertility and where water may be a limiting factor. With rootstocks that are very dwarfing, such as the apple EM IX and the Quince EM C, spacing can be quite close (3 to 6 feet); whereas with the apple EM XIII, trees must be a farther apart (20 to 30 feet). Some varieties, such as the Gallia and Cortland apples, are smaller-growing trees and will bear closer planting than will such large-growing sorts as the Northern Spy. And the type of training and culture will control spacing to considerable extent, such as espalier-type trees trained on wires, dwarf pyramids, and hedgerow plantings.

The suggested spacings that follow are general in nature and must be adjusted to meet these and other factors. They are all at minimum distances, and are dependent upon good pruning practices to keep the trees in hand. If less-intensive operations are intended, as where modern mechanical equipment is used, the spacing between rows should be at the maximum distances given, or even wider. This applies primarily to bush trees, pyramids, and trellised trees; and it applies less to cordons and espaliers:

Single cordons (trained as single stems)
    Apple
        EM IX: 2 feet, in rows 6 to 10 feet apart
        EM I, II, VII: 2 to 2½ feet, in rows 8 to 10 feet apart

Pear on Quince
    EM C: 2 feet, in rows 6 to 10 feet apart
    EM A: 2½ feet, in rows 8 to 10 feet apart
Palmettes, Espaliers (trained in one plane)
  Apple
    EM IX (small forms): 3 to 4 feet, in rows 10 feet apart
          (large forms): 10 to 12 feet, in rows 6 to 10 feet apart
    EM VII and IV: 12 to 15 feet, in rows 6 to 10 feet apart
    EM II: 15 to 18 feet, in rows 6 to 10 feet apart
  Pear on Quince
    EM C: 3 feet apart, in rows 10 feet apart
    EM A (small forms): 3 to 4 feet, in rows 10 feet apart
          (large forms): 10 to 12 feet, in rows 6 to 10 feet apart
Fan-shape (trained in one plane)
  Apricot, peach, plum: 14 feet apart
Pyramids (trained in low-headed, pyramidal form)
  Apple
    EM IX (weak varieties): 2½ to 3 feet, in rows 6 to 9 feet apart
          (strong varieties): 4 to 6 feet, in rows 8 to 12 feet apart
    EM IV, VII, II: 4 to 6 feet, in rows 10 to 12 feet apart
  Pear on Quince
    EM C: 3 to 4 feet, in rows 6 to 12 feet apart
    EM A: 4 to 6 feet, in rows 8 to 10 feet apart
Trellised trees (as employed in commercial fruit production)
  Apple
    EM IX (weak varieties): 4 to 6 feet, in rows 10 to 12 feet apart
          (strong varieties): 8 to 10 feet, in rows 10 to 20 feet
             apart
    EM VII: 15 feet, in rows 25 feet apart
    EM II: 20 feet, in rows 30 feet apart
  Pear on Quince
    EM C: 8 to 10 feet, in rows 10 to 12 feet apart
    EM A: 10 feet, in rows 10 to 20 feet apart
Bush trees (low-headed, little-pruned)
  Apple
    Dwarf
      EM IX (weak varieties): 6 to 8 feet, in rows 10 to 12 feet apart

(strong varieties): 8 to 10 feet, in rows 10 to 20 feet apart

Semidwarf
    EM VII: 12 to 15 feet, in rows 20 to 25 feet apart
    EM II: 12 to 18 feet, in rows 20 to 30 feet apart
Semistandard
    EM XIII: 20 to 30 feet, in rows 25 to 30 feet apart
    EM XVI: 25 to 30 feet, in rows 30 feet apart
Pear on Quince
    EM C: 6 to 8 feet, in rows 10 to 12 feet apart
    EM: A: 12 to 15 feet, in rows 15 to 20 feet apart
Cherry
    Sour: 15 to 20 feet, in rows 15 to 20 feet apart
    Sweet: 20 to 25 feet, in rows 25 to 30 feet apart
Peach
    15 feet, in rows 15 to 20 feet apart
Plum
    (Japanese and American varieties): 15 to 18 feet, in rows 15 to 18 feet apart
    (European varieties): 15 to 20 feet, in rows 20 to 25 feet apart
Apricot
    15 to 18 feet, in rows 20 to 28 feet apart
Hedgerow planting (bush trees)
Apple
    EM IX: 4 to 6 feet, in rows 10 to 12 feet apart
    EM VII, II: 10 to 12 feet, in rows 25 to 30 feet apart
Pear on Quince
    EM C: 4 to 6 feet, in rows 10 to 12 feet apart
    EM A: 10 to 12 feet, in rows 20 to 25 feet apart

There are many variations from these suggested planting distances. For example, a common planting distance for semidwarf apple trees (EM VII) in the United States is 20 × 25 feet. But if the planting is made with trees still closer in the rows and with rows farther apart, such as 15 × 30, it will better suit modern mechanical equipment, which favors closer spacing in the row and wider spacing between rows. Further, the number of trees per acre is increased by this distance from 87 (20 × 25) to 96 (15 × 30).

A spacing that has been found suitable to apple trees on EM VII rootstock in southwestern Michigan is 16 × 24, or 113 trees per acre. This approaches a solid hedgerow planting. When spaced in this way it has been found useful to break the row by omitting every 31st tree. This provides areaways at 500-foot intervals through which equipment can be moved without going to the end of the row.

Interplanting

It is sometimes advantageous to interplant the main planting with some other variety or kind of fruit or with trees that are smaller or earlier in bearing. Peaches, cherries, pears, or plums, for example, were at one time frequently interplanted in American apple orchards, being removed when six to ten years of age. By that time, sufficient fruit had been produced by the peaches, cherries, or plums to pay for some of the cost of bringing the main block of apples into production.

The difficulty with this practice was, however, that as control of insects and diseases became more difficult, spray schedules often conflicted in timing or were injurious to the one fruit or the other. Accordingly, a solid planting of one class of fruit, such as apple or pear, was deemed much more desirable from this point of view. This has become the accepted procedure in commercial plantings.

In this situation the dwarfing rootstocks are admirably suited. Not only can a single class of fruit be interplanted with smaller and earlier-bearing trees of the same class, such as the apple, but also the same desired variety of apple may be used throughout, allowing for pollinators.

Further, if an orchardist is uncertain as to the performance of a variety or of a rootstock, he can plant two different varieties or rootstocks in the same block, expecting one or the other to be removed when the orchard reaches bearing age. He can thus watch the performance of the two kinds for several years before the decision must be made.

Such a plan is diagramed in Fig. 82, prepared from suggestions by Preston of the East Malling Research Station in England. In this case, the question is which of two rootstocks will be better, namely, MM 104 or MM 111. The planting is made with trees 17 feet apart on the square, using diagonal rows of the two different rootstocks. The block is predominantly of a single variety, with a row of a pollinating variety placed every third or fourth row.

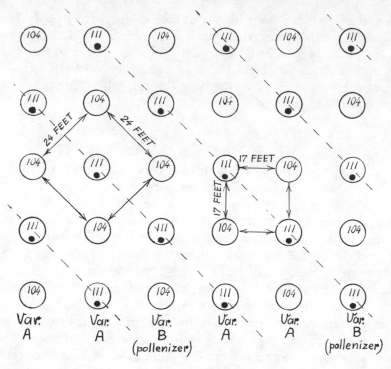

Fig. 82. Planting plan at 17 × 17 feet, involving two rootstocks (EM 104 and EM 111) and providing for pollenizers. Trees on either the EM 104 or the EM 111 may be removed as the trees get older, leaving permanent trees at 24 × 24 feet. (After A. N. Preston)

When the decision has been made as to which of the two rootstocks to retain, the other is removed on diagonal lines, leaving the permanent trees at 24 feet on the square. It will be noted from the diagram that either one of the two rootstocks may be removed.

The principle applies to any other two rootstocks that are somewhat similar in vigor, as EM II and MM 111, MM 106 and EM VII, and EM 26 and EM IX. If the grower is familiar with the performance of the older rootstocks (EM II, EM VII, and EM IX) on his land, he is taking less of a risk with some of the newer ones (MM III, MM 106, and EM 26) if he plants in this way. If he becomes dissatisfied with the performance of the newer rootstocks, he can always remove them as shown, leaving the permanent planting on EM II, EM VII, or EM IX, as the case may be.

Interplanting of dwarf and semidwarf trees with semistandard or standard trees is also possible. This system is perhaps better suited to the fruit garden or small suburban plantings than to large-scale commercial ventures. Such a scheme is shown in Fig. 83. The planting illustrated consists of permanent trees on the EM XVI rootstock (semistandard) set at 34 feet on the square. In the middle of the square is a tree on the EM VII rootstock (semidwarf), and between the permanent trees on the square are those on the EM IX rootstock (dwarf). This means that the orchard at planting consists of trees 17 feet apart in rows 17 feet apart.

The trees on the EM IX rootstock may be expected to produce five or six crops of fruit before the planting begins to crowd. At this time, all the trees on this rootstock may be removed, leaving a planting of semistandard (EM XVI) and semidwarf (EM VII) at a distance of 24 feet on the square. This planting, in turn, may continue for another four or five years, during which time the semidwarf trees in the center of each square will have fruited. They may then be removed, leaving the permanent trees at 34 feet on the square.

While seemingly somewhat complicated, the system has been used successfully. The feature is, of course, that the planting begins to fruit early in its life, so that the cost of developing a permanent orchard of larger trees is carried more economically. One great difficulty is in removing the trees successively before the planting crowds and injures the permanent trees. There is always the temptation to leave them "one more season." On the other hand, the nature of the rootstocks used will hold the interplants in check so that the competition between trees and the possibility of injury to the permanent trees is greatly reduced.

Fig. 83. Planting scheme involving dwarf (**EM IX**), semidwarf (**EM VII**), and semistandard (**EM XVI**) apple rootstocks. Trees are originally 17 × 17 feet, reduced to 24 × 24 feet when dwarf trees (**EM IX**) are removed, and ending with a permanent planting 34 × 34 feet (**EM XVI**) when semidwarf trees (**EM VII**) are removed.

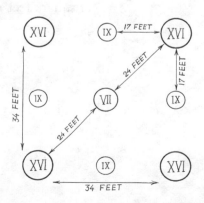

There are, of course, various modifications of interplanting. One of these is the planting of one or two dwarf (EM IX) trees or semidwarf (EM VII or EM II) trees between the permanent trees in the row. Another method is to plant a solid row of dwarf or semidwarf trees spaced closely and parallel with the permanent rows and between them.

In this connection it should also be noted that the newer spray materials available are more likely to be suited to interplantings of different classes of trees, such as apples and peaches, than were older spray materials. The newer chemicals are less harsh than older ones such as lead arsenate and lime sulfur; also, they have a wider range of usefulness. In consequence there is some disposition to return again to mixed plantings, such as peaches and apples or cherry and apples. Dwarfing rootstocks are especially useful for this kind of planting.

The Size of the Planting

A word of caution must be inserted at this point for those who may be inclined to rush into plantings of closely planted dwarfed fruit trees on a large scale without first giving consideration to the costs and labor involved in intensive culture.

The capital investment, for example, on five acres of pyramids set 3×9 feet, or 1,613 trees per acre, is high. Bush trees at 18×18 represent only 134 trees per acre, which means a substantial reduction in outlay. To offset this, of course, pyramids may yield fruit the second and third year, whereas, bush trees may not yield appreciably until the fourth or fifth year.

Two acres of closely planted trees is usually a large-enough venture for a man and wife to handle alone on the part-time basis. Four acres can be cared for by a man and wife provided there is additional help for harvest, as in the "pick-your-own" system. Ten acres is probably the limit for a couple with no outside help.

Number of Plants Required to Set an Acre of Land at Given Distances

The following table is computed by dividing 43,560 (the number of square feet in an acre) by the product of the two distances in feet.

For example, 4 feet by 8 feet equals 32 square feet, which when divided into 43.500 equals 1,361 individuals:

Distance	Plants	Distance	Plants	Distance	Plants	Distance	Plants
1 × 3 =	14,520	5 × 5 =	1,742	9 × 18 =	268	16 × 24 =	113
1 × 4 =	10,890	5 × 6 =	1,452	9 × 20 =	242	16 × 30 =	91
1 × 5 =	8,712	5 × 7 =	1,244	10 × 10 =	435	17 × 17 =	151
1 × 6 =	7,260	5 × 8 =	1,089	10 × 12 =	363	17 × 20 =	128
2 × 3 =	7,260	5 × 9 =	968	10 × 15 =	290	17 × 24 =	107
2 × 4 =	5,444	5 × 10 =	871	10 × 18 =	242	17 × 30 =	85
2 × 5 =	4,356	5 × 11 =	792	10 × 20 =	217	18 × 18 =	134
2 × 6 =	3,630	5 × 12 =	726	10 × 22 =	198	18 × 20 =	121
2 × 7 =	3,111	6 × 6 =	1,210	10 × 24 =	181	18 × 24 =	100
2 × 8 =	2,722	6 × 7 =	1,037	10 × 25 =	174	18 × 30 =	80
2 × 9 =	2,420	6 × 8 =	907	10 × 30 =	145	18 × 36 =	67
2 × 10 =	2,178	6 × 9 =	806	12 × 12 =	302	20 × 20 =	108
2 × 11 =	1,980	6 × 10 =	726	12 × 15 =	242	20 × 22 =	99
2 × 12 =	1,815	6 × 11 =	660	12 × 18 =	201	20 × 24 =	90
3 × 3 =	4,840	6 × 12 =	605	12 × 20 =	181	20 × 28 =	78
3 × 4 =	3,630	7 × 7 =	889	12 × 24 =	151	20 × 30 =	72
3 × 5 =	2,904	7 × 8 =	777	12 × 30 =	121	20 × 36 =	60
3 × 6 =	2,420	7 × 9 =	691	14 × 14 =	222	24 × 24 =	75
3 × 7 =	2,074	7 × 10 =	622	14 × 16 =	194	24 × 30 =	60
3 × 8 =	1,815	7 × 11 =	565	14 × 18 =	173	24 × 36 =	50
3 × 9 =	1,613	7 × 12 =	518	14 × 20 =	156	25 × 25 =	69
3 × 10 =	1,452	8 × 8 =	680	14 × 22 =	141	25 × 30 =	58
3 × 11 =	1,320	8 × 9 =	605	14 × 24 =	130	30 × 30 =	48
3 × 12 =	1,210	8 × 10 =	544	14 × 30 =	104	30 × 36 =	40
4 × 4 =	2,722	8 × 11 =	495	15 × 15 =	193	32 × 32 =	42
4 × 5 =	2,178	8 × 12 =	453	15 × 18 =	161	34 × 34 =	38
4 × 6 =	1,815	9 × 9 =	537	15 × 20 =	145	35 × 35 =	35
4 × 7 =	1,556	9 × 10 =	484	15 × 24 =	121	36 × 36 =	33
4 × 8 =	1,361	9 × 11 =	440	15 × 30 =	97	36 × 42 =	28
4 × 9 =	1,210	9 × 12 =	403	16 × 16 =	170	36 × 48 =	25
4 × 10 =	1,089	9 × 14 =	345	16 × 18 =	151	38 × 38 =	30
4 × 11 =	990	9 × 15 =	322	16 × 20 =	136	38 × 40 =	28
4 × 12 =	907	9 × 16 =	302	16 × 22 =	124	40 × 40 =	27

Providing for Pollination

A prospective planter should not overlook pollination requirements. Varieties should be carefully chosen, and plantings should be arranged to take care of this need. Peaches, sour cherries, and most apricots are self-fruitful. They do not require special pollenizers, and may be set in solid blocks. For other tree fruits, solid blocks are not advisable. In fact, there is some evidence that all fruits benefit from cross-pollination.

For best results a pollenizer should be placed adjacent to every tree. The minimum number of pollenizing trees in this arrangement is one tree in nine, a pollenizer being set every other tree in every other row. While very useful for small plantings, such a design presents problems in large orchards at harvesttime. Accordingly, the preferred practice is to plant solid rows of pollenizers at regular intervals in the block. Every fourth or fifth row is usually sufficient. Some orchardists prefer two or three pollenizers rather than a single variety. Certainly the evidence is clear that mixed plantings, or those in which good pollination is provided, are the most fruitful over a period of years. The entire matter of pollination is more fully discussed in Chapter 23.

THE PLANTING OPERATION

Time of Planting

Where severe winters are not a problem, and where nursery stock that is mature can be secured, fall planting is best. The trees are then dug and transplanted promptly. New roots are favored by the cool soils of fall days, so that the trees become established by spring, and take hold well.

On the other hand, if nursery trees are not thoroughly mature when dug and if the fall season is short, followed by severe winter, losses to young trees may be very high. Under these conditions, spring planting is to be preferred. If spring is selected, the planting operation should be done just as soon as the ground can be worked. Delay often means reduced growth and a poor stand of trees.

A satisfactory system, especially for both sour and sweet cherries, is to secure the trees in the fall and then "trench" them outdoors in the fall, to be planted in spring. The method is simple, a trench being dug sufficiently large to accommodate and cover the roots in opened bundles.

The trenching is done in a protected spot, such as near a farm building, and the tops are slanted away from the prevailing wind. Straw may be used as additional protection, but great care must be observed to prevent damage from rodents.

Depth of Planting

A special consideration with dwarfed fruit trees is the depth of planting. It will be remembered from previous discussion that the dwarfing influence of the rootstock may be lost if the scion becomes rooted and becomes established on its own roots. Accordingly, the union between the stock and the scion must be placed above the ground level so that the scion cannot form roots (Fig. 84).

It has been the opinion of the author that the matter of scion-rooting has been overemphasized. Only occasionally, with the materials with which he has worked, has the scion taken root. When this does occur, the surge of new growth is so striking as to be unmistakable. It has been his feeling that the union need be only slightly above the ground line. This position also lessens the chance of winter injury to the crown. Data collected after the fall freeze of 1955 in Hood River, Oregon, showed injury to several of the Malling rootstocks planted with the bud of the scion variety more than 2 inches above the ground. With the bud at not over 1 to 2 inches above the ground, as frequently recommended, little injury occurred. Ground cover in the form of mulch or snow reduced injury.

On the other hand, other persons have had different experiences and have encountered considerable scion-rooting with dwarfed trees. The reasons are not clear. Perhaps in these cases the union has been set substantially below the soil line. Or perhaps certain varieties were used that root fairly easily, such as the Northern Spy.

At all events, from these experiences and differences of opinion it

Fig. 84. **The graft union must be slightly above the soil level so as to avoid scion rooting, which destroys the effect of the rootstock. (Left) Graft union properly placed. (Right) Graft union placed too deep, with the result that scion has rooted and the effect of the dwarfing rootstock has been lost. (After Raymond Bush)**

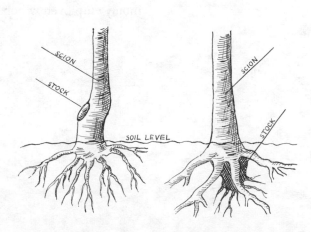

would seem the wise precaution to place the union 1 or 2 inches above the soil line. Some planters prefer even 3 to 4 inches.

Another consideration in depth of planting is the fact that some of the dwarfing rootstocks are inclined to be shallow-rooted. It has been suggested that the resulting tendency of dwarfed trees to lean, necessitating staking and bracing, might be reduced by planting the tree deeper than usual. Though the point has not been fully proved, there seems merit in the idea. Some nurserymen bud trees at 5 to 8 inches above the ground line. Such a tree can then be planted 4 to 6 inches deeper than it stood in the nursery and yet have the union 1 or 2 inches above the ground level. Certainly, the depth of planting of a tree on a dwarfing rootstock is a matter that should receive special attention.

Method of Planting

One of the features of some dwarfing rootstocks is their ability to regenerate roots readily, as has already been pointed out. This is especially so with many of the clonal apple dwarfing rootstocks, such as most of the Malling and Malling-Merton series. As a result, the planting of trees on these rootstocks usually meets with considerable success and satisfaction. It is not uncommon to have 1,000 trees survive from a planting of 1,000. Some loss is usually expected from trees on standard seedling rootstocks.

This does not mean, however, that good planting methods can be neglected,. When digging and planting are done by hand, care should be taken to dig a hole of sufficient size to accommodate the tree roots without excessive pruning. Awkward and broken roots may be cut off or trimmed back. A hole two feet across and eighteen inches deep is usually sufficient. Some good growers insist upon holes three feet across, from which the soil is removed, to be replaced with garden soil when the tree is planted. Another proven method is to mix a 12-quart pail of wet granulated sphagnum moss into the soil as the tree hole is filled. This is especially beneficial on heavy soils.

For large orchard plantings, holes are dug mechanically with a large soil augur mounted at the rear of a tractor. The tree is merely placed in the center of the hole. For smaller plantings, a planting board is commonly employed which serves to place the tree exactly where the marker stake stands. This is made from a 1 by 4-inch piece of wood, about 5 feet long. A notch is made in the board at the center on one edge, and either holes or notches are made near the two ends.

The planting board is then placed so that the stake that marks the position of the tree lies in the center notch (Fig. 85). Small end stakes are tamped through the end holes or notches into the soil. The board is then removed, leaving the end stakes as markers, and the tree hole is dug around the original stake.

The planting board is now replaced, being positioned by the two end stakes, which pass through the end holes or notches. The tree is placed in the tree hole, with the trunk fitting into the center notch to position it exactly where the marking stake stood, and with the union held at the desired height above the ground line, as previously discussed.

A small amount of soil should be placed around the roots and tamped down firmly to hold the tree in place. The planting board may then be removed and the tree hole filled and firmed, being careful not to tamp the roots down along the stem but rather to lift them gently and progressively as the hole is filled so as to place them in a natural position for better growth and anchorage.

It may be well to mention at this point that no manure or chemical fertilizer should be placed in or around the tree hole. There is always the temptation to "help the tree along" in this way, but the results are often fatal to the tree. Young roots are easily destroyed by excessive concentrations of salts. After the tree has begun growth and has become established is time enough to consider adding nutrients to the soil.

A method the author has used with very great success in planting is to stake out the orchard in the fall of the year, placing a forkful of straw, hay, or other mulching material over the stake. In the spring the mulching material is drawn back from around the stake, and the tree is planted. The mulch is then left around the tree to serve in conserving moisture and suppressing weeds. The method has proved very satisfying insofar as planting and subsequent tree growth are concerned.

Fig. 85. The tree is placed snugly in the center notch, which marks the position of the planting stake; the tree is planted firmly, and the planting board is removed. (After W. H. Upshall)

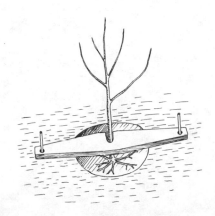

18

STAKING, BRACING,
AND TYING

A N Y discussion of dwarfed trees sooner or later moves to the topic of whether or not they require staking or bracing. There is no universal answer one way or the other. Various factors are involved, such as soil, variety, rootstock, and prevalence of wind. Certainly it is true, however, that unlike standard trees on vigorous seedling rootstocks, many situations with dwarfed trees do require attention to bracing, at least until the trees have become well rooted and well established. For those who are concerned about the necessity for bracing, the rejoinder has been well made that there should also be some concern about the necessity for a 26-rung ladder for standard-size trees.

In any event, dwarf apple trees on the EM IX rootstock require special consideration. The root system is not extensive, and the roots are brittle and easily broken. Further, the trees come into bearing early. They may carry fruit the first year set, before the tree is firmly anchored, so that the weight of fruit may make the tree top-heavy and cause it to lean.

Trees on the EM VII rootstock are less likely to require steadying, but they, too, are early fruiting and need watching. Many English growers stake EM VII as a matter of routine practice. The EM IV rootstock has a high top-to-root ratio, and is early in fruiting. As a result, it is likely to lean or even to blow over and pull from the soil, especially in heavy winds accompanied by heavy rains that reduce root anchorage. On the other hand, trees on the EM II, EM I, and EM XVI seldom require bracing if common precautions are taken. Nevertheless, many growers prefer to steady all trees on dwarfing rootstocks, as a matter of insurance.

316

Trees on good soil will take hold better and become established earlier than trees on poor soil where growth is poor and the trees do not become well anchored by themselves.

A vigorous variety, such as the Northern Spy, with upright growing branches and strong terminal growth, may require bracing where a less vigorous variety with wider angles of branching does not. The Fenton and Lodi apples develop into low-growing, symmetrical trees with wide angles of branching. Even on the EM IX rootstock they seldom require bracing. On the other hand, the Northern Spy commonly requires bracing on EM VII or even on EM II. These are some of the factors involved.

Reducing the Necessity for Staking

At all events, much can be done to reduce the requirements for staking and bracing. Trees may be headed low and encouraged to develop a compact shape. If trees on the very dwarfing rootstock (EM IX) are headed at a height of 12 inches they can be developed into a low bush form that will frequently require no staking. The dwarf pyramid form usually does not require staking.

Again, if blossoms and fruits are removed the first two seasons, there is less danger from leaning caused by a top-heavy weight of fruit. The early-bearing feature of dwarf trees is too often accepted, with the result that trees lean, become poorly established, and remain a problem for many years.

Much can be done by giving careful attention to pointing the newly planted tree slightly into the prevailing wind and to favoring the development of branches on that side. Not only does the weight of the branches help to balance the tree in the direction of the wind, but what is more important, root development is encouraged on that side. It is commonly observed that a vigorous branch lies directly over a vigorous root. If a vigorous branch can be induced to develop on the side facing the prevailing wind, a strong root system will form similarly on that side and will steady and anchor the tree. Just the reverse is true if shoot development is predominantly on the leeward side of the tree. A weak root system forms on the windward side, and the tree is easily tilted by the prevailing wind.

In Oregon it has been observed that clean cultivation results in leaning and blowing over, especially in exposed areas and on light soils. Mulching or chemical control of weeds without disturbing the soil proved better in this respect.

All of this refers primarily to the young tree as it is becoming established. As the tree becomes older, it tends to form branches and roots on all sides, but it is in the early life of the tree that the problem of bracing is most important. The young or "childhood" period of a dwarfed fruit tree is most important.

Steadying the Young Tree

In many instances all that is required is a steadying of the tree until it has become well established. This statement refers to trees that have the potential to become well anchored if steadied for a few years, such as apple trees on the EM VII, II, I, and IV rootstocks. One staking at the time the tree is set is sufficient for the period required, namely five to seven years. This does not apply to trees that are always going to require support, as some varieties of apples on the EM IX rootstock.

A 3½ to 4-foot length of old ¾-inch or 1-inch gas pipe or water pipe makes a good stake, driven to a depth of 18 to 20 inches. It requires no pointing, drives well, does not disintegrate, and can be taken up and used again.

Old iron fenceposts are also useful. Usually 7 or 8 feet in length, they can be left 5 or 6 feet aboveground and can be used for supporting low fruiting branches with strings during the early fruiting years. Such a post is especially suited to very dwarf trees, as apples on the EM IX rootstock.

Wood stakes, 2 × 2 inches, are also used effectively, but are more difficult to drive and to maintain. They should be boiled in creosote to protect the below-ground part. Chestnut is preferred in England.

For somewhat better support from a single stake, the stake is driven into the ground at an angle of about 60 degrees, with the head of the stake pointing into the prevailing wind (Fig. 86). In this position, the side of the stake is against the tree and it is easily tied in place. The tree tends to lean against the stake, and is given very good anchorage. With

Fig. 86. **Oblique staking against the prevailing wind for maximum support.** (**After D. Macer Wright**, *Dwarf Fruit Trees*, Faber & Faber, London, **1953**)

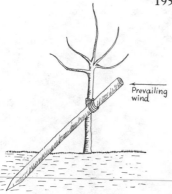

Prevailing wind

this method a shorter stake may be used and it need not be driven so deep, since the pressure on the stake tends to firm it rather than loosen it. Lengths even as short as 18 to 24 inches, placed close to the tree and only 6 inches in the ground, have served to steady trees until they have become well established. A disadvantage of this method is that the foot of the stake or pipe is something to catch onto or trip over in any orchard operation.

Still another method, devised by Frank Green, of Allen, Michigan, is to brace the tree with a short guy wire, not unlike the manner in which a telephone pole is braced. To avoid encircling the trunk with a wire, a $\frac{3}{16}$-inch hole is drilled through the trunk at a height of about 24 inches. Some 30 inches from the trunk, and on the windward side, a 2×2-inch stake or peg, 18 to 24 inches long, is driven firmly into the ground. A length of 10- to 12-gauge galvanized wire is securely fastened to the stake at one end, while the other end of the wire is threaded through the hole in the tree and drawn taut with a twist or knotting of the wire on the opposite side of the tree. A modification is to use a metal eye-screw. The system is simple to install, avoids any possibility of girdling the tree, and requires a minimum of attention. A disadvantage is, again, the hazard of snagging or tripping over the diagonal wire or the short stake. This method has proved excellent for apple trees on the EM VII, I, II, and IV rootstocks. Once in place, the matter of steadying the tree can be all but forgotten.

An interesting method of support has been developed by Henry Miller, of Paw Paw, West Virginia. Salvaged piping is cut into 3-foot lengths. These are driven half their length into the ground, where the tree is to be set. At planting time, the tree hole is dug in the lee of the pipe, away from the direction of the prevailing wind. The tree is planted as close to the pipe as possible. Both the tree and the pipe are then wrapped together with a piece of heavily galvanized, closely woven wire cloth or screening. This should not be confused with wide-mesh ($\frac{1}{4}$-inch) hardware cloth. The screening is secured from salvage operations and cut into squares 12 inches wide by 18 inches high. The 12-inch width represents the material for wrapping and fastening and will vary according to the size of the stake, the size of the tree, and how close the tree can be planted to the stake. The wire wrap is set like a mouse guard. If part of it is placed below ground, so much the better. The upper corners of the wire screen are pulled together, "dog-eared," and fastened with a hand stapler. The band can now be easily stapled three or more places along the seam. This is a simple and inexpensive operation, and provides mouse protection as well as bracing for the few years necessary for the tree to become established.

Permanent Supports

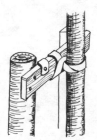

Where prevailing winds are quite strong, and where the tree requires a permanent support, two stakes and a crosspiece are often used. The stakes are heavier than for temporary support, being 3 inches in diameter. They are set on each side of, and about a foot away from, the trunk of the tree. The crossbar is nailed to the stake on the lee side of the tree, so that the tree leans against it. For still more substantial and permanent support, the stakes should be spaced 2 feet from the trunk on each side (Fig. 87).

Instead of being driven and set in place after the tree is planted, supports of this kind are set in dug holes and the tree then planted in position. All this is a laborious and expensive process, requiring considerable upkeep and replacement. The length of life of the stake is, of course, greatly improved if the stakes are treated with creosote or copper sulfate before being set in place.

Wire trellising has for the most part replaced elaborate double staking. In this, horizontal wires are strung at various desired distances on iron, wooden, or concrete posts, to provide support of varying degrees of elaborateness. Some of the trellis supports used in America provide minimum requirements, as described in Chapter 21. They may consist of no more than a single 10- or 12-gauge wire strung 4 feet from the ground on 6-foot posts set 20 feet apart. On the other hand, some of the trellises found in parts of Europe are more substantial, employing 6-inch concrete posts 10 to 12 feet tall, set in 18 inches of concrete, with wire strung at 3 feet, 5 feet, 7 feet, and 9 feet (Fig. 88).

Fig. 87. (Above) A useful tie for young trees, using a strip of rubber inner tubing, providing good protection against rubbing. (After Raymond Bush)

Fig. 88. (Below) Tie of twisted straw or burlap, fastened with a piece of wire back of the stake, providing good protection against rubbing. (After Raymond Bush)

While supports of this more elaborate kind may be essential for palmettes trained to special forms, and under special conditions, they are too expensive for modern commercial plantings where economic competition is keen.

Tying Materials and Tying Methods Used in Bracing the Tree

Materials that are somewhat bulky and that will not cut the tree are used for tying. Pieces of burlap sacking, tarred twine, or old rope are good. Sections of rubber inner tubing are also useful (Fig. 89). One of the best is old nylon hose. Some growers insist that whatever is used must be 2 or 3 inches in width. The material may be brought snugly around the tree, twisted once, and then brought around the stake or pipe and the ends tied tight with a twist of wire. This is a good method (Fig. 90). It holds the tree snugly and at the same time prevents rubbing against the stake in the event that the tie becomes at all loose. Further, merely by loosening the wire and retightening it, the tie can be tightened or repositioned at any time. Trees must be re-tied each year, done usually at blossom time, and the tie loosened to accommodate the increase in girth by fall.

If twine or cord is used, there should be a 2- to 3-inch width of sacking or some other similar material placed around the tree before tying. In the tying operation, three or four strands of the cord or twine should be wound around both the tree and the stake, and should then be wound around the tie itself between the tree and stake, as shown in Fig. 91. This last step in the tying operation is often overlooked, but is most important. Not only is the tie made snug but a cushion of the tying material is also placed between the tree and the stake. Old spray hose and garden hose are also useful, in which a wire is run through a section of the hose and tied in a figure eight around the tree and stake.

There are a number of special, manufactured ties that are available,

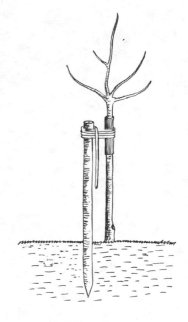

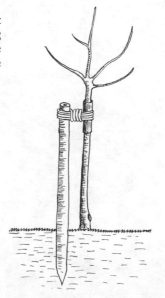

Fig. 89. **First step in tying (above), in which the tying material is first wound around the stake and the tree, and then tightened by winding around the connecting strands. Complete tie (below). Note protective banding around tree. In actual operation the stake should be closer to the tree. (After D. Macer Wright)**

Fig. 90. Commercial dwarf apple orchard on the EM IX rootstock in England, with trees tied against a crossbar supported by two upright stakes.

and some of them are quite good. One of these is in the form of a plastic belt with buckle, furnished in several lengths. A plastic collar is slipped onto the belt between the tree and the stake so that rubbing is prevented. A feature of this tie is that it can be drawn up from time to time if need be, and it can be easily removed and reused.

Finally, the method of bracing previously described, which employs a guy wire, may be again mentioned.

Tying Materials and Tying Methods Used on Trellises and for Training

For tying developing branches of trees to a trellis or other support, there are many suggested materials. Light string, cord, or willow shoots are commonly used. In the training of trees to special forms on wire trellises, bamboo canes are first fastened securely to the wire trellis where the various branches of the tree are to be trained, and the young shoots are fastened in turn to the bamboo canes. Wire is used for fastening the bamboo canes in place.

In order to prevent rubbing against the wire, the tie material for the branches should be first looped over the wire or bamboo or other support to which the branch is to be fastened, and then looped over the shoot and tied.

Special knots have been suggested. The common clove hitch is useful to make around a bamboo or stake, since it will not slip while the rest of the tie is being completed. The square knot is best with which to complete the fastening.

An excellent and inexpensive tie is made from old rubber tire inner tubes. Slice the tube into rings about one-half inch wide, and cut these rings open. This will provide rubber strips of just about the right length

Fig. 91. Dwarfed fruit trees may be trained to special forms for commercial fruit production. Portion of 140-acre orchard of pears and apples grown as Verrier palmettes with four vertical branches, planted one meter apart in the row and three meters between rows (Chaptainville, France). Note permanent metal posts set in concrete.

and width for tying. Make one turn around the shoot and another turn around the wire or stem. In this way, there is no rubbing of the shoot on the support. Either tie the rubber in a knot or fasten the ends together with a hand stapler.

Because of the labor and time in fastening with string or cord, a number of mechanical fasteners have appeared that have proved very useful. One of these is the common wooden spring clothespin. These can be clipped on easily when desired and will not slip or rub. They are used to fasten young shoots to wire trellises or to other shoots. They are especially useful in fastening and bending of young branches, where shoots are conveniently fastened to each other. In order to avoid injury to the tender growth, the fastening should be placed several inches back from the tip.

Another manufactured fastener is one used for grapes, but it is also effective for fastening fruit trees. It is essentially a strap of lightweight galvanized metal, in lengths of approximately 12, 18, and 24 inches. The metal strap is placed around both the tree or branch and the support to which it is to be fastened. One end of the strap is then passed through a slot in the other end of the strap, pulled as tight as desired, and bent back upon itself. When bent in this way the metal is stiff enough to hold in position. If at any time it is desired to loosen and retie, the metal tongue is merely straightened back and the strap loosened to a new position and rebent to fasten.

Still another product, from France, is not unlike the rubber-band fastener used to hold children's braids. It consists of a rubber strip a quarter-inch wide, notched at one end and slit at the other. After the band is in place around the branch and the support, the notched end is slipped through the slit and held in place by the elastic contraction. Fasteners are said to last four or five seasons and able to accommodate to tree growth and expansion.

PROTECTING TREES FROM RABBITS AND MICE

Some of the dwarfing apple rootstocks seem particularly attractive to mice. The EM IX has a relatively thick, succulent cortex, and trees on this rootstock are very subject to mouse injury. Pears, plums, cherries, apricots, and peaches are much less likely to be attacked, but protection from possible injury cannot be entirely ignored.

Dwarfed trees are also very subject to injury from rabbits. The trees are headed low, and the branches are in easy reach after even a moderate snowfall. It is essential to gain complete control over both mice and rabbits if there is to be any satisfaction with dwarfed fruit trees.

Control of Mice

Protective guards of half-inch or quarter-inch galvanized wire mesh are effective against mice for 5 to 7 years after planting. The wire mesh is cut 18 to 24 inches wide and 12 to 18 inches high, depending upon the height of heading and the lowest branches. This will provide space between the guard and the tree for up to seven years of tree growth. The wire should be set into the soil an inch or two to protect against burrowing. The guard is crimped at the top and caught at several places along the seam to hold it together.

A combination support and mouse guard has already been described under the discussion of supports for dwarfed trees. It consists of heavily galvanized wire screening (not hardware cloth) cut into 12 × 18-inch squares; these are wrapped tightly around a pipe support placed close to the tree, and stapled along the seam. Here also, the lower edge should be embedded an inch in the soil.

Some growers place gravel or cinders around the tree as a deterrent to mice. One successful grower in eastern New York has for years placed broken glass and broken bottles both in the tree hole and around the crown of the tree.

The use of baiting stations is another recommended practice. Bales of mulch, a forkful of hay, a 12-foot square of tarpaper, or a slab of wood may be placed out under the ends of the branches in late summer. Mice will develop runways under these covers. In early fall or winter, the cover may be carefully lifted, and cracked corn or apple cubes treated with zinc phosphide placed in the runways.

By far the best treatment is 2 per cent zinc-phosphide-treated cracked corn or oats spread broadcast throughout the orchard. Ten pounds are used to the acre, making an application the first part of October. A second application may be made two or three weeks later where the ground cover is heavy and the mouse population high.

There are other chemical poisons available for mouse control, but they are extremely poisonous and must be used with great caution. It is questionable whether they should ever be used in gardens or orchards.

Control of Rabbits

Since dwarfed trees are headed low, the low bronches are subject to severe injury, often when least expected. In late fall or even in early spring, extensive damage may be done at a time when all danger of injury seems past. In fact, even during the summer rabbits will sometimes strip bark from just above the union. Newly planted trees seem especially attractive. The writer has experienced damage to apple trees on the EM IX rootstock, by chewing and stripping in this way a few inches above the union at odd times during July and August. It would seem that this section of dwarf trees is especially savory to rabbits, in preference to such tender morsels as carrots and lettuce. It is a serious mistake to leave newly planted, dwarf apple trees unprotected even for a single day after planting.

The best way to get rid of rabbits is to exterminate them by hunting them down. Since this cannot always be done, for a few individual trees chicken-wire petticoats may be fashioned 4 feet high and strung on three stakes set in a triangle in such a way that the fastened wire will be about 6 inches from the branches, so as to avoid rubbing of tender shoots.

There are a number of chemical rabbit repellents available that may be sprayed or painted onto the trees. They cannot be guaranteed to give complete protection, but if used according to the directions of the manufacturer they give considerable protection.

As good a material as any is a rosin and alcohol mixture. To one gallon of ethyl alcohol (not methyl alcohol), 7 pounds of crushed rosin are added and allowed to stand in a closed container overnight in a warm room. It may be sprayed onto the tree or applied with a brush, taking care to cover the trunk and all branches to a height of 3 or even 4 feet.

19

PRINCIPLES
OF
PRUNING AND TRAINING

P R U N I N G, in simplest terms, is the removal of a shoot or branch or other plant part. In fact, nature conducts pruning operations of her own, as anyone knows who has observed the branches that accumulate beneath the trees on the lawn or in the woods. Many of these branches were low ones that were shaded out and literally "pinched off" as the tree grew.

Pruning, as practiced by fruit growers, implies a rational, purposeful cutting. It is done with two ends in view: (1) forming and shaping the tree so that it can bear maximum amounts of the best fruit, and (2) regulating the bearing of the tree so that it produces regular annual or nearly annual crops of fruit with good color, size, and uniformity.

Pruning is not something that can be considered by itself. It must be balanced with growing conditions, with climate, with nutrient supply, with varietal characteristics, with rootstock, and with the vigor of the tree. What is effective in a mild climate may be disastrous in a region of severe winter cold. What is desirable for one variety may be undesirable for another. Throughout our discussion, these factors must be borne in mind. The gross structure of a fruit tree is shown in the accompany diagram (Fig. 92).

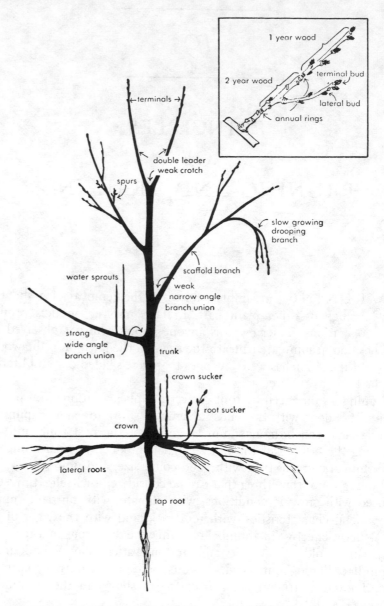

Fig. 92. The gross structure of a fruit tree. (From *The Pruning Manual* by
E. P. Christopher, The Macmillan Company, New York, 1957)

GENERAL RULES IN PRUNING
AND TRAINING

Prune with a Purpose

Too often, pruning is done with no clear plan of what is desired. This is seemingly an elementary point, yet it is the first rule. An orchardist should have a definite image or plan in mind before he begins. Thus, in training the tree, does he wish an open-center tree? a modified leader tree? a low-headed tree? Is he interested primarily in developing a sturdy framework for long years of fruiting? Or is he concerned solely with early fruiting and a short-lived tree? Does he desire some special artificial form of tree? What is the bearing habit of the variety? Is it a terminal bearer? Does it bear latterly on one-year wood, or does it bear on spurs? Is it inclined to develop bad crotches that are weak? Is it an annual bearer or a biennial bearer? Does it come into bearing early in its life? Is it a late bearer?

As the trees come into fruiting, will they be too thick for easy spray penetration and the development of good color? Do the trees need opening to sunlight? Are the trees in the "on year" and are they likely to be overloaded with fruit unless some of the fruit buds are removed in a heading back or thinning process? Do the trees need heading back to keep them in hand? What are the problems in relation to soil, climate, and rootstock? What will the effect be upon vigor and fruiting? (Fig. 93.)

These are some the questions that must be answered before the shears or the saw are taken in hand. And these questions are more frequent and of greater variety with dwarfed fruit trees than with free-growing standard fruit trees. With standard trees, the pattern has been very largely set, and the variation from the existing pattern is not great. But with dwarfed trees, the diversity of forms of training are many. In fact, this is one of the features of dwarfed fruit trees. They can be trained and controlled, by virtue of the type of rootstock employed, to suit the desires of the grower. They can be trained as early-bearing, short-lived filler trees that are to be removed when the permanent trees in the planting are in full fruiting. They may be grown in special artistic forms. They may be supported on wire trellises. They may be grown as dwarf pyra-

Fig. 93. Varietal bearing habits must be taken into consideration when training the fruit tree. (Left) Gallia, which bears terminally and tends to produce a low, spreading willow tree which is not suited to the very dwarf EM IX rootstock and in which a central leader should be encouraged. This tree is on the stronger growing EM I rootstock. (Right) Coxe's Orange Pippin, which bears largely on spurs and tends to develop a sturdy tree with wide angle of branching, well suited to the very dwarfing EM IX rootstock. This tree is on the very dwarfing EM IX rootstock.

mids or as bushes or in a manner similar to that of standard trees. The same variety can be grown in many different ways.

Prune as Little as Possible

A second general rule in pruning is to prune only as much as is necessary to accomplish the desired results. It is better to underprune than to overprune. When a branch has been removed, as has been explained in an earlier chapter, the photosynthetic or food-manufacturing capacity of the tree is reduced by just that much. Heavy pruning tends to obscure the differences in rootstock effects, and the desired value from the use of a particular rootstock may be largely lost.

Obviously, the removal of a branch from a large fruiting tree is proportionately less significant than when one is removed from a small, young tree in full vigor. In fact, a vigorous young tree can be weakened and prevented from coming into bearing merely by constant removal of a major part of the branch system each year. One the other hand, with a thick, overcrowded older tree, the removal of a few branches so as to reduce the crop and to let in more light is often most beneficial. In general, apple, pear, sweet cherry, apricot, and plum trees require less pruning than do peach and sour cherry trees.

Pruning trials in England with three styles of pruning have shown highest yields from little pruning. Comparisons were made between (1) severe pruning to develop spurs freely, (2) moderate pruning to contain the tree and to develop renewal shoots, and (3) little pruning other than to shape the tree and to thin out weak and crowded branches. The severely pruned trees were considerably the smallest at seven years of age, followed by the moderately and the lightly pruned trees in that order. The blossoming of the severely pruned trees was heavy, but the "June drop" was also heavy. Over a fourteen-year period the yield per tree was generally highest on the lightly pruned trees (15¾ to 25½ bushels) and lowest on the severely pruned trees (11½ to 14¾ bushels).

The Young Tree and the Older Tree Compared

With the young tree, the purpose of pruning is to shape the tree. A basic framework of scaffold branches is required upon which future crops will be borne. And this shaping should be accomplished with as little pruning as possible, for the reasons previously stated. Happily, a tree can be pruned heavily at the time it is planted, since the removal of some of the top is essential to balance the loss of roots that occurs in any digging and transplanting operation.

As the tree becomes older, the purpose in pruning is largely to keep the tree open for entry of light, for spray penetration, and for air circulation. It is also to keep the tree in hand, remove bad crotches and interfering branches, as well as to reduce potential overloads of fruit. This is essentially a fruit-thinning operation, but one done during the dormant season. The very dwarfing rootstocks dispose the tree to excessive fruitfulness, and unless the fruiting branches are cut back or thinned out so as to reduce the number of bearing buds and to produce new shoots, the tree may become weakened and strongly biennial.

Some General Principles for Pruning Dwarfed Fruit Trees

It has frequently been said that dwarfed fruit trees require less pruning than do vigorous trees of standard size. But there is little evidence to support this view. On the basis of a single tree, less pruning is obviously required by virtue of the fact that a dwarfed tree is smaller; also, the nature of the pruning and of the cuts made may be different. But observations in Canada indicate that the amount of prunings from a tree is related to its size and that the weight of prunings from a dwarfed tree is proportional to that from a large tree.

The expectation in some quarters has been that dwarfed fruit trees can be handled like standard trees, except that the trees are smaller, earlier-fruiting, and more closely planted. Unfortunately, this is not entirely true. Insofar as the general principles of pruning are concerned, there is no difference between dwarfed and standard trees, yet the operation usually becomes more involved, with more consideration to details. It becomes a more refined, less gross, operation.

For one reason, in a dwarfed tree the rootstock appreciably controls the vigor, the type of growth, and the fruiting. Accordingly, a variety that is pruned by experience in an accepted standard fashion when growing on standard rootstocks now becomes a problem. The orchardist is now confronted with the difference in behavior of this variety on different rootstocks.

It is at this point that the expression "stion" takes on really significant meaning, in which the combination of Northern Spy/EM IX, for example, is quite different from Northern Spy/EM XIII and both are different from standard trees. Each combination now becomes a different individual, requiring special attention. Pruning practices for dwarfed trees must be adjusted with regard to the rootstock as well as the scion variety, something which need not be considered for standard trees.

Thus, as a general principle, apple trees on a very dwarfing rootstock must be trained so that the lateral or scaffold branches arise quite low on the trunk, say between 12 and 24 inches from the ground. Otherwise the tree becomes leggy, top-heavy, and inclined to lean or to blow over; also, every care must be taken to develop a modified central leader and a compact form. Long, thin, willowy branches that bear fruit early, usually at their tips, should be discouraged. Blossoms and fruit should be removed the first year; especially should they be removed from the leader and the main scaffold branches. Early-bearing varieties, as Jonathan, Golden Delicious, and McIntosh, need to be watched closely. Too often the potential leader becomes weighted with fruit and is bent over and downward. This, as has been shown, still further encourages premature fruiting, with the result that a scraggly, runted tree develops that never realizes its full bearing potential (Fig. 94).

Furthermore, trees on dwarfing rootstocks reach adulthood quite early. This fact must be recognized and adjustments made in pruning and in other practices at an earlier date than for standard trees. Thus, scion varieties that tend to a biennial habit of fruiting exhibit this condition while still relatively young in terms of years. To meet this situation, dwarfed trees of this type must be kept in vigorous growing condition, with sufficient detailed pruning to renew the top with vigorous wood. A regular renewal system must be followed as trees become older. Old fruiting shoots must be systematically cut back or thinned out, and new shoots must be induced to develop that in turn will form fruit buds and spurs for continuation of good crops of quality fruit.

Another major difference between dwarfed and standard trees is that the former may be pruned more heavily at an earlier age, if so desired, with less adverse effects. This is because the dwarfing rootstock checks the growth of the tree, so that it does not remain in an overvegetative condition of unfruitfulness. In fact, relatively heavy pruning may be essential to develop a compact tree with the fruit carried on short branches close in to the center of the tree.

The significance of the rootstock in relation to vigor and pruning is illustrated by an experience the writer once had with Delicious apple trees on EM II rootstocks, grown as Verrier palmettes. In order to maintain the desired trained form of the trees, severe pruning was required. In this instance the EM II was not sufficiently restricting to top growth. As a result, over a period of thirteen years, these trees never blossomed

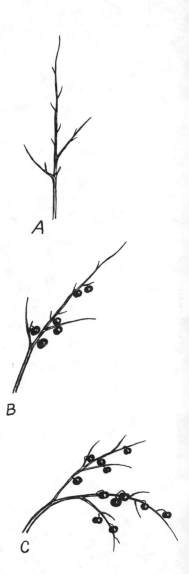

Fig. 94. This is what happens when the central leader of a dwarfed tree is permitted to carry fruit. The terminal (A) is bent somewhat by the weight of fruit (B) and is bent still further in subsequent years by the heavy fruiting that is induced in turn by the bending (C). Such trees are often severely dwarfed and even stunted. If a taller and more substantial tree is desired, the fruit should be removed from the terminal of the central leader for the first two or three years. (After E. P. Christopher in *The Pruning Manual*)

and never fruited. On the other hand, trees of this same scion variety on the more restricting EM IX rootstock were grown and fruited successfully as Verrier palmettes, with the severe pruning required to keep them in proper form. The balance between root growth and top growth is the significant point. It is fully explained in the discussion of carbohydrate-nitrogen relationship in a previous chapter.

Still another difference between dwarfed and standard trees is the possibility of summer pruning and pinching. The trees are small and within reach, and the response is often quite satisfying, provided one understands what he is doing.

THREE COMMON TREE FORMS

Although fruit trees may be pruned and developed into a great variety of shapes and forms, there are three major natural categories; namely, the leader type, the open-center type, and the modified leader, or delayed open-center type (Fig. 95). These types are not restricted to dwarfed trees. On the contrary, they are the ones also used almost universally for free-growing standard fruit trees.

It must be repeated here that the term "standard" as used in America refers to the typical system, or the "standard of behavior." It should not be confused with "standard" as used in Europe to designate the length of the stem or trunk between the ground line and the lowest branches. Thus a "standard" tree in Europe is a tree trained with a relatively high head, in which the lowest branches are 5 to 7 feet from the ground line. A "half-standard" tree is one with a clean trunk or stem 3½ to 5 feet in length. A "bush" tree has a short stem, 3½ feet or less in length.

The Leader Tree

This is the natural type of tree for free-growing pears and for many varieties of apples if left unpruned (Fig. 95A). The central shoot of the young tree is permitted to grow upward uninterrupted year after year as an extension of the trunk. From this central shaft, or axis, twenty or more lateral branches develop into scaffold limbs, with the oldest and largest at the bottom.

This is the type of tree found on free-growing rootstocks when the trees are started in this way and then pruned no more. But large trees of this kind are no longer grown purposefully on standard rootstocks. The tree is very strong and there is little breakage, but the fruit is mostly in the tops of the trees where it is difficult to spray and to harvest, and lower potential bearing areas are bare and unproductive.

Various modifications of the central-leader type, are, however, grown successfully on dwarfing rootstocks. Vertical cordons and dwarf pyramids may be thought of as central-leader trees. The EM IX apple and the Quince EM A rootstocks are used for this purpose. Such trees are not difficult to develop, and are among the most successful forms in which dwarfed trees are grown.

The Open-Center Tree

In this type, there is no central leader (Fig. 95B). The tree is started by selecting four or five scaffold branches, spaced only about 3 to 5 inches apart along the trunk, the lowest being 20 to 24 inches from the ground. The general effect is that of a whorl of branches comparatively close together. Any central leader is removed.

The branches that are to serve as the scaffold branches for the tree are headed back in early years. This causes them to branch and also helps to keep them the same size.

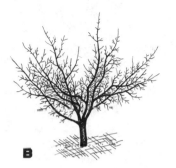

Such trees are relatively low and flat-topped. They have the disadvantage of breaking rather easily if the crotches are narrow-angled. Peach trees develop into this form rather naturally. The apricot, cherry, and plum can be trained this way with modest attention. The apple and pear can also be grown quite successfully in this manner, although more attention is required to keep any potential central leader from developing. The semidwarfing rootstocks are well suited to this type tree, including Quince A for pears and EM VII, I, II, and IV for apples.

The Modified Leader Tree

This type occupies a position midway between the two just described (Fig. 95C). A central leader is permitted to grow to a height of 5 or 6

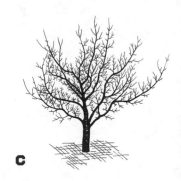

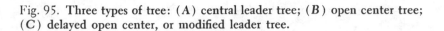

Fig. 95. **Three types of tree: (A) central leader tree; (B) open center tree; (C) delayed open center, or modified leader tree.**

feet, and then headed at that height. Lateral branches are selected for scaffold limbs at various intervals along this modified leader. The lowest scaffold may be at 24 to 30 inches from the ground, and the uppermost at 5 or 6 feet, usually 6 to 12 inches apart. This type of tree differs from the two- or three-story tree in that the branches are more or less evenly distributed along the central modified leader rather than in groups or whorls. Such trees combine many of the advantages of both the central leader and of the open center. The crotches are strong, breakage is less than with the open center, and the trees are lower and more open than with the central-leader form. This type is well suited to the apple, especially when worked on semidwarf and semistandard rootstocks, such as EM VII, EM II, EM I, EM IV, EM XIII, and EM XVI.

HEADING BACK AND THINNING OUT

There are two major ways in which a branch can be treated in the pruning operation. Either it can be cut back partway, or it can be removed entirely. If it is cut back partway, it is said to have been "headed back" or "leader tipped." If it is removed entirely back to the limb from which it arises, it is said to have been "thinned out" (Fig. 96).

While these statements may seem at first to be obvious and overly simple, they are not without significance. Because when a branch is "headed back," the terminal bud of the branch is removed, thus removing the hormonal dominance of that bud, which in turn affects the remaining buds. The result is that lateral buds are induced to develop. On the other hand, when a branch is thinned out or removed entirely, there is no such influence on lateral buds, since they are removed along with the terminal bud that controls them.

In the pruning of dwarfed trees, heading back by one means or another, such as by pinching or summer or dormant cutting back, is practiced much more than with free-growing standard trees on strong rootstocks. With the latter, the result of excessive heading back is often merely to thicken the tree and delay fruiting. The value of heading back

Fig. 96. **Two methods of pruning compared: (Left) cutting back, and (Right) heading back to an outside lateral branch and heading back of the remaining branches. (From *The Pruning Manual* by E. P. Christopher)**

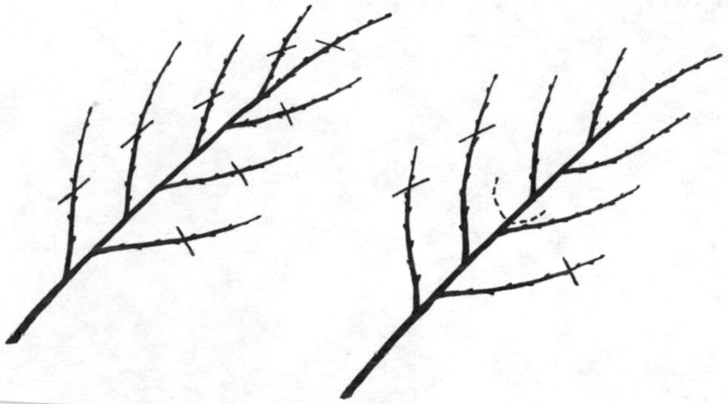

as a means of inducing fruiting has been very much overestimated for trees of this kind. The idea of its value in this respect may have come from Old World practices involving trees growing on dwarfing root-stocks, in which vigorous shoot growth is balanced and restrained by the rootstock.

At all events heading back is a practice that can be employed to advantage with trees on dwarfing rootstocks. Properly used, it induces blossom-bud formation along with keeping the tree well in hand. The three rules that follow will be found helpful in any heading-back opera-tion:

Rule 1. If two branches are competing, the one cut back the most grows the least. Thus, in practical operation, if two branches are coming out close together and are of about equal size, the one cut back the least will make the most growth during the following season. This greater growth does not always appear as new shoot growth; it may be in the form of short spur growths or in increase in diameter of the cut shoot. This technique is very helpful in preventing weak "Y" crotches from forming, which they are certain to do if two competing branches are cut back evenly.

Rule 2. The bud nearest the point of cut is likely to assume the lead and outgrow the buds below it. If this bud is injured, the next one below will take over. The situation frequently occurs when the cut is made close to the bud or is made early in the winter, so that the bud dries out. The shoot arising from the bud that takes over is likely to follow the same general direction as the terminal part of the shoot that was removed.

Rule 3. When a branch is headed back, the remaining buds are en-couraged into growth, thus developing more branching than when no heading back is done. This is especially the case with young, vigorous trees. Heading back thus thickens the top and makes a more dense and a sturdier tree.

TIME OF PRUNING

There are two principal times for pruning, namely, (*a*) summer, and (*b*) winter. Summer pruning involves cutting of the current season's growth while it is still growing and in full leaf. It includes the pinching of young shoots, as well. Summer pruning is a topic to itself, and will be discussed in a separate section as well as in connection with the development of special forms and methods of growing dwarfed fruit trees.

Winter pruning, more commonly called dormant pruning, is done during the dormant season, as the name implies.

In regions of mild winters, pruning may be done at any time during the dormant season. In regions where very severe freezing temperatures occur, it is best to postpone the operation until danger of winter cold has passed. Trees that have been pruned are much more subject to injury than are unpruned trees. If trees are pruned early, as in December, and a low cold spell follows, injury to the tree may be great. Then, too, exposed pruning wounds offer the hazard of entry of disease organisms. The closer to the time of active young growth that the operation can be done, the shorter period of time before the wound begins to close over. This is a very important matter with peach trees where the Valsa disease occurs.

Further, with trees that are subject to bud killing in winter, such as the peach, it is best to postpone pruning until the degree of bud injury has been determined. Then the pruning can be adjusted so as to remove more wood if there has been little injury, and less wood if injury has been appreciable.

The same consideration applies in northern Europe or other regions where bacterial canker (*Pseudomonas mors-prunorum*) and silver leaf (*Stereum purpureum*) are prevalent. In fact, pruning is best done during the growing season, usually after blossoming, so as to minimize disease entry into the pruning cuts.

In large orchard operations where pruning must often of necessity begin earlier than one would like, the hardiest sorts are pruned first. Apples are pruned first, beginning with the hardiest varieties, and are followed by pears, cherries, plums, apricots, with peaches last. In the small planting, or where only a few trees are involved, best results will always follow from the delay of pruning until just before spring growth commences, or even until it has just begun.

Pruning of the same type as dormant pruning, and involving large branches and older wood, can also be done in spring or summer. It tends, however, to devitalize the tree, and is seldom employed. Pruning of this type during the growing season must not, however, be confused with the technique known as "summer pruning." The latter refers not only to the time of pruning but also to special methods, as noted above.

Making the Cuts

Pruning tools should be sharp. If you will take a magnifying glass and examine the severed ends of a branch or a twig, you will better appreciate why this is so. Dull tools will mangle the bark and the wood, destroying more living tissue than is necessary, delaying healing, and favoring disease entry. Sharp tools make a clean cut from which callous tissue quickly arises to cover the wound.

A sharp pruning knife in the hands of an experienced operator is probably the best pruning tool for small cuts. Unfortunately, to use a pruning knife effectively requires special skill and continued practice. There are few individuals who acquire it.

A sharp hand shears is the most used pruning tool. It will take care of most of the cuts that are required on dwarfed trees, especially those grown on trellises or trained to special forms. A pair of larger lopping shears will take care of branches an inch in diameter. If cuts larger than an inch are to be made, a sharp saw is to be preferred. There is a great temptation to use a lopping shears for cuts larger than should be made, with resulting tearing and bruising of tissues.

The first branches to be removed are those that are diseased, broken, or interfering. Water sprouts and strong vertical branches that grow straight up through the tree should be taken out. Suckers from the roots or the base of the tree should also be removed. With dwarfed trees the matter of sprouts and suckers commands special attention, since dwarfing rootstocks sucker more frequently than do free-growing standard rootstocks.

When a branch is headed back, the cut should be made fairly close to a bud, but not so close as to result in drying out and death of the bud. If the cutting is done several weeks or months before growth is renewed in spring, considerable dying out may occur unless the cut is made a half-inch above the bud.

In order to encourage a branch to spread laterally rather than to grow in too upright a fashion, the cutting back of the terminal growth should be to a bud on the outside or underside of the branch. As the end bud now starts into growth, the shoot that develops grows outward and in a somewhat horizontal direction, rather than inward and upward as it would be encouraged to do if the cutting was done back to a bud on the inner or upper side of the branch. This is an important point, and should be carefully adhered to. The practice is especially valuable on the windward side of the tree, where the tendency is for the prevailing wind to cause new shoots to grow upward or even somewhat back into the tree.

As a tree becomes older, this same general principle of cutting to outside buds should be applied to pruning to outside lateral branches. That is, a branch that is growing vigorously and is making perhaps two feet of new growth will have sent out lateral shoots from near the end of last season's growth. These shoots will arise in various positions on the branch, pointing in different directions. One of these should be selected that arises on the outerside or underside of the branch, and the branch should be cut back to this point. The result is, as with cutting to an outside or underside branch, to encourage the development of low-spreading growth.

TRAINING THE YOUNG TREE

As with a child, the early years of training a tree are the most important in its life. Given a good start, and properly formed, much of what can be done for a tree has been done. This seems especially true for dwarfed trees. With standard trees it is often possible to plant poor trees, grow them with a little attention for several years, and still secure good crops following remedial treatment. On the other hand, if the nursery stock of dwarfed trees is weak or indifferent, no amount of training and subsequent care will correct the situation.

This requirement for good trees of good vigor, off to a good start by proper training and attention, cannot be overestimated for dwarfed trees. It is associated with the fact that dwarfed trees tend to come into bearing at an early age. If not given proper framework and sturdiness before they begin to bear, they can never be remade. Much of the dissatisfaction that is sometimes experienced with dwarfed trees is traceable to this misunderstanding.

Varietal Characteristics

It is often difficult, if not impossible, to train a variety of fruit exactly as one desires. Each variety has a characteristic growing habit of its own. Each rootstock and each stock-scion combination has its own peculiarities. Climate and nutritional factors also have a bearing on how a tree can be trained. It is necessary to understand these habits, and to work with them rather than against them in any plan of training.

Peaches, for example, are well suited to the open-center form of tree. They also perform well when trained to a fan shape and grown against a wall. On the other hand, pears form spurs freely and are typically strong central leaders in habit. They are well adapted to the leader and modified leader types, and to forms in which their strong vertical-growing tendencies are controlled yet not eliminated, as with the Verrier palmettes and with cordons.

Apples have a wide range of natural characteristics. Some are terminal bearers, with willowy branches, such as the Gallia and Cortland (Fig. 93). These cannot be grown as central-leader trees no matter how hard one will try. The best that can be done is to try for a modified leader tree, and the open-center type is probably most easily realized. Neither does this type of variety do well when trained to cordons or Verrier palmettes. It will, however, perform well on low trellises where branches can be supported (Fig. 97).

On the other hand, varieties that tend to form spurs early in the life of the tree are well suited to the modified leader or even central-

Fig. 97. **Pruning of a tip-bearing apple variety (Jonathan and Cortland): (Left) branch showing where cuts are to be made; (Right) same branch after pruning. The leader has been shortened by about half. One of the laterals from the main branch has been shortened to three buds; most of the shoots from laterals have been cut back to one bud; a number of shoots have been left entirely unpruned so that they may form blossom buds and fruit at the tips. (After *The Fruit Garden Displayed*, The Royal Horticultural Society, London, 1951)**

leader tree, and they perform excellently as cordons and espaliers. Golden Delicious, Early McIntosh, Coxc's Orange Pippin, and Stark-rimson are excellent examples (Fig. 93). They are easily trained to special forms where spurring close to the framework is essential.

Pruning the Young Apple and Pear Tree

The young tree received from the nursery is often an unbranched stem or "whip." This is especially likely to be the case in America. If it is a one-year-old tree, it will usually be of this type. Some varieties, however, typically tend to develop lateral branches on one-year trees, and two-year-old trees will have several such branches.

On the other hand, in Europe it is not uncommon to grow and train a tree for three or four years in the nursery, so that the form is already attained at the time of planting in the orchard. This is a good plan for an individual who desires only a few trees and who can induce his neighborhood nurseryman to train the trees for him to the desired shape. Unfortunately, this has not caught on well in America, and a planter may be forced to train the trees himself. Most orchardists prefer to do so, anyway.

Pruning at planting. Dwarf apple trees on the EM IX rootstock should be headed low. The scaffold branches should be developed in a region between 12 and 22 inches from the ground. If a few branches arise even as low as 6 inches, they should be permitted to grow, since they help to form a sturdy trunk; they can always be removed later.

Unfortunately, nurserymen frequently rub off all buds below 24 inches while the tree is still in the nursery. This is poor practice for dwarfed trees, and should not be done. If the scaffold branches are to be developed between 12 and 22 inches from the ground, and all buds have been removed from this region, the tree must regenerate new growth from this area.

But if buds remain on this portion of the main stem, it can be cut back to a height of 24 to 30 inches as the first step in developing the head. With trees that already have lateral branches, such as most two-year-old trees, any of these that can be used may be cut back one-third to one-half their length, and retained (Figs. 98 and 99). More often there are no branches that can be retained, or at best only one.

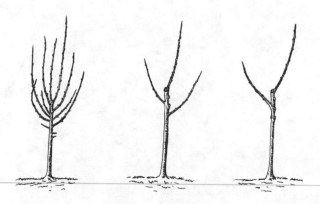

The undesired ones must be removed. The net result is that most trees on a very dwarfing rootstock are started from a single stem so as to force the development of the low head.

This low heading is all-important. A compact tree must be formed. If a leggy tree is developed, with a central leader, it becomes top-heavy, and tends to lean or to blow over. In addition, the root growth is weak and anchorage is poor. Further, if it is a terminal bearing variety, it may carry a few fruits the weight of which may bend the branches and cause the undesirable premature fruiting that has been noted. All blossoms and fruits should be removed the first year, except as one may have a tree or two in his planting that he is growing simply for the novelty of the thing. This may be done by hand or by spraying with one of the chemical plant regulators used for blossom removal.

Semidwarf trees, as on the EM VII rootstock, may be cut back to a point somewhat higher than the EM IX, such as 30 or 36 inches, and the head may be formed between 18 and 30 inches. With slightly less dwarfing rootstocks, as EM I, EM II, and EM IV, the tree may be cut back to 40 inches, and the head may be formed between 24 and 36 inches.

While these figures may seem excessively low, it is the writer's experience that the greater error is in starting the tree insufficiently low. Branches that prove later to be too low can always be removed, and in the meantime they help to develop a sturdy, compact, well-anchored tree, which is what is desired.

Pruning the first winter. In the first winter, four, five, or six good scaffold branches should be selected. They should be well spaced around

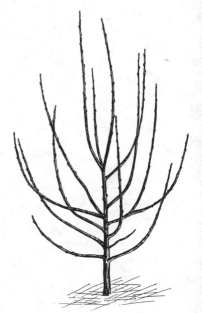

Fig. 98. (Opposite) Two-year-old apple tree at planting, to be trained to the modified leader system. This is typical of American trees and American practices. Compare with Fig. 99. (Left) Before pruning, and (Center) after pruning. Two wide-angled laterals have been selected to form the lower scaffold branches. The leader is left longer than the lateral shoots. (Right) After pruning. Often it is advisable to leave only one wide-angled lateral.

Fig. 99. Very vigorous two-year-old apple tree at planting time before pruning (Above, right) and after pruning (Below, right). This type of vigorous, well-grown tree is seldom seen in America but is found in the more equable climate that parts of Europe provide. Compare with Figure 98.

the tree, and 4 to 5 inches apart up the trunk. If the tree is to be formed as a modified leader, the central leader should also be retained. This is the ideal, and will not always be realized, but the principle remains and can be worked toward during subsequent years. Other branches should be removed entirely, and those that remain should be cut back one-half to two-thirds their length.

If a good selection has been made for scaffold branches, and if growing conditions are favorable so that a good head is developed, pruning the next few years is aimed at keeping the tree in hand and in inducing the formation of spurs and blossom buds close in on the tree. Any sucker growth should be removed, and branches that cross or interfere should be eliminated.

Pruning the second winter. For dwarf trees on the EM IX rootstock, the new growth may be cut back one-third its length again the second year. This will force buds into development toward the base of the shoot that might otherwise not start. Some of these buds will form into spurs and blossom buds, with the result that the fruiting wood will be close to the center of the tree. This same procedure may be carried out in subsequent years until a heavily spurred, compact tree has been formed.

The central leader should not be permitted to grow beyond 4 or perhaps 5 feet. Where sharp crotches develop, one or the other of the branches involved should be cut back severely so that the other may dominate.

Semidwarf and semistandard trees are handled in somewhat similar manner except that the new shoot growth is not cut back after the second year, or only tipped back modestly if this seems necessary to keep the tree compact and in hand. The effort is largely to remove suckers and crossing branches and to prevent poor crotches from forming as described above. For a modified leader tree, the central leader should be headed at the height desired. This is commonly at 6 or 7 feet. Any cutting beyond what is necessary to form the tree will only delay fruiting.

Pruning dwarf pear trees. Dwarfed pear trees on quince rootstocks are handled much as has been described for apple trees. The pear is, however, inclined to develop strong terminal-shoot growth, and may require more severe cutting back in some cases while the framework is being formed. On the other hand, excellent results have been secured with Comice on Quince A with no pruning at all, not even the year of planting.

Pruning the Young Peach Tree

Peach trees are received from the nursery commonly as one-year-old "whips." They should be headed low, at 18 to 24 inches from the ground, and trained in an open-center form with three or five scaffold branches. No central leader is permitted to develop (Fig. 100).

In the first winter, the new growth may be cut back one-third to one-half so as to induce branching. From here on, the procedure is to keep strong branches in hand by cutting back, to head back branches so as to keep the tree compact and induce branching. Awkward crotches should be prevented from forming, as has been described for the apple. The peach, because of its strong growth, requires more heavy pruning than the apple and the pear during the formative years.

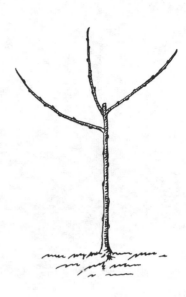

Pruning the Young Sour Cherry Tree

The sour cherry naturally develops into a central leader or modified leader tree. As received from the nursery, the trees frequently have developed several lateral branches. Four or five of these branches, if well spaced, may be selected for the head, the lowest one being 18 to 20 inches from the ground. The central leader should be cut back to 40 inches. The laterals should be cut back to half their length.

Pruning in subsequent years should be light. It will consist primarily in balancing the growth of scaffold branches by cutting back the terminal and in thinning out shoots that tend to make the tree too thick.

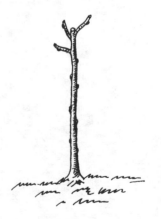

Pruning the Young Sweet Cherry Tree

The sweet cherry is inclined to grow upright. Trees received from the nursery may be either one-year whips or two-year-old trees with some lateral branching. One-year trees are cut back to 60 inches from the ground, and the head is developed at 24 to 40 inches. As for two-year-old trees, some of the lateral branches may be retained for scaffold branches if found to be in the proper position. Four, five, or six scaffold branches

Fig. 100. **Newly planted peach tree, pruned to the three-scaffold open-center type. When three branches of sufficient vigor and uniformity cannot be found, the branches should be cut back to short stubs.**

should be developed, and the central leader cut at 5 or 6 feet from the ground.

Terminal growth of scaffold branches and of the leader is likely to be quite strong. The new growth may be cut back one-half the first winter. Thereafter, pruning consists in thinning out unwanted branches and heading back lightly if this becomes necessary to keep the tree in hand.

Pruning the Young Plum Tree

There are a number of different species of plum, each requiring different treatment. The Japanese varieties grow much like the peach, and may be handled similarly. They should be headed at 18 to 24 inches, and developed into an open-center form.

The Damson or Insititia varieties grow quite upright, and may be treated in a manner not unlike the sweet cherry. The European or Domestica varieties are intermediate between these two types. They are best grown in a modified leader form, with four or five scaffold branches. A number greater than this tends to develop too thick a tree.

The major problem with plum trees is to keep them from becoming too thick. After the first year, it is best not to cut back the terminals unless it is necessary to balance the tree. Cutting of this type induces lateral shoots and a thick, twiggy habit. It is usually necessary to thin out the slender shoots that form, so as to keep the trees open.

PRUNING THE BEARING TREE

Pruning the Bearing Apple and Pear Tree

Little pruning is required for the well-formed young bearing apple and pear tree. The tree should be kept open to permit light to enter, and for spray materials to penetrate, but that is about all. The cutting will be mostly to remove or thin out undesired branches. Little or no heading back is necessary (Fig. 101).

At the same time, one need be less concerned about the amount of cutting except as it may remove fruiting wood and fruiting branches and reduce yield. Once the tree has reached the balance of fruiting, the fear of delay in bearing from excessive removal of top growth is no longer a concern. In fact, some of the corrective pruning that may have been overlooked can now be done more or less with impunity.

The problem with dwarfed trees, as they reach bearing age, is largely one of keeping new wood coming along to replace old wood, to keep the tree in regular annual bearing, and to keep the tree in hand. This is especially so with the very dwarf forms, such as apple trees on the EM IX rootstock. Trees on this rootstock may tend to overload, and while excess blossoms may be thinned off by hand or by chemical means, they can also be removed by judicious pruning. This will often mean that older branches may be cut back severely or entire branches removed, carrying excess blossom buds with them. This is good practice, since new shoots will be induced to develop that will spur up and replace the older wood. Thin and weak wood should constantly be removed; also, if the tree is too heavily spurred, as often happens with pears, some of these may be cut off, or the shoots cut back heavily. This will cause lateral shoots to arise, which in turn will develop young fruit spurs and blossom buds.

The tree must also be kept in hand by thinning out, tipping, or cutting back growth that is too vigorous or out of balance. However, this is not a difficult problem, and can be handled with a little judicious pruning.

As the trees become still older, some of the old leaders may be replaced by well-placed new laterals. In this way the tree is kept from becoming senescent, by inducing new shoots here and there to renew the fruiting wood.

For semidwarf and semistandard trees the pruning system to be followed is much like that for a standard tree. That is, the tree is kept open by the removal of entire branches. Weak wood should be cut out. Strong new shoots may develop. These should be removed entirely if not needed, or cut back to an outside lateral. Such branches may be used to build new fruiting wood, replacing some of the old fruiting branches, which should then be removed.

Fig. 101. **Dwarf pyramid apple, a form that has proved very useful for small gardens in England and is well adapted for America. (After *The Fruit Garden Displayed*)**

Much that has been written here will become more intelligible as the orchardist observes and prunes his trees each year. One of the first lessons that will be learned is that dwarfed trees require constant attention. It is not so easy, as with standard trees, to set a pruning pattern and then put unskilled men to work following that pattern. But skilled men who have learned how to handle dwarfed trees do not find the operation difficult. The cutting for the most part deals with relatively small branches that are easily removed, and the major operations can be done from the ground or from a short ladder. This is really one of the features of dwarfed fruit trees, once the system is learned.

Pruning the Bearing Cherry Tree

As the trees come into bearing (five to seven years of age), little pruning is required other than to remove branches that cross and rub, and to thin out any excess branches so as to open the tree to light and spray entry. When the trees become older, they may slow down in growth, and the fruit may be carried at or near the tips of the branches. Such trees can be reinvigorated by cutting the terminal branches back severely to an outside lateral growth. The cut should not be made to a stub, or an excessive number of bushy shoots may arise that are tedious to thin out. Dead and injured branches occur more frequently on cherry trees than on apple and pear trees, and special attention must be given to removing these regularly each year.

Pruning the Bearing Peach Tree

Bearing peach trees require heavier pruning than do apple trees of the same age, and they require annual pruning. It must be remembered that peaches are borne on one-year wood. Accordingly, pruning should be aimed at maintaining new shoot growth well distributed throughout the tree. Terminal growth of 12 to 16 inches is considered desirable in a standard peach tree eight to twelve years of age. At the same time, if all the potential blossom buds on a tree were to set as fruit, the tree would be seriously overloaded. So the procedure is to head back or cut back the new shoot growth by about half, and to thin out any excess shoots. This is especially necessary if pruning has been back into two- and three-year-old wood, which may induce more shoot growth than is desirable. At all events, thinning out and heading back in the peach are essentially fruit-thinning operations, and are employed with this in mind.

As the peach trees get older—and this is especially true with dwarfed trees—the terminal growth may slow down to only a few inches of spur-like growth. These shoots also fruit, so that the crop is eventually borne at the ends of long, leggy branches. Such branches should be cut back into two- or three-year-old wood as a rejuvenation process. The branches should not merely be stubbed back; rather, the cuts should be made to outside lateral branches. This will reduce the formation of thick brushy shoot growth near the point of cut.

Pruning the Bearing Plum Tree

Once plum trees have reached bearing age, they require little pruning other than thinning out excessive thin shoots that tend to make the tree thick; it is also necessary to remove dead or diseased branches.

Dwarfed trees soon slow down to a regular fruiting habit with little strong shoot growth. As long as cropping is satisfactory, this condition should be maintained. If it becomes necessary to reinvigorate the tree, this can be done by cutting back into two- and three-year-old wood, or older, making the cut to an outside lateral rather than to a stub. The new shoots that result from such treatment of the plum will need to be thinned out so as to prevent a bushy growth.

20

SUMMER PRUNING

SUMMER pruning is a special system of pruning. It is not merely the cutting out of water sprouts and dead limbs in the summertime. It is not winter-type pruning done in the summer, such as removing large limbs or heading back main branches. Nor is it merely rubbing off buds or making minor corrective cuts to shape the young tree.

Rather, it often becomes a detailed and complicated pruning procedure aimed at the early and regular production of quality fruit, and performed during the growing season. To be sure, summer pruning may be supplemented by dormant pruning, or vice versa. And there may be more of the one than of the other in a particular system of summer pruning. Moreover, there are any number of variations or special forms of summer pruning, such as the Lorette System and the Modified Lorette System, to be described later in this chapter.

Summer pruning is essential for trees that are to be grown in a restricted space, such as cordons and espaliers (Fig. 102). With trees that are not to be restricted, there is no absolute necessity for summer pruning. Further, it has no place with very vigorous trees or with extensive orchard operations. In such situations the usual result is the development of thick, branchy trees, added labor costs, and delayed fruiting.

On the other hand the operation is useful for dwarfed fruit trees when they are being trained and when they are beginning to crop. Under

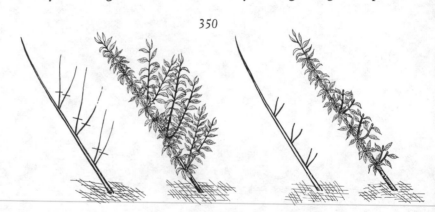

ideal conditions the practice removes some of the foliage, checks root growth, slows shoot growth, keeps the tree compact, and favors development of fruit buds at the bases of the shoots that are cut back. Further, light and air are admitted to the tree more easily, transpiration losses are reduced, and fruit size and color are improved (Fig. 103.)

It seems to the writer that one of the main features of summer pruning has not been sufficiently emphasized. That is, by judicious summer pruning the gardener can cause fruit buds to develop just about where he wants them to be. He can position them on a branch or shoot to suit his desires. It is this control in the position and development of specific individual fruit buds that appeals to the writer as the most significant contribution of summer pruning. This is in contrast to generalizations regarding total growth, total yield, and general orchard performance. It is the refinement, control, and specificity of summer pruning that is significant.

Not uncommonly the generalization is made that summer pruning favors fruit production and that winter pruning favors shoot growth. While there is some truth in the general idea, this is not an accurate statement. Summer pruning improperly performed can be devitalizing to a tree, and may result in poor crops of inferior fruit. It is only by a thorough understanding of the principles involved and the methods of procedure that good results will be obtained.

Fig. 102. (Opposite) Summer pruning of cordons: (Left) oblique cordon ready for summer pruning, in late July or early August, after wood has hardened for three-fourths of its length. The leader should not be touched. (Right) Same cordon summer-pruned by cutting back to three leaves those laterals that arise from the main stem, and by cutting back to one leaf those laterals that develop from existing side shoots. In counting leaves, the cluster at the base of each shoot is ignored. (After *The Fruit Garden Displayed*)

Fig. 103. (Below) Summer pruning of branch of bearing apple tree: (Left) branch ready for summer pruning, in early August, when shoots of the current season's growth have become woody at the base. (Right) Same branch summer-pruned by cutting back to five inches those shoots of the current season's growth that have become woody at the base. The leaders should not be pruned. (After *The Fruit Garden Displayed*)

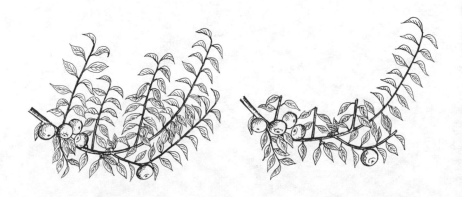

General Principles of Summer Pruning

There are many different individual ideas as to how and when summer pruning should be done. These differences in viewpoint are brought about largely because of special conditions encountered by the person concerned, such as variety, rootstock, soil, and climate. Thus, one person may be growing a variety of apple or pear that is a terminal bearer. Another may be concerned with a spur-type variety, which forms spurs readily. Still another person may be growing a variety inclined to produce fruit buds laterally on the current season's wood, much as a peach does. Or the variety may be inclined to bear too heavily, so that a procedure favoring fruit-bud development may only aggravate the problem. Obviously, the most favorable summer pruning technique will vary according to the fruiting habit of the variety.

Again, the tree may be growing on a very vigorous rootstock, so that very severe pruning must be resorted to if the trees are to be restricted. But such pruning under these conditions merely makes matters worse, and will keep the tree in a vegetative condition and will delay fruiting. On the other hand, trees on very dwarfing rootstocks will be checked in growth by that rootstock. The resulting growth is predisposed to fruit-bud formation. Summer pruning can be used in such conditions to control fruit-bud formation still further. More especially, summer pruning will be found to help shape the trees and to position the fruit buds where desired and where they will be most efficient in the production of quality fruit.

Likewise, soil and climate play a part in determining what system of summer pruning is most effective. Trees on very fertile soil that promotes vigorous growth will respond differently from those on less fertile soil. And a location with high summer rainfall will promote tree growth quite different from what will be found in locations of characteristically low rainfall or even summer drought. In addition to these combinations there are complicating factors of length of growing season, severity of winter cold, and hazards of spring frost. These are all factors to be considered in summer pruning.

Accordingly, it should not be an occasion for great surprise that there are a number of different conceptions about the value of summer pruning and the techniques to be employed.

There are, however, certain general principles that prevail and that may help to explain some of the differences that are observed. First of all, the reduction in foliage and in the efficiency with which it functions tends to weaken the tree. It makes no difference whether the reduction is due to an insect, a fungus, or artificial cutting off of the foliage. Insofar as the foliage is reduced, the ability of the tree to carry on the process of photosynthesis is reduced. This is especially noticeable with trees that are in a vigorous growing condition and have an ample supply of nitrogen and moisture. Both top and root growth are reduced, and the tree is weakened.

Since an accumulation of carbohydrates is essential to fruiting, as has been explained in the section dealing with the carbohydrate/nitrogen relationship (Chapter 4), fruiting is delayed and the operation is termed unsuccessful. However, if the root system is one that is dwarfing and restrictive by nature, the supply of nitrogen to the tree is limited. Under such conditions, carbohydrates tend to accumulate in the tree in relation to nitrogen intake, and fruit-bud formation is promoted. The reduction in photosynthetic activity and accumulation of carbohydrates may be exactly what is needed to balance growth.

The time of the pruning operation is also a factor in success. The earlier in the season the pruning is done, the less restrictive it becomes. In fact, pruning done just as growth starts in spring is scarcely different from dormant pruning. Experiments have shown that midsummer pruning is the most devitalizing and restrictive on general tree performance. This is because by this time of year the tree has expended its reserves in new growth. It is just beginning to accumulate carbohydrate materials. New growth must now be developed, with further drain on the tree. On the other hand, if pruning is done late in the summer, the operation may be less devitalizing. This is because substantial reserves may have been accumulated by this time, yet new growth is not stimulated into development.

While what has been said may help to explain some of the general responses from summer pruning, it leaves much explaining yet to be done. It does not explain some of the unusual specific responses that occur. The Rhode Island Greening variety of apple may be taken as an example. It is a vigorous growing variety when young. Trees with a vigorous terminal growth of 36 to 48 inches do not typically form blossom buds. Yet the writer has seen trees of this variety on the EM I rootstock that not only made this much growth but also developed blossom buds in the axils of the leaves on the current season's growth.

From examples such as this—and there are many—it would seem that the processes involved in response to summer pruning are much more complex than anything that has been mentioned. It would not be surprising that some of the mechanism is hormonal in nature. Surely, the downward flow of auxin from the growing tip of a shoot, and its control in bud development are suggestive. The stage of development of both the parts removed and the parts that remain after pruning may be important in relation to one another. Much that is often passed off about summer pruning being an "art" merits careful scientific study.

Typical Summer Pruning

While there is no accepted pattern for summer pruning as such, nevertheless there are practices that are more or less characteristic and that may be accepted as typifying the method.

To begin with, the terminal or extension growth is cut back early in spring at a time when the new growth has reached a length of 2 or 3 inches. The cutting back is by one-third to one-half the extension growth. The operation is aimed at balancing the main branches of the tree and keeping them in hand. The more vigorous branches receive the heavier cutting back (one-half) and the less vigorous ones are pruned less heavily (one-third). One of the buds on the shoot, usually the one nearest the cut, will assume the lead, and develop as the terminal growth. It is permitted to grow unchecked the remainder of the season.

Lateral shoots, or secondary branches, may be expected to develop from the buds in the axils of the leaves. These and all other such shoots are cut back to three, four, or five leaves (not counting small basal leaves) when they have reached the proper stage of development (Fig. 104). This is when the shoots are the diameter of a pencil, usually a foot or more in length. Depending upon the climate and the season, the time will be in late June or early July. If there is any question about the proper maturity, the error should be on the side of delaying the pruning a week or two. Any individual shoots that have not reached the proper stage of maturity should be untouched. They may be cut back later when they have reached the desired stage.

The degree of cutting back seems to be a matter of opinion. Some gardeners insist upon leaving three leaves, others four, and others five. As will be seen under the discussion of the Lorette System later in this chapter, there are still other opinions. The less-severe cutting back is preferred where trees are vigorous and where the growing season favors late shoot growth. The shorter pruning is best suited to less vigorous trees and a season that does not favor late growth.

A second cutting is done about a month later, to young shoots that have developed as a result of the first summer pruning. This is about the last of July or the first part of August, depending upon the climate and the season. Again the cutting is done when the shoots are a foot or more long, about the diameter of a pencil, and woody for half their length. Any individual shoots that have not reached the proper stage should not be cut, but can be cut later when they have developed further. The cutting is likewise done back to three, four, or five main leaves, not counting basal leaves, depending upon the choice of the gardener.

Finally, a third cutting may be necessary a month after the second, owing to new growth that may have been forced similarly from the earlier pruning. This is usually about the last of August or the first part of September. This may also be the time to cut back some of the shoots that had not attained the proper stage of maturity at the previous cutting back in August.

While this brief description may seem an oversimplification, it nevertheless gives the general features of summer pruning as commonly employed. The refinements, the detail, and the personal likes and dislikes of different gardeners go far beyond this, and often seem to approach the state of idiosyncrasy. One need only read the numerous

Fig. 104. **Young shoot after summer pruning.** (After C. R. Thompson in *The Pruning of Apples and Pears by Renewal Methods*, Faber & Faber, London, 1949)

books and articles on the topic to appreciate the point. There are no two that treat the subject exactly alike. And there are few gardeners who will agree in all detail.

Summer Pruning of Vigorous Young Trees in America

Summer pruning has been tried in a small way in America with vigorous young trees, beginning with the second growing season. It differs from methods commonly accepted as typical of summer pruning in that only one cutting is made. Trees have been brought into bearing earlier than trees given dormant pruning only, and they have been maintained in good form. The experience does not represent any wide acceptance of the method, but it does add one bit more of information as to how trees may be handled in America by summer pruning.

In early or mid-July, depending on the locality, the new wood of the current season's growth begins to toughen and become woody from the base toward the tip. When the hardening has progressed to or about the length that would be left if winter pruning were to be done, and when the leaves on the part to be retained are about as far apart as they would be in the fall, the soft upper portion is clipped back. From the more vigorous upright branches, sometimes a foot or more is cut off. From the less vigorous laterals it may be necessary to remove only 4 to 6 inches. Competing branches should be cut unevenly in summer pruning, just as in winter pruning. It is unnecessary to cut the smaller, weaker branches at all.

Within a few weeks new branches will start in the angles of the upper leaves, just below the cut. Varieties behave differently, but ordinarily two to four shoots may be expected to start. These new shoots or branches usually grow from 3 to 18 inches by fall. Thus, instead of a long, unbranched shoot that otherwise would have developed, there is a sturdy basal part, 12 to 18 inches long, from which arise two to several less sturdy shoots ranging from mere spurs to branches of considerable length.

When winter comes, if too many of these secondary branches have been produced, it is well to thin them out moderately; and if growth has been exceptionally long it is well to tip them back lightly. Most of these branches require no winter pruning whatsoever, and this is where the system gains over repeated winter heading-in. During the coming season these branches behave as do ordinary ones that were not headed back, and these are also the ones that tend to become fruitful.

This type of pruning can be continued for several years if the trees continue to grow vigorously. The need for winter pruning can be virtually eliminated. At the same time, nearly as many water sprouts and undesirable limbs can be produced as by winter pruning. And heavy pruning of older bearing trees in summer is questionable.

Summer Pinching

The pinching of soft terminals or tips of branches during the summer season is considered by some as a form of summer pruning. More often it is treated as a minor supplement to regular pruning practices.

Nevertheless, young fruit trees can be shaped and brought into bearing by the use of this method almost exclusively. It was more commonly employed in the mid-1800's, particularly for pears and for peaches. Pear trees on the quince root are reported to have been handled effectively by summer pinching in New England about 1850, having been brought into bearing at four to five years of age.

Summer pinching aims to anticipate. It is designed to save the loss of plant energy expended in shoot growth that is later removed in remedial pruning. It consists in merely pinching off the soft end or tip of the shoot with the finger and thumb. If a small part of the remaining shoot is bruised in the process, there need be no concern. Some growers feel that a slight bruising has a greater effect in checking shoot growth than if a clean cut is made.

The time of pinching will depend upon what is to be accomplished. If it is a case of regulating and balancing growth, then the operation should be performed as soon as the need is observed. For example, if a branch has been headed back during the dormant season, a number of lateral buds below the cut will be forced into growth. Any of these that tend to outgrow the terminal bud from which the leader is expected to be developed should be pinched as soon as it is seen that they are gaining the ascendancy. This may be when the shoots have made only 3 or 4 inches of growth.

The leader and any branches it is desired to extend will not be pinched. The others will be pinched when they have reached a length of 3 or 4 inches. If lateral shoots develop from these pinched shoots, the new growth should be again similarly treated. A third pinching may be necessary. The objective is to balance the tree as the shoots develop. Gardeners who practice this method feel that they thus avoid much dormant pruning and also promote fruit-bud formation.

THE LORETTE SYSTEM OF PRUNING

The Lorette system of pruning is named for M. Lorette, curator and professor at the Practical School of Agriculture at Wagnonville in northeastern France. He developed his system about the turn of the twentieth century, and published his experiences in 1925.

The principal feature is that no pruning of trees occurs during the dormant season. All is done during the summer as summer pruning. There is one pruning period in early spring and three periods more occur during the summer. The method seems especially suited to the pear, but is said also to be useful for the apple and for espaliers on walls and for cordons of cherries, plums, and apricots. It is most useful with trees that are not overly vigorous, and where the climate does not favor late wood growth.

The preferred forms of trees employed are the cordons, the single U palmette, the double U palmette, and the winged pyramid. The less vigorous varieties, as the Beurré Clairgeau and the Passé Crassane are grown as cordons and single-U palmettes. The more vigorous varieties, as the Beurré Hardy and Doyenné du Comice are grown as double U palmettes and winged pyramids.

The system consists essentially in cutting back the new shoot growth during the summer as it reaches a length of about 12 inches and is half-woody and the thickness of a pencil. The shoots are cut back severely to within ¼ to ⅜ of an inch from their bases. The process is repeated at approximately monthly intervals during early summer, midsummer, and late summer.

The system attempts to induce "dards" and fruit spurs to form from "stipulary" buds or "eyes" instead of from wood buds. The claim is that growth developing from stipulary eyes is more readily brought into fruiting condition; also, dards can be developed rapidly into fruit buds.

Special terms

Lorette uses certain terms (Fig. 105) in describing his method of pruning:

1. *Eye*—a leaf bud and potential wood shoot, which is commonly found in the axil of a leaf.

2. *Stipulary eyes*—incorrectly named. Really secondary buds in the axils of scales on either side of a main bud. The misnomer has arisen because of their position near the stipules of the subtending leaves.

3. *Latent eyes*—buds that are hidden among the folds and wrinkles in the bark at the base of a shoot.

4. *Bourgeon*—a developing shoot.

5. *Brindille* (translated "brindle"), a thin lateral shoot, about 3 to 12 inches long, which usually terminates in a blossom bud.

6. *Lambourd*—fruit spur.

7. *Coursonne*—small, short, fruiting laterals.

8. *Bourse* (translated "knob" or "cluster base")—the thickened or swollen stem at the base of the inflorescence in apples and pears. Called "cluster base" in the United States. It is persistent, and may carry fruit buds or wood buds.

9. *Bourse bud*—a fruit bud or wood bud arising on a bourse, either directly or on a short stem (up to ¼ inch).

10. *Bourse shoot*—a shoot growing from a bourse.

11. *Dard* (translated "dart")—an eye or bud surrounded by a rosette of small leaves, which may develop into a shoot but which may also develop into a fruit spur.

12. *Extension shoots*—wood shoots that terminate the framework branches or leaders and produce the annual extensions of growth.

Fig. 105. Special terms used in pruning: (1) Dard, which has developed into a shoot terminating in a fruit bud; (2) Bourse, with two bourse buds; (3) Spur (three-year-old) with two bourses, left-hand one with a bourse bud; right-hand one with a very short bourse shoot; (4) Dards, which may develop into fruit buds; (5) Brindille, ending in a fruit bud. (After "A glossary of terms used in pruning fruit trees," *Scientific Horticulture* 11: 67–74. 1952–1954)

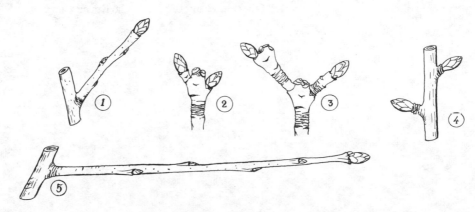

13. *Extension growth*—the annual terminal wood growth of a framework branch or leader.

14. *Secondary growth*—shoots that arise from summer-pruned leaders and laterals during the same season in which these were pruned.

15. *Pruning back to the base*—cutting a branch or shoot back to within ¼ to ⅜ of an inch of the branch or shoot from which it arises, so as to retain the cluster of small leaves often found there.

Directions for the Lorette System of Pruning

1. In winter there is no pruning to do.

2. In early spring, when the new growth of lateral buds is about 2 inches in length, the terminal growth of branches is cut back (Fig. 106). This is done back into the last season's growth (extension growth). If the branches are vigorous, about one-fourth of last season's extension growth is removed. If branches are less vigorous, they are cut back more severely, by half their extension growth. This checks terminal growth, and aids in the development of adventitious buds and stipulary buds on this short stublike growth, which will later develop into blossom buds. At the same time, any superfluous shoots may be removed so that fruiting laterals will be 3 to 4 inches apart.

3. The new extension growth is left undisturbed to continue growth during the season, excepting that the extreme tips of extra-strong, branches may be removed to maintain balance in the tree. The lateral shoots, however, are severely cut back (Fig 107). This is done when they have reached 12 inches in length, are the diameter of a pencil, and are half-woody. This is about the middle to the last of June for pears, and the middle to the last of July for apples. They are cut back to about ⅜

Fig. 106. Lorette pruning, Step 1. Last season's terminal or extension shoot was cut back in early spring to the point indicated by the topmost bend in the branch. As a result, four shoots have developed; the central one is allowed to grow as the new extension growth, and will be cut back next spring. The three other shoots, just below, are marked by bars for cutting back in mid- to late June for pears, and early to mid-July for apples. Still lower down are three other shoots produced from older wood, and these are similarly marked for cutting back. At the base is a developing fruit, carried on a "bourse" or cluster base. (After *La Taille Lorette* by Louis Lorette, 6th Ed., Bibliothèque de la Revue "Jardinage," Versailles, 1926)

of an inch, leaving, if possible, one or two small leaves. From this short stub, especially from the axils of the small leaves, typically short, pointed buds will develop into what are called *dards*, translated "darts" (Fig. 108). Special attention must be called to these buds. Much dependence is placed on the development of dards into blossom buds in the Lorette system. Some of the dards are forced in this way into forming fruit buds. Some others will continue as dards until another year, and still others will be forced into forming small branches.

4. About a month later, toward the middle or end of July for pears and the middle to the end of August for apples, any other shoots that may have reached a similar length of 12 inches, the thickness of a pencil, and a half-woody condition, are similarly cut back to about ⅜ of an inch (Figs. 109 and 110).

5. Again, about a month later (about the middle of August for pears and the middle of September for apples), toward the middle or the end of August, the operation is repeated on any shoots that have reached the characteristic stage described.

Fig. 107. (Above right) Lorette pruning, Step 2. Shoot similar to those on the branch in Fig. 106, which is marked for cutting back to the basal leaves or to ¼ to ⅜ of an inch, as indicated by the slant line. This is done when the shoot has reached at least a foot in length and the diameter of a pencil, and is half-woody. From the axils of the small leaves or from the "wrinkles in the bark" one or more dards (D) will form. (After Louis Lorette)

Fig. 108. (Above center, right) Lorette pruning, Step 3. Dards (D) produced as a result of cutting back of shoot shown in Fig. 107. Some of these will develop into fruit buds, but the majority will continue as dards throughout the year, unless they are forced into growth as wood shoots. (After Louis Lorette)

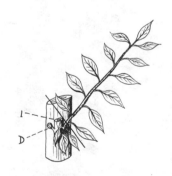

Fig. 109. (Below center, right) Lorette pruning, Step 4. New wood shoot that has developed as a result of cutting back (1) now must be cut back in turn (indicated by slant line) when it has reached a length of a foot or more and the diameter of a pencil and is half-woody. This will be the middle or end of July for pears, and the middle to the end of August for apples. D is a dard.

Fig. 110. (Below, right) Lorette pruning, Step 5. The first cutting back (1), and the second cutting back (2) have resulted in the formation of dards (D).

Although the steps given above are the essentials of the Lorette system, there are numerous additional refinements. For example, weak shoots that will not form dards, and that are known as *brindilles* (translated "brindles") are not cut back. They are bent round and tied with their tips downward. In addition, they may be pierced and slit longitudinally with the point of a knife between the third and fourth leaf. This tends to induce dards and fruit buds to form at the base of the brindille.

In the final analysis, the system promotes the formation of very short fruiting growths that arise from the main branches of the tree (Figs. 111, 112, 113, 114).

Additional Considerations

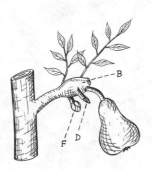

The Lorette system of pruning is suited to the garden rather than to the large orchard; also, it is seemingly adapted to northern France, where daylight is long in summer and where the climate is equable. Thus in the preface to Lorette's book, the point is made that gardeners having as many as 50 or 80 pear trees are sometimes disappointed in their crop of fruit. Continuing, it is remarked that when garden owners have their trees pruned by professional gardeners in winter, they almost invariably omit pruning and pinching during the summer season, with unhappy results. "Since pruning by the Lorette system must be carried out in June," the argument continues, "the amateur is, in a way, forced to carry out all the operations. . . ." Further: "One of the advantages of the Lorette system is that it enables amateurs to take entire charge of their trees. . . . [They] find a real difficulty in carrying out winter pruning because in the winter, either by necessity or from preference, they are separated from their trees. On the other hand, in summer, they take up their abode in the country, where they have the mornings and the evenings of the longest days in the year in which to garden."

Fig. 111. (Above, left) Typical growth that may develop from the bourse or cluster base (*B*) upon which the fruit is borne. *F* is a fruit bud. *D* is a dard. The shoots are bourse shoots or bourse brindilles.

Fig. 112. (Below, left) Bourse shoots must be cut back to the base, as shown.

Trees must be kept in good vigor. The soil should be of good depth and texture, well drained, and high in fertility. The climate where the Lorette system has been most successful, it will be observed, is equable. Temperatures seldom go much below the freezing point in winter, and the summers are free from drought and excessive heat.

Dwarfing rootstocks are a requirement. For the pear, the quince is used. For the apple, both the Doucin types, such as EM II, and the paradise types, such as EM IX, are used.

Those who have observed the trees under M. Lorette's system at Wagnonville claim that the number of blossoms that develop is remarkable. For example, it is said that 30 inches of branches may carry 100 clusters of flowers, each cluster comprising 10 to 13 flowers. To prevent overloading, the fruit must be thinned. A spacing 3 inches apart is considered about right for the developing fruit. And to keep the trees in good vigor, they must be well supplied with moisture and nutrients.

The Modified Lorette System of Pruning

The very reasons that make the Lorette system of summer pruning so acceptable to gardeners make it unacceptable to large-scale commercial orchardists. Thus, the point is made that the gardener is more likely to be in his garden in the summer than in the winter, and finds detailed summer pruning an interesting occupation. On the other hand, the large orchardist finds his summer sufficiently heavily occupied with-

Fig. 113. (Above, right) A brindille, terminating in a flower cluster. A brindille is a thin shoot that does not reach pencil diameter. Some are 7 to 12 inches long, others are 12 to 18 inches long. The shorter ones usually produce a terminal blossom bud, as shown. The longer ones seldom produce a terminal blossom bud. They should be pruned back to three buds in early summer because they are too weak to be cut back to the base as is done with other shoots. See Fig. 104.

Fig. 114. (Below, right) A pear that has developed from terminal flower bud. The shoot is bent and tied in this way so that dards will form at the base. The process is accelerated by piercing longitudinally with a knife, as shown. After fruiting, the brindille is cut back to two or three buds.

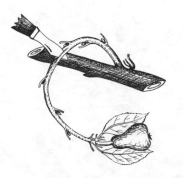

out recourse to summer pruning. He welcomes the distribution of labor that winter pruning provides.

The Modified Lorette system, as developed in England, is a combination of summer and winter pruning, with less severe summer pruning. In fact, it may be thought of as one summer pruning and a winter pruning, although some growers practice two summer prunings and a winter pruning. The system, because the cutting is less severe than with the Lorette system, is better suited to more vigorous trees and to a climate that favors late summer growth.

The procedure in the Modified system is to cut lateral shoots back to three leaves above the basal cluster at a time in summer when they have reached proper maturity. This is when they have reached about the diameter of a pencil and are woody for at least half their length. The time will vary according to the climate and the growing season, but it will be late June to early July for pears, and mid to late July or even early August for apples. If the cutting is done too early, considerable new shoot growth will develop that may be late in maturing, and susceptible to winter injury.

In England it has been observed that with accurate timing, that is, when the basal portion of the young shoot has started to lignify, it is possible to omit further pruning four years out of five. If, however, a wet August occurs, secondary growths that develop must be cut back to one fully developed leaf.

It is better to err on the side of being a week or two late than to prune too early. Even then, if a period of ample rainfall follows in late summer, new growth may be stimulated more than is desirable. Trees on too vigorous a rootstock and in too strongly vegetative a condition are also hard to handle by this system because of renewed late growth. This is one of the real hazards of summer pruning in America, as has already been noted.

Not all shoots will reach the proper stage of maturity at one time. They should be untouched until they reach that stage.

In the winter, the leaders are cut back as desired to balance the tree. Laterals are cut back to three good buds. Brindille-type growths are left to bear fruit terminally or are otherwise also cut back to three buds. All wood shoots are removed completely from older spurs that are bearing dards and blossom buds.

Obviously, the directions that have been given are schematic and general. What will succeed in one section may not succeed in another. For example, if exceptionally vigorous growth is induced, as in regions of wet summers, summer cutting can be more severe. Instead of cutting back to three leaves, lateral shoots may be cut to two leaves; also, varieties that are terminal bearers, such as Jonathan and Cortland, must be treated so as to retain the terminal fruiting brindille-type growth. On the other hand, the varieties that spur readily, such as Wealthy and Golden Delicious, can be handled very well by following the directions rather closely.

OBSTACLES TO SUMMER PRUNING AND SUGGESTIONS FOR THE AMATEUR

There are several serious obstacles to summer pruning in America. The first concerns maturity of wood and buds. In good growing seasons with plenty of moisture, new growth may develop vigorously late in the season, and the wood may not ripen properly. The sudden onset of winter or a severe fall freeze in late October and early November may be disastrous. Not only may immature tips be injured, but also buds, crotches, and crown may be severely damaged.

Injury to crotches and to crown may result in trees so weakened and so susceptible to disease that they never fully recover. Such experiences are not uncommon, even in the more favored horticultural sections of America, excepting for limited areas on the Pacific Coast west of the Cascade Mountains where generally mild winter temperatures prevail.

Added to this are problems associated with summer drought, high summer temperatures, and high light intensity. For greatest success with summer pruning it is important that growing conditions be generally favorable throughout the season. Extremes of temperature, the unpredictability of the climate one year as compared to another, and the general harshness of the climate do not lend themselves well to summer pruning.

It is not uncommon, for example, for periods of drought to occur during midsummer in America, followed by rainy spells. If summer pruning is practiced just prior to the drought period, new shoot growth from lateral buds is retarded. Then, when adequate moisture arrives during a rainy period, the new growth is activated too late in the season for proper ripening prior to the onset of winter. As a result of periodic disasters of this kind over more than a century, American horticulture has generally condemned summer pruning. Over and over again it has been tried and has been found disappointing.

It must also be recognized that there has been no clear understanding in America of just what summer pruning really is. As implied in the opening of this discussion, summer pruning has been confused with removal of water sprouts in summer, with disbudding and minor training practices performed in summer, and even with dormant-type pruning conducted during the growing season.

Finally, American fruit growing has been mainly concerned with large trees and extensive, low-cost operations. Most of the summer pruning that has been tried has been a direct transfer of European ideas and practices. Perhaps the time and attention have not yet been given to developing a system that might succeed in America. It may very well be that the American climate and tradition are not suited to European methods in detail. Yet it may also be that some suggestions may be secured from a better understanding of those methods that can be applied in some degree and in some instances. Perhaps also a system or systems of summer pruning can be eventually developed that are specifically adapted to certain areas and regions of America.

Special Suggestions for the Amateur Gardener

Pruning may be developed into a hobby or avocation with as much satisfaction to the operator as any sport or game. Someone has said that one of the features of tree fruit culture is that "the gardening can be done standing up." Be that as it may, many amateur orchardists have attained remarkable skill in studying how a tree grows and in training and pruning it with governable finesse. The intricacies of treatment and of response can become fascinating, and the entire performance can become truly personal.

It has been said that a skillfully trained tree is the product not of *labor* but of *attention*. Unremitting watchfulness is essential in developing trees into desired forms and shapes.

Daladier, a former Premier of France, secured relaxation from his governmental duties by daily visits to his dwarfed-tree garden, where he was thoroughly absorbed in a not too strenuous examination and training of his trees. Fifteen or twenty minutes a day, or several times a week, will usually be sufficient to observe every shoot in a small fruit garden. The eye very soon becomes trained so that, as Patrick Barry has said, "a glance at a tree will detect the parts that are either too strong or too weak, or that in any way require attention. We are never allowed to forget them. This is one of the most interesting features in the management of garden trees. From day to day they require some attention, and offer some new point of interest that attracts us to them." As Daladier said, "Those trees which enjoy the sympathetic presence of the gardener every day are sure to fare best."

In somewhat the same vein, Frank Waugh has pointed out that "the objective with dwarfed trees is to maintain equilibrium between vegetative growth and fruit bearing. This relationship should be established early, and maintained thereafter. Although this balance is also sought in a standard tree, it is sought and secured earlier in the life of a dwarfed tree, and should be more accurately maintained.

"The tree must make a certain amount of growth each year, but this must be only enough to keep the tree in good health, and to supply sufficient foliage to develop and mature the fruit. Beyond this, there is no need for encouraging wood growth and foliage. Annual bearing becomes not just an ideal but a rule. A proper system of summer pruning and winter pruning, combined with judicious cultivation and fertilizer application, must be maintained to balance vegetation and fruiting. It is a delicate business, like courting two girls at once, but it can be carried out successfully."

21

SPECIAL SYSTEMS
OF TRAINING AND TRELLISING
FOR COMMERCIAL PRODUCTION

GENERAL CONSIDERATIONS

WHEN one thinks of growing tree fruits on trellises, he immediately thinks of trees trained to special forms for use in the garden, as described in another chapter, such as Verrier palmettes, U forms, double-U forms, fan shapes, and various cordons —horizontal, vertical, and oblique. While it is true that when fruit trees are grown on trellises they are perforce trained in special forms and manners, yet the methods discussed in this chapter are designed with regard to profitable commercial fruit production rather than to artistic form. This does not mean that there are not some artistic forms that are suitable for fruit production. As a matter of fact, most of the forms now considered artistic arose functionally in fruit production; and they employ methods of bending and arching and shaping that will promote production of quality fruit. Most of them were devised because of their value in this respect. But much of the emphasis on artistically shaped trees has turned toward refinement of the art rather than toward economic tonnage production of fruit—with some exceptions, as will be noted.

At all events, the emphasis in this chapter is upon commercial production of fruits from trees grown as espaliers and supported on trellises, principally apples and pears. Properly speaking, as discussed more fully in Chapter 26, trees trained in the open on trellises are called "contre-espaliers," in contrast to the trees trained on a wall, which are called "espaliers." In fact, the French also designated trellis-trained trees "espal-ier-aere," which indicates a feature of the method in that the trees are open to the air. The distinction between "contre-espalier" and "espal-ier" has seemingly been lost to English-speaking people, and both kinds are by them called "espalier."

There are many methods and styles of training. Each has its expo-nents. One method seems adapted to one area and another to another. Undoubtedly there are peculiar circumstances that have brought these forms into use where they are found. In one instance it may be the need to hold a desirable but overly vigorous variety in check. In another it may be to facilitate entry of sunlight so as to secure fruit of high color and quality. In another instance it may be the ease of pruning and han-dling and harvesting, in which available women and children may be employed. But in the final analysis one is led to believe that the adoption of a method in a given area is likely to follow the success some fruit grower and leader has had with it—either his innovation or his use and development of some already available form or method. It would be interesting to study and determine how the various methods arose and how they came into general practice.

As for the trellising methods, they are all similar in that they try to take advantage of the dwarfing rootstocks now available, for early and high yields of quality fruit. They utilize, also, the best information in modern nutrition and plant physiology, in which the tree is strictly con-trolled by rootstocks, fertilizers, bending, pruning, and position and angle of branch training. Further, the amount of fruiting wood per tree is easily adjusted to the vigor and bearing habit of the variety, thus tending toward uniform fruits of appropriate size and quality.

Coupled with this is the necessity for some form of bracing of trees on very dwarfing rootstocks, such as the EM IX apple. Individual stakes are expensive and difficult to handle. The erection of a modest trellis, with appropriate wires, is a good answer; also, modern orchard equip-ment is adapted more to row operations and mass operations than to individual tree operations. The hedgerow and the trellis, with their two-dimensional aspect, meet these requirements. Pest control from the air as well as by row spraying and dusting is favored. Trees are open; and light, air, and pesticides enter the planting easily.

Finally, the high cost of hand labor has forced the development of efficient operations. There is evidence that small, low trees, supported by trellises, are well adapted to the new needs.

USE OF CORDONS AND PALMETTES FOR COMMERCIAL FRUIT GROWING IN EUROPE

Cordon Forms

As discussed in a previous chapter, cordons are single-stemmed trees, as the name implies. This form is seldom found in commercial fruit production, having been replaced by the dwarf pyramid. It is more common in gardens as a catalogue of varieties or to provide an interesting succession and variety of fruit in a small area. Occasionally a small planting of vertical cordons or of oblique cordons for commercial fruit production may be seen in Europe, but to the writer's knowledge they have never been tried commercially in America. Obviously a few acres are all that one man can handle.

There is an instance on record in England of several acres of cordons set 1 foot by 3 feet, or 14,527 trees to the acre. Most plantings of cordons, however, are at distances of 2 to 3 feet in the row, with rows at least 6 feet apart. The closer distance is for apples on the EM IX rootstock and for pears on Quince EM C. The wider distance is for apples on EM VII or possibly EM II and EM IV, and for pears on Quince EM A.

Strong yearling whips are used for planting stock. The oblique planting, in which trees are set at a 45-degree angle, is preferred to the vertical planting, especially for the stronger-growing varieties. In the oblique position, most of the buds on the main leader are likely to break, whereas in the vertical position only the buds on the upper half of the tree may start. This leaves a bare leader for the lower half of the tree, where the buds have remained dormant.

The trellis is made of three wires, spaced on 10- to 12-foot posts that have been set 2½ to 3 feet in the ground. The top wire may be a few inches below the top of the post; the center one, 5 feet from the ground; and the bottom one, 3 feet from the ground.

If the oblique method is used, 7-foot lengths of bamboo are fastened in position onto the trellis wires with a piece of 4-inch wire made into a "scaffold knot." The tree is fastened with soft string, not to the wires but to the bamboo cane, using a clove hitch and a reef or square knot. As the tree grows, the angle at which it was planted may be reduced about 5 degrees each year until the terminal has ceased growth.

Cordons must be summer pruned. In fact, in all systems of culture where growth must be severely controlled, summer pruning is said to be a necessity. As practiced in Europe, the Modified Lorette system is commonly used for cordons, in which all mature laterals from the main stem are cut back to three leaves in mid or late July. Any lateral shoots that have arisen in turn from these laterals are cut back to a single leaf. The operation is done after the first flush of growth is completed, and when the lower half of the shoot is woody and the thickness of a lead pencil. Any shoots not sufficiently mature for pruning are passed by. Later in the season, as in early September, these shoots and any others that may have developed are treated as were the others in July. The leader is not touched.

Yet, with all that is said and written about the necessity for detailed summer pruning and for meticulous training, there are experiences that indicate much of this to be unnecessary. For example, the writer has seen a thirty-five-year-old, five-acre planting of dwarf apples in England that was grown successfully with no summer pruning. The trees were set 6 feet apart in rows 10 feet apart and grown as oblique cordons supported on a 3-wire trellis. By European standards the planting was overgrown and extremely untidy and was seldom shown to visitors. The owner, however, maintained that the block had proved much more productive and profitable than some of his more conventional plantings.

The cordon form represents probably the quickest method of cropping, with the production of a pound or more of fruit the second year set. At three or four years of age the yield may be 10 pounds or more, with an average of 3 to 5 pounds. Although of interest to American growers, the cordon system as described is not well adapted to American climate and management. The extremes of winter cold and summer heat, plus the wide variations in growth from season to season, make the system unreliable. It may succeed for a few years and then be utterly destroyed by adverse conditions.

Palmette Forms

Palmette forms, as described in Chapter 26, dealing fully with trees trained to special forms, are fundamentally all similar in that they are either fan-shape or have a vertical central leader or leaders from which scaffold arms may arise to the right, to the left, or in both directions, but always in the same plane. The tree thus has a two-dimensional aspect, namely, height and width but not thickness or depth, in contrast to the natural habit of a tree in which the branches spread in all directions in a bush or columnar form.

Scaffold branches may arise horizontally from the main leader, when it is called a palmette with horizontal branches or a "horizontal palmette." Or the scaffolds may arise at an angle (obliquely), when the tree is called an "oblique palmette." If two vertical leaders are developed so that the tree is U-shaped, it is called a "U-form palmette." Sometimes each of these two leaders may be again formed into a "U" so that the tree is then called a "double-U palmette."

The Verrier Palmette

The most common form found for commercial production, however, is the Verrier palmette, named for the man who developed it. It is really a series of tiers or "U's," the one above and within the other. The simplest form of Verrier palmette is composed of two "U's" and is similar in this respect to the double-U palmette. The Verrier palmette differs, however, in that each of the two U's is derived from the central leader, the one being inside and above the other. Such a four-arm Verrier palmette thus has four vertical stems, which are spaced about a foot apart. The number of vertical stems may be increased variously to four, five, six, seven, or eight, by the development of additional U's. In each case, the increase in number of stems is by the formation of successive tiers of bent U-shaped arms one above the other, derived from the central leader.

The four-arm Verrier palmette is the form most likely to be met with. It seems especially well adapted to pears. The vertical stems are spaced at about 30 cm. (12 inches) apart. The main horizontal stem or bottom of the "U" from which they arise is about 35 to 40 cm. (15 inches) from the ground line. The number of vertical stems, and therefore the amount of fruiting wood, is adjusted to the vigor and fruiting habit of the variety. Thus, the strong-growing Beurré Hardy pear is trained to eight vertical stems, while weaker varieties are developed with four, five, or perhaps six. Favorite varieties of pear for this mode of culture are Passé Crassane, William's (Bartlett), Louise Bonne de Jersey, Doyenné du Comice, Passé Colmar, Duc de Boudeau, and Beurré Hardy.

Extensive plantings of pears as Verrier palmettes are often found in France. One orchard with which the writer is familiar comprises 150 acres. The trees were set 1 meter (40 inches) apart in rows 3 meters (10 feet) apart. Supports were from posts 10 to 12 feet long, set 2½ to 3 feet in concrete. Horizontal wires were strung about 30 cm. (12 inches) apart. Seven wires were used for moderately vigorous varieties and as many as nine wires for very strong growers.

The trees bore some fruit at four years of age and were in full production at eight years. At this time they yielded 30 to 40 pounds of fruit per tree. At 1,452 trees to the acre, this means roughly 900 to 1,200 bushels of choice fruit per acre. Varying with the variety, yields of 1,000 to 1,500 bushels were expected annually, and were highly profitable.

Fan-shape Palmettes

In Italy the fan-shape palmette has met with considerable success. On the fertile soils of northern Italy, under the long growing seasons which prevail there, young fruit trees on vigorous rootstocks will make strong growth—often five to eight feet and of exceptional diameter and strength. Under such conditions it is possible to develop a strong central leader and two strong oblique arms the first season, and to form an excellent fan-shape palmette very quickly. The more vigorous the growth, the more wide the angle of branching, suggestive of the Delbar system described later in this chapter. Similarly, the weaker the growth, the more acute the angle of branching.

The arms are not headed back but are permitted to make long extension growth, relatively free from laterals, to serve as the scaffold branches. Tipping or heading back will induce lateral-shoot growth, which is undesirable.

The arms are fastened to wires for training. Considerable bending and spreading is done with branches that arise in subsequent years. Superfluous growth is thinned out; that is, it is removed back to the branch from which it arose, rather than being tipped back. The net result is a relatively large fan-shaped palmette produced at a very early age, namely, the second or third growing season, with good anchorage and heavy early production.

Since the trees are well anchored, the wire trellis is not needed for support and is used primarily for training the tree branches in the desired fan-shaped palmette. Accordingly, the trellis need be neither expensive nor substantial. Some growers use heavily pruned live poplars as end posts several hundred feet apart, and depend upon the fruit trees themselves to support the wires in the row. The number of horizontal strands of wire depends upon the height of the palmette, usually spaced about two feet apart.

The system is used for apples, pears, and peaches. It has been tried with some success with sweet cherries, plums, and even olives. It is claimed that the method provides low-cost production of quality fruit with a minimum of labor.

SPECIAL SYSTEMS OF TRAINING

The Bouche-Thomas System

This method, found to only a limited extent in France, relies almost exclusively upon the slanting, bending, and weaving of the branches, with a minimum of cutting (Fig. 115). In fact, descriptive literature notes that "branches in the vertical position promote wood growth; branches in the oblique or arched position promote fruitfulness."

Strong yearling whips on seedling rootstocks or on EM VII and II and on quince rootstocks are used for planting stock. Where dwarfing rootstocks are used, they are often planted deep to induce scion-rooting eventually. They are set in pairs in the row at a pronounced angle—

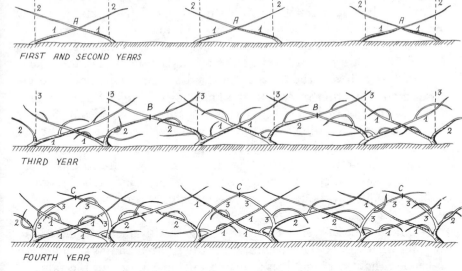

FIRST AND SECOND YEARS

THIRD YEAR

FOURTH YEAR

Fig. 115. Bouche-Thomas system. Designed as a low, early-fruiting hedge-like form. First year: Vigorous one-year-old trees planted in pairs at 30-degree angle in opposite directions (1), and fastened where they cross (A). Second year: Vigorous vertical shoots encouraged (dotted line 2). Third year: Shoots bent in reverse direction (2), and fastened where they cross (B); vertical shoots again encouraged (dotted line 3). Fourth year: Shoots again bent in reverse direction (3), and fastened where they cross (C).

30 degrees with the horizontal—and slanting toward each other in opposite directions so as to form a low "X." The trees are tied together at the point where they intersect.

In the second year the trees are induced to produce strong shoots or suckers close to the base and on the upper side. To aid in this, competing buds are reduced by pinching. The resulting shoot is bent and trained in a direction opposite to the main branch from which it arises, as shown in the diagram. The shoot may be held in a bent position by tying it with a long string to a peg placed in the ground.

In the third year these selected strong shoots of the previous season from adjacent trees are crossed in the direction of the row and tied at the point where they meet. Bending the shoots backward in this way serves further to suppress terminal growth, and induces fruit-bud formation. Adjacent trees are now joined one to another in a relatively firm hedgerow that requires little trellising or artificial support. Other branches that arise are bent downward and arched and interwoven in the hedgerow. During the third year, other strong-growing sucker shoots are induced to develop from the bases of the last series of bent suckers.

In the fourth year these strong shoots are again similarly bent backward in the direction reverse to the branches from which they sprang, and are similarly tied where they cross. In this way a series of tiers of crossed and recrossed branches is developed that has the appearance of an interwoven hedge fence. It may be developed to 5, 6, or 8 feet in height. Superfluous and conflicting shoots may be thinned out. Very vigorous shoots, which constantly arise, may be bent and arched severely, and so restrained without cutting. Less-vigorous shoots may be removed or bent less severely.

While not extensively employed, this method is of special interest in showing what can be done to promote early fruiting and to keep growth in hand without pruning. It is a practical demonstration of the value of slanting, bending, and arching. It seems especially useful for apricots, nectarines, and peaches, but may also be employed with apples and pears.

The Delbard Three-Crossarm System

George Delbard of Paris, France, has devised a system of training apple and pear trees that is not unlike the so-called Belgian fence and the Losange system. It consists of two main arms or scaffold branches for each tree, off which arise single arms at almost right angles to the primary arms. The trees are planted in such a manner that the arms overlap so as to give a three-crossarm appearance, as shown in Fig. 116. The EM IX rootstock is used for the apple, and the quince rootstock is used for the pear.

A feature of the system is that the length of the scaffold branches, which constitute the bearing wood, is adjusted to the vigor and the bearing capacity of the variety. Varieties are grouped into five classes, according to vigor, namely, weak growers, moderate growers, rather strong growers, strong growers, and very strong growers. The total length of scaffold branches is greater for the stronger-growing varieties and less for the weaker ones. This is accomplished by wider planting distance within

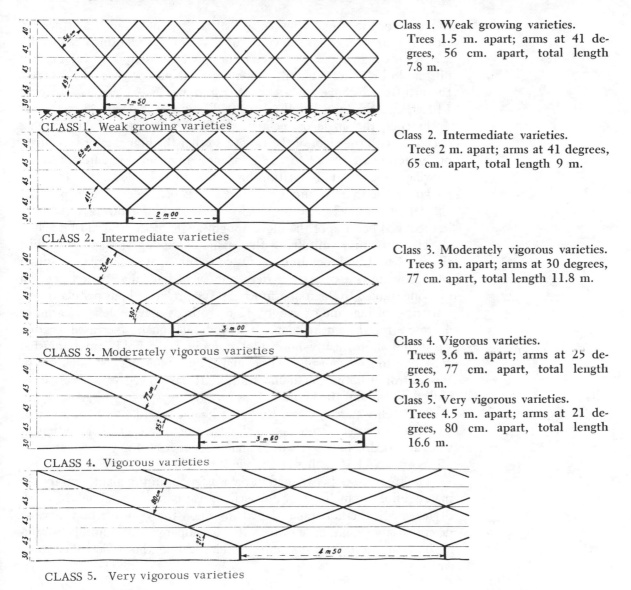

CLASS 1. Weak growing varieties

CLASS 2. Intermediate varieties

CLASS 3. Moderately vigorous varieties

CLASS 4. Vigorous varieties

CLASS 5. Very vigorous varieties

Class 1. Weak growing varieties.
Trees 1.5 m. apart; arms at 41 degrees, 56 cm. apart, total length 7.8 m.

Class 2. Intermediate varieties.
Trees 2 m. apart; arms at 41 degrees, 65 cm. apart, total length 9 m.

Class 3. Moderately vigorous varieties.
Trees 3 m. apart; arms at 30 degrees, 77 cm. apart, total length 11.8 m.

Class 4. Vigorous varieties.
Trees 3.6 m. apart; arms at 25 degrees, 77 cm. apart, total length 13.6 m.

Class 5. Very vigorous varieties.
Trees 4.5 m. apart; arms at 21 degrees, 80 cm. apart, total length 16.6 m.

Fig. 116. The Delbard three-crossarm system (*Tricroisillons Delbard*), adjusted to the vigor of the variety.

the rows for the strong growers, so that the scaffold branches of bearing wood may be longer. As shown in the diagram, the planting distance for a weak-growing variety is 1.5 meters, or about 5 feet, and the total length of scaffold branches is about 7.8 meters, or about 25½ feet. In contrast, the planting distance for a very strong grower is 4.5 meters, or about 14 feet, and the total length of scaffold branches is about 43 feet.

Further, the angle or inclination of the primary scaffold branches decreases with the vigor of the variety, so that the branches of very vigorous varieties are closer to the horizontal than are branches of weak-growing varieties. This relationship tends to hold down the terminal growth of the more vigorous varieties, as has been indicated in previous chapters. Conversely, the terminal growth of weak-growing varieties is less reduced because of the more upright angle of the scaffold branches.

The system is remindful of the balanced pruning of grapes, in which the number of buds per arm is adjusted to the vigor of the plant. With a weak-growing vine or grape variety, only eight buds may be left per arm of a vine trained to the four-arm Kniffin system, or 32 buds per plant. For a strong-growing vine or variety, 12 to 16 buds may be left per arm, or 48 to 54 for the entire plant. The combination of increased bearing wood and more nearly horizontal training of scaffold arms, adjusted according to the vigor of the tree, makes an interesting combination for the control of both growth and fruiting. It has proved very effective for both the pear and the apple.

The trellises are built to the same height regardless of the vigor of the variety, namely, 2 meters, or slightly more than 6 feet. Five tiers of wires are used. The lower wire is about a foot from the ground, and the others are about 16 inches apart, one above the other.

The trees are either trained in the nursery or in the plantation to two main scaffold branches that are inclined variously from about 20 to 50 degrees, with the horizontal depending upon the vigor of the variety as described above and as shown in the accompaning diagram. The training is done according to common classical methods described elsewhere.

The Lepage Method

The Lepage system of training (L'arcure Lepage) was devised by Henri Lepage of Angers about 1925. More extensive plantings were made about 1933, which proved their worth. The method has reached some commercial importance in the Angers area.

In this system the scaffolds are trained in a series of arches or arcs suggestive of the "arcure" method of espalier training often seen in France (Fig. 136).

In commercial operation, the management is not so intense, however, as with the usual well-trained espalier of similar design. Instead, the bending, tying, and pruning are done as simply as possible, and even somewhat grossly. In fact, it is the simplicity of the method and the fact that it can be quickly taught to inexperienced help that gives it its widest appeal.

Vigorous yearling apple trees are planted 4 to 7 feet apart in rows 8 to 10 feet apart. The distance varies with the vigor of the variety and the fertility of the soil. The rootstock for apples is EM IX and possibly EM VII, EM IV, and even EM II for weak-growing varieties. The quince root is used for the pear. Early plantings were made 2 feet apart but have been found to be too close.

The trellis is made with posts 7 feet above the ground, using three wires strung at 2-foot intervals, the lowest wire being 16 to 18 inches above the ground line.

The young trees are bent and tied to the lowest wire in an arched position. At the top of the bend, or slightly beyond it, a strong sucker-like shoot is induced to arise from the main scaffold branch. After the tree is established, usually the next year, this shoot is bowed and arched back in the opposite direction, being similarly tied to the next wire and thus forming the second tier. The third year, a similar strong shoot is developed from near the tip of the second scaffold, which is again arched and fastened to the third wire. In some instances, as with strong-growing varieties or with stronger rootstocks than the EM IX (such as EM II), a fourth tier of arches is developed.

Golden Delicious/IX has proved especially well suited to the method. One planting of 50 acres of Golden Delicious/IX averaged 400 bushels of fruit at four years of age, 600 bushels at five years, 800 bushels at six years, and between 800 and 1,000 bushels at full production annually. The Jonathan variety is also successful for this method, as are other varieties whose wood bends easily. On the other hand, Delicious is not well suited, since the angle of branching is sharp and the wood does not bow easily.

The method is of further interest as illustrating the value of bending in promoting early fruiting. It has not been used in America.

The Marchand Method

In the Marchand system, developed in France, yearling whips are planted at 30- to 45-degree angles, preferably pointing north and in the direction of the row, which should run north and south. The EM IX is used exclusively; EM II has proved too strong a grower. The trees are set about 6 to 8 feet apart, depending upon the vigor of the variety and the fertility of the soil. Rows are about 10 feet apart. The trellis consists of three or four stretches of wire, about 16 inches apart.

By virtue of the slant position of the tree, practically every bud will start into growth the first year. Strong shoots develop which are trained obliquely at right angles from the trunk and fastened to the wires. The result is a lattice effect which virtually covers the trellis.

Before the next spring, the main shoot should be cut back about one-third the growth of the preceding season. The cut is preferably made to a bud on the underside of the shoot. The laterals should be cut back lightly, and any shoots should be removed that do not arise vertically from the main shoot, leaving about one shoot every 6 inches, or about seven arms. The lowest shoot should be no less than 10 inches above the ground line.

During the second year, the main branches should be pinched back in April. In August, later growth must be suppressed and tips cut back. Winter pruning consists in thinning out undesired shoots.

This method calls for considerable detail pruning and special management. Success depends upon keeping a proper balance between vegetative growth and fruiting. Yet with all this, the method has much to recommend it. For example, the planting stock requires no special training in the nursery. The trees are merely yearling whips. They can be easily and rapidly planted before the trellis is set in place, a trench being opened and the plants set rapidly at the 45-degree angle that is required.

Moreover, it is possible further to simplify management by omitting some of the details of pruning and tipping. While the plantation thereby loses some nicety of appearance, the performance is still the criterion.

Records show plantations that have yielded 80 to 100 bushels of fruit per acre the second year set, 200 to 400 bushels the third year set, and 500 to 600 bushels at six years. In full production, claims are made of 1,000 to 1,500 bushels at ten years of age. Golden Delicious/IX has proved very successful with this method. One twenty-year-old apple planting of four acres is reported to have produced 1,000 bushels to the acre in 1959. Another planting of 200 acres of apples averaged 200 bushels per acre at two years of age, 250 bushels at three years of age, and 2,000 bushels at six years. An 80-acre planting of pears on Provence quince is said to have yielded 300 bushels per acre the third year set, and 2,500 bushels per acre at ten years.

Another interesting feature of this method is that if the young trees appear to be too vigorous, the angle or slant of the tree can be easily altered by dropping the tree slightly closer to the horizontal. Not only does this check growth and promote fruiting, as has been shown; it also permits a greater area of bearing wood, and so balances fruiting and wood growth.

The Spindle Bush (Spindelbusch)

The spindle bush (Spindelbusch) has been a feature of the Rhineland. It has spread into Holland and has become popular there as the "Vrie-spiel" or free spindle bush. It is really a modification of the dwarf pyramid; or it may be thought of as intermediate between a vertical cordon and a bush form. It differs from the dwarf pyramid in that it has no specific arrangement of scaffold branches; it differs from the vertical cordon in that the fruit is borne on short branches rather than directly on the main stem or trunk (Fig. 117). However, the most important feature of the spindle bush is the tying down of lateral shoots in a horizontal position, with little or no summer pruning. By choosing early-cropping varieties, such as Jonathan, Golden Delicious, and even Coxe's Orange Pippin, the weight of early fruits will bring the branches into this position without recourse to tying down.

Trees will bear fruit at two years or three years after planting; and

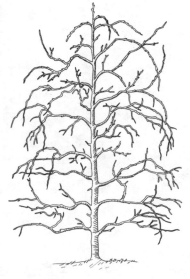

Fig. 117. **Spindle bush.** *Courtesy of V. H. Goldschmidt and A. V. Delap and the Royal Horticultural Society, London*

because of small size, the trees may be set 10 feet on the square, or 435 trees to the acre; or four to six feet apart in rows 10 to 12 feet apart. A common spacing is 8×13 (417 trees per acre). The EM IX, MM 26, or EM IV are used for the apple, and Quince C is used for the pear, although the latter is none too hardy against winter cold.

One-year-old or two-year-old trees are planted, headed at about two feet, with the lowest shoots arising at 12 to 14 inches above the ground. Any lower branches are removed because they only interfere with general operations. The leader is encouraged and maintained. Stakes may be planted with the tree, being sufficiently long to stand 5 to 6 feet above the ground line. Or a wire trellis may be used to steady the trees, employing either a single wire at 3 or 4 feet, or two wires at 2 and 4 feet. With some varieties, no bracing is required. Varieties with wide angles of branching, such as Beacon and Lodi, are especially suited to this method.

Any lateral shoots from the newly planted tree are cut back to about four buds. This is done about the middle of June or the first part of July, before the wood has become too stiff to handle. The branches will stiffen and remain in this position, and fruiting will still further bend the branches down.

A very important point is that the horizontal position of the branches must be truly horizontal, and not bowed in a sharp arc with the tips lower than the origin of the shoot from the main trunk. If the branch is severely arced, vigorous shoots will arise at the bend, as illustrated in Chapter 5. If properly done, the horizontal shoot will form spur growth rather than shoot growth.

Raffia and light string have been used for tying the branches, holding them in the horizontal position long enough for them to become firm and yet not so as to girdle the shoot. Paper-covered and cloth-covered wire is also good. In addition, various metal clamps have been devised, which are especially useful for the amateur and home gardener. Wire-sprung clothespins are inexpensive and simple to use. Success depends upon placing the clamp at just the right time, which is usually when the shoot is about ten inches in length and not yet woody.

No summer pruning is done. In the dormant season, the terminal and all laterals are cut back to about eight buds. Weak and interfering shoots are removed entirely. The next summer, and thereafter, laterals are again spread and fastened in the horizontal position.

When the tree has reached the desired height of 5, 6, or even 8 feet, the central leader is cut to a lateral, or it may be left to develop fruit spurs, depending on the weight of fruit to bend it and check its further elongation.

Yields of Coxe's Orange Pippin in the Meckenheim district of Germany have been recorded as 4.4 pounds per tree the second year, 6.6 pounds the third year, 9 to 11 pounds the fourth year, and 28 to 33 the eighth year. Records for other varieties have shown averages of 33 to 40 pounds in the eighth year, with an occasional yield as high as 74.8 pounds per tree.

Modified Spindle Bush

In the lower Loire Valley of France a method of growing apples and pears on dwarfing rootstocks has been promoted locally to a limited degree. It is not important economically but it is included here because it illustrates nicely another variation of the many different systems of pruning and training that can be developed.

It is essentially an adaptation of the vertical cordon. Or it may be thought of as a modified spindle bush trained in one plane like an espalier. The EM IX is used for apples, and the quince for the pear. It is fairly simple and easy to develop.

A trellis is constructed with poles 5 to 7 feet long set 25 to 30 feet apart, the height of the trellis depending on the fertility of the soil. Four rows of wires are strung, the lowest one being about 14 inches from the ground level, and the other three being about a foot apart, one above the other.

Vigorous yearling whips are selected and planted about 2 feet apart —farther apart for vigorous varieties—in rows 6 to 8 feet apart, again depending on the fertility of the soil.

Immediately after planting, the yearling whip is pruned back to about a third its length, or about 2 feet. Generally, three main branches will develop. Two of these should be arched downward and fastened to the lowest wire, while the third branch continues as a central leader. This operation will usually be done in late July or early August. During the winter, the central leader should be cut back to about 15 inches.

During the second year, lateral branches will again develop, and should be similarly arched downward and tied to the second wire in late July or early August, when they will have reached ¼ inch in diameter, as was done the preceding season. The terminal shoot should be untouched until winter, when it may again be cut back to about 15 inches. Further, the lateral branches, which should now carry many fruit buds, should be cut back to prevent overbearing, and to induce the production of large, quality fruit.

In this way the vertical cordon is developed into a palmette about 28 to 30 inches in spread. Branches that are not favorable for tying downward to the wire should be removed, leaving fruiting branches about 12 inches apart.

The Dwarf Pyramid

The dwarf pyramid, found frequently in England and on the Continent, is essentially a low-headed, compact, central-leader tree, with the lowest branches at 12 to 14 inches from the ground level and with successive branches radiating at intervals along the main leader, gradually diminishing in length from bottom to top. This gives a somewhat pyramidal effect, hence the name. As grown in England, dwarf pyramids are less severely and less exactly trained and the tree is less stocky than those grown on the Continent (Figs. 118 and 119). Although used largely in small gardens, trees of this type have commanded attention for commercial planting, and are found quite desirable. If set 3 feet × 9 feet, the number of trees per acre is 1,613. Pyramids are reputed to produce an average of 5 to 8 pounds per tree the third or fourth year after planting, or in the neighborhood of 200 bushels to the acre. Mature trees may produce 20 pounds per tree. A good average is 500 bushels per acre, although yields of 1,000 to 1,500 bushels have been reported. Good production is estimated for a period of twenty years.

Fig. 118. (Below) Development of a dwarf pyramid apple tree (English style). Pruning cuts made at the end of the season's growth as indicated: one-year-old, two-year-old, three-year-old. Compare with more severe, compact form (Fig. 119) typical of the Continent. (After S. B. Whitehead, *Fruit from Trained Trees*, J. M. Dent and Sons, London, 1954)

The system is especially useful for the apple but may also be used for the pear, the cherry, and the plum. EM IX rootstock is principally used for the apple, with some EM IV and EM VII, and even EM II, for weak varieties. EM 26, MM 106, and MM 104 are other possibilities. Quince EM A is used for the pear, and St. Julien C for the plum.

For the apple, trees are planted 3 feet apart in rows 6 to 9 feet apart, with the tendency being to widen the row to as much as 10 feet. Some successful growers recommend 4 × 9 feet for apples and 4 × 6 feet for pears. This latter distance permits the use of efficient modern machinery and other orchard equipment. Where the EM IX or EM IV is used, support must commonly be provided, although there are exceptions. Support is accomplished by running two strands of wire, one above the other down the rows, on posts set 15 feet apart. The lowest wire should be about 18 inches above the ground level and the upper wire about 3 feet. The trees are fastened to the wires. Some growers have found a single wire at 18 inches to be sufficient for the dwarf pyramid. Others prefer 3 feet, since the trees are better steadied by a wire at that height. The second wire can be added, if necessary, when trees come into full bearing, but it is much more difficult to do at that time. The cost of a second wire is not great in comparison to the entire cost, and may as well be strung in the first place. Trees on EM VII and EM II may be grown as dwarf pyramids without support. Less bracing is employed on the Continent, where trees in this form are more compactly and sturdily trained.

Both yearling whips and two- or three-year-old nursery stock may be used for planting. The latter will come into bearing a year or two earlier, but such trees are difficult to locate, so that reliance must more often be placed on yearling whips or two-year-old trees.

The central leader should be cut back to about 20 inches at planting.

Fig. 119. (Below) Development of a dwarf pyramid apple tree (German style). Compare with less severe, less compact form (Fig. 118) typical of England. (Left) Newly planted one-year-old tree, to be headed back to point indicated; (Center) growth the next year; (Right) pruning of the pyramid the second year to develop the first tier of branches. (After Fr. Lucas, *Die Lehre von Baumschnitt*, Stuttgart, 1909)

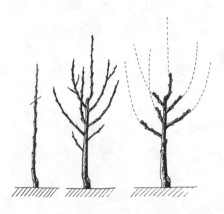

The maintenance of this leader is a most important matter. If there are lateral shoots, any over 6 inches in length should be reduced to about five good buds.

The next winter, the leader is reduced to about 8 to 10 inches of the last season's growth. Laterals are cut back to 6 or 8 inches of the last season's growth.

In Europe, summer pruning is considered an essential adjunct to success with this method. The Modified Lorette system of summer pruning, described in Chapter 20, is used successfully. Here again the summer pruning is performed at a time when most of the shoots are about the size of lead pencils and when they are woody and mature along half their length—firm and resistant to bending. Immature shoots are not touched.

At this time, then, which is about mid-July, the mature shoots are cut back to five or six leaves, or about 6 inches of the new growth. The central leader is left untouched. Later in the summer, perhaps mid-August or early September, any shoots that were not mature in July and were not therefore cut back, are similarly cut back.

The second winter, and thereafter, the central leader is reduced to about 8 to 10 inches of new growth, and any secondary growth that may have resulted from summer pruning is cut back to a single good bud. Undesired branches are removed entirely.

The central leader is not allowed to reach beyond a height of 7 feet, for the life of the tree. Any blossoms arising on the central leader in early years are removed. If allowed to develop into fruit, they may cause the central leader to bend under their weight, and so destroy the strong central-leader effect essential to a good pyramid. In the ideal pyramid, the lower branches should attain a length of 3 feet, the middle branches should be 2 to 2½ feet, and the upper branches should be about 18 inches.

Although the dwarf pyramid has been well received in Europe, and is apparently very useful for sizable commercial operations, it becomes a question whether it will succeed in America under the conditions that prevail. One of the large questions is the requirement of summer pruning, which does not succeed well in America. However, there are suggestions from this discussion that should be of real value in working out a system of training dwarf trees on wires that is adapted to America.

The Dwarf Bush

The dwarf bush form scarcely needs or deserves a name. It is the small, short, modified leader or open-center tree most commonly seen. It appeals to many growers because they are familiar with developing a tree of this type in larger size, and they can adapt to it very well. It most nearly resembles the form of dwarfed tree that is grown in America and is employed in the dwarf hedgerow system.

Yearling or two-year-old trees are set at about 6 to 8 feet apart in rows 12 to 15 feet apart. The central leader is suppressed at about 4 feet so that a low tree is developed. Trees are handled as any bush-type tree is handled, using conventional pruning and training methods. A single wire at a height of 4 feet is usually sufficient. In fact, with the proper variety and with care to head the tree low, no support may be required. For example, trees of Lodi and Fenton are characteristically and naturally well balanced, with wide angle of branching. Jonathan, too, can be grown without support.

The Pillar System

The Pillar system has been developed by Gordon A. Maclean of Abingdon, Berkshire, England, as a modified system of vertical cordons, without the requirement of support (Figs. 120, 121).

The trees are grown on well-anchored semi-dwarfing (vigorous) rootstocks, such as EM II, MM 104, and MM 109. They are planted 4 to 6 feet apart in rows 12 feet apart, or 907 and 605 trees to the acre, respectively. The 6 × 12 planting is preferred. This arrangement provides a planting comparable to a hedgerow type, with spraying done between the rows in the one direction. Mowing and other orchard operations are similarly performed between the rows.

A distinctive feature of the system is the simplicity of a standardized method of pruning that provides regular renewal of new wood. This is accomplished by developing a series of so-called "fruiting units" or "production units" on each tree, about 30 to 35 in number on a central axis not over 10 feet high.

Each "production unit" consists of three shoots or laterals; namely,

Fig. 121. (Above) Old pear tree pruned to Pillar System, showing old stubs and one- and two-year-old laterals after pruning. *Courtesy of Gordon Maclean*

Fig. 120. (Opposite) Steps in training and pruning according to the Pillar System: (A) (Above) Step 1—Cutting out third-year fruiting laterals to short stubs; (B) (Center) Step 2—Cutting off extension growth on second-year laterals; (C) (Below) Step 3—Cutting out surplus one-year laterals. Note growth of two strong one-year laterals from the stub of a third-year lateral. Approximately thirty shoots of each of the above three age groups are maintained on a mature fruiting tree. *Courtesy of Gordon Maclean*

(1) a one-year-old shoot, (2) a two-year-old shoot upon which spurs and blossom buds are formed, and (3) a three-year-old shoot that bears fruit. Each year, after fruiting, the shoot that has fruited is removed. This again leaves a unit comprising a new one-year-old shoot, a two-year-old shoot upon which spurs and blossom buds will form, and a three-year-old shoot that will bear fruit. In this way the "production units" are steadily renewed each year. The renewal is not unlike that used for gooseberries or currants, in which old wood is regularly removed and is succeeded by new wood. Since the maintenance of new wood growth by some system of renewal is one of the problems with dwarfed trees, the Pillar system seems, theoretically at least, to have important advantages.

The procedure for developing such a planting begins with the planting of a yearling tree, which is cut back to 30 inches, and from which all laterals are removed.

The following winter the central leader is cut back half its new growth, the top lateral is cut back to a half-inch stub, and three laterals are left untouched. Any other laterals are cut back to a half-inch stub. The untouched laterals are the first in the series of "production units" that are to be formed.

In the winter of the next year, the central leader is again cut back by one-half of its new growth. The topmost lateral is cut back to a half inch stub. Three new laterals are left untouched. The remaining one-year laterals are thinned out and cut back to half-inch stubs. The two-year laterals are cut back by the amount of the last season's growth. The three-year laterals, which have now borne fruit, are removed. So the system is repeated, year after year, and a series of "production units" is built up one above the other at approximately 1-foot intervals, until 30 to 35 units are formed on the 10-foot vertical axis that is the tree. Three units are added each year.

Each unit is said to carry perhaps 10 fruits, for 40 to 60 pounds per tree and 750 to 1,000 bushels per acre. Since the trees are open to light, the ratio of leaves per fruit is relatively low, being about 25 to 1 as compared to 35 or 40 to 1 for modified leader trees. Coloring of fruit is favored, and spray coverage is effectively done.

Those who encourage this system point out that the pruning is light work. Even inexperienced persons can be easily trained to perform the routine work, or "drill." Simply stated, the "drill" is as follows:

1. Cut back all three-year laterals to short stubs, whether they have fruited or not.

2. Cut back all two-year laterals by the amount of new growth they made the previous season, leaving the fruit buds intact.

3. Thin out the one-year laterals, leaving about 30 to 35 shoots well distributed over the entire tree and selecting those shoots that are not unduly vigorous and that grow out or away from the main stem. All pruning can be done from the ground, using a long-handled pruning shears for the highest cuts. Four minutes is about average for a tree.

Both Jonathan and Coxe's Orange Pippin apples have been found well adapted to this method. It is felt that the Pillar system is well suited to mechanical harvesting and other mechanization proceedures.

It should be borne in mind that the amount of cropping under this system is dependent very largely upon the length of one-year laterals which will carry the crop two years later. Thus, the rootstock and the growing conditions that will give strong laterals, are essential. The system is best suited to regions of long growing season and equable climate, such as England and the Pacific Northwest; and free from the severities of summer heat, winter cold, drought, and early fall freezes. One severe season of fire blight, for example, with blossom infection on spurs close to the main trunk can wipe out a plantation in a single season. Hot summer sun can cause ruinous trunk sunburn, and sudden fall cold or severe winter low temperatures can result in extensive spur and body damage.

AMERICAN EXPERIENCES WITH TRELLIS METHODS

American emphasis has always been largely on function. The questions asked are, typically, Does it work? What does it cost? How is it any better than what we now have? The niceties of symmetry and artistic form that have featured trained trees in Europe do not carry over into American commercial fruit production. Shortage of labor, high costs of labor, and the absolute requirement of efficiency for survival in a highly competitive industry are controlling factors in any management practice. Accordingly, by European standards, methods used in America for training trees on trellises often seem lacking in refinement and precision, and are even regarded as crude.

There is some question in the writer's mind whether there has not been too much of an attempt in America to follow European practices with trellised trees without due consideration to the differences that exist in growing conditions. Only on the Pacific Coast are conditions somewhat comparable. In eastern America, seasons are shorter than in Europe and in the Pacific Northwest; the climate is more rigorous. The result is that trees do not make so strong a growth in the East and are not so easily trained to special forms.

The writer has had such an experience in Michigan. After visiting the trellised plantings in Angers, France, an attempt was made to grow Golden Delicious/EM IX in the Marchand system which looked so attractive there. Unfortunately, the strong lateral shoots did not develop. Instead of a growth of lateral shoots three feet in length, the growth was eight inches to a foot. In fact, after four growing seasons in Michigan the trellises were not well covered with wood, whereas one or possibly two seasons would have sufficed in the climate and soil of Angers.

To meet this situation, it would seem that more vigorous rootstocks should be used in the East in comparison to those used in the Pacific Northwest or in Europe.

It would seem that EM II, EM IV, EM XVI, appropriate interstocks, and even standard seedling roots might be found more successful in the East when trees are trained in the palmette form on trellises, with appropriate bending and training of shoots at wide angles. Only an inexpensive trellis would be required, since the trees would be sufficiently well rooted to stand without support; and the trellis would be required only for training the branches in the desired form.

Nevertheless, attempts have been made to grow trees commercially in America on trellises as cordons, espaliers, pyramids, semibushes, and as free-growing trees in hedgerows. This has been accomplished with a considerable degree of success, confined mostly to apples, but including the pear to some degree.

Rootstocks

The rootstocks used are almost exclusively the EM IX for apples and the EM A quince for pears. Some EM VII, EM IV, and EM II have also been used for apples, and there is some disposition to employ more of these two. Trees formed with dwarfing interstocks, such as Clark Dwarf, EM VIII, and EM IX, have also been used, although in a small way only. The bulk of experience is with the EM IX apple rootstock. For the pear, Quince A has proved a bit too vigorous for strong varieties. For these it has been suggested that Quince C might be an improvement, using an interstem of Old Home or Beurré Hardy to overcome incompatibilities.

Spacing

In keeping with requirements for modern tools and equipment, the distance between rows is generally wider than in Europe. Ten feet is the minimum, although small operations and part-time orchardists close to markets may employ closer plantings. Distances as wide as 20 feet have been preferred by some large-scale operators.

In the row, spacing for apples has varied from 4 feet to 6, 7½, 8, and even 10 feet—all on the EM IX rootstock. Successful plantings have been made at 6 × 12 feet and at 8 × 15 feet. Distances of 10 × 20 feet may seem excessive by European standards, but have not proved so in the hands of growers who prefer a fairly free-growing type of tree.

Trellising

Trellises have been as simple and inexpensive as possible, and not so high as they often are in Europe. In some instances they scarcely deserve the name "trellis." In fact, trellises serve often for no more than modest bracing and steadying in place of staking. Some growers have used 2 × 2-inch wooden stakes for the first year or two, to be replaced later with a wire. In such cases, a single wire has been found very successful, strung at a height of 3 to 4 feet from the ground. In others, a wire has been placed at 1½ feet from the ground, with the possibility of stringing a second wire later at 3 or 4 feet. But the difficulty of placing a second wire after the trees have made some growth does not encourage this procedure. The single wire at 3 to 4 feet is better.

A successful method has been to erect a 5-foot trellis with four wires, at 2, 3, 4, and 5 feet from the ground. Posts are set 30 feet apart. End posts are set 3 feet deep, and intermediate posts at 2 feet deep.

Iron posts and pipe are good but they must be set firmly, especially the end posts. They are more easily pulled out and tipped than are wooden posts. Substantial trellises have been made with 8-inch end posts and 4- to 6-inch intermediate posts. Wire is at least 12-gauge.

In the state of Washington, where trees grow more vigorously, one commercial planting of apples on the EM IX rootstock has been supported on a higher trellis erected on 10-foot treated cedar posts, 5 or 6 inches in diameter, 35 feet apart. They are set 2½ feet in the ground, leaving about 7½ feet above ground. The wire is 9-gauge used telephone wire. The first wire is placed 24 inches above the ground line, and three additional wires are placed above the first wire at 20-inch intervals. Another successful fruit grower in Oregon has used trellises made of 9-foot, treated poles, set 2 feet in the ground. Anchor posts are 10 feet in length, set with 1¼ cubic feet of concrete. Three strands of No. 9 wire are used per row and posts are set 40 feet apart in the row.

General Management

As with any intensive crop, trees planted as close as has been described require soil and orchard management effective under such conditions. Further, trees on the EM IX or on quince roots do not have extensive root systems under these conditions.

Accordingly, special attention must be paid to keeping up the fertility of the land as well as maintaining adequate moisture. Soils that are relatively heavy and have good moisture-holding capacity are superior to light and sandy soils. A buildup of humus prior to planting is good practice, and anything that will add organic matter to the soil will help. With modern fertilizer practices, plus supplemental irrigation with light-weight portable equipment, trees have been grown successfully even on light soils. Supplemental irrigation has proved very effective as the trees have come into production and as water requirements have increased.

Weed growth has been controlled in trellis plantings by use of a grape hoe and some of the automatic mechanical hoeing devices. There are suggestions that some of the weed-control chemicals can be used successfully for eliminating weeds in the row and close to the trees.

At the wider spacings of 12 to 20 feet between rows, regular orchard equipment can be used for spraying and for management. There is much to be said, however, for distances of not greater than 10 feet, with small sprayers, dusters, and other orchard equipment adapted especially to the trellis system. Airplane dusting and spraying have also proved effective.

Tying

There are many special ties and tying materials for training trees on trellises. Willows, soft string, and cloth strips have been used for tying the trunk to the wires. A 6-inch section of garden hose slipped over the wire for each tree has been found useful in protecting the tree from rubbing against the wire.

Wooden spring clothespins have become almost standard for fastening laterals to wires or to other laterals. They are an interesting American development—not at all artistic but extremely functional. They do not choke or constrict the branches unduly, and can be shifted and used over again. One branch can be fastened to another branch or to a wire very quickly and with very little effort or special instruction. The clothespins should be fastened about 2 to 4 inches back from the tip so as to prevent breaking the tender terminal growth.

Pruning and Training

The general principles of pruning apply to trees grown on trellises, as discussed in Chapters 19 and 20. However, there are special pruning and training operations that are especially suited to the different methods of training trees in America on trellises and as espaliers. These are described in the paragraphs that follow.

Cordons

Cordons, which are essentially single leaders well fitted with spurs, have not been used for commercial fruit production in America. It should be repeated that American climatic conditions, with hot summer sun and cold winter temperatures, often produce sun scald and serious bark injury on this form of tree. They are not well adapted. However, limited trials have been made with apples as cordons, which are worth mentioning (Fig. 122). The appearance is not attractive as judged by European standards, but at least the trees have grown and produced fruit.

A vigorous yearling tree 5 to 6 feet in length on the EM IX rootstock is planted either at a 45-degree angle or erect, and later tied at a 30- to 40-degree angle to a wire trellis. The trellis is of two wires, at 2 feet and 4 feet. The cordon is tied to the wires by any of the conventional methods.

The central leader is left unpruned until it has reached the desired height. The lateral shoots are cut back to about 5 inches in mid-July when they are the thickness of lead pencils and are woody for at least half their length. The shoots are again cut back in winter to 2 to 3 inches. The result is often only a crude likeness of the delightful cordons seen in Europe, but trees have borne a few apples the second season, a pound or two the third season, and 10 pounds at full bearing.

Espaliers (Contre-Espaliers)

Commercial plantings of both apples and pears have been successfully grown as primitive espalier (really contre-espalier) forms that might be likened to horizontal palmettes. In one system, the trees are grown as grapes are grown in the Kniffen system. The trellis is 5 feet high with either two wires at 3 feet and 5 feet or four wires at 2, 3, 4, and 5 feet. It is easier to train the tree on four wires than on two (Fig. 123).

There is considerable difference of opinion as to planting distances. Plantings of 10 feet apart in rows 12 feet apart (363 trees per acre) have been successful, as have plantings 12 feet apart in rows 10 feet apart (363 trees per acre), as well as 8 by 15 (363 trees) per acre, and

Fig. 122. (Left) Apple tree in British Columbia trained as a functional oblique cordon supported on two wires. Trees 15 years old, on the EM IX rootstock. *Courtesy of J. H. Harris*

Fig. 123. (Right) Four-arm espalier apple tree, in British Columbia; supported on a two-wire trellis. Trees 15 years old, on EM IX rootstock. *Courtesy of J. H. Harris*

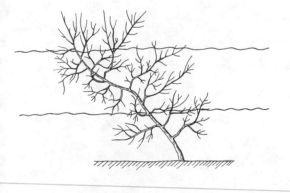
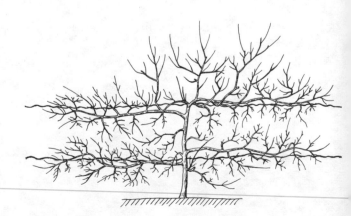

6 by 12 feet (605 trees per acre). The differences are seemingly due to variations in growing conditions, soil, fertility, and available orchard equipment. These spacings are for apple trees on the EM IX rootstock. Pears on Quince EM A have succeeded at 10 × 12 feet.

Either vigorous yearling trees at least 5 feet in height or two-year-old trees with strong central leader are selected for planting. Yearling trees are a bit more manageable, although two-year-old trees can be brought into bearing earlier. Any lateral shoots are cut back to about two buds, and the leader is headed at the top wire. All other laterals are removed. The trees are tied loosely to at least two of the four wires.

With good stock and good growing conditions it is possible to select eight branches, four on a side, to be tied loosely to the four wires. The branches need not come at the exact point of the wire, but they may be fastened up to a wire somewhat above or down to a wire somewhat below, as is done with grape vines in the Kniffen system. Wooden spring-clip clothespins are ideal for this operation. As the lateral branches extend, they need be fastened near their tips only, and the trunk needs no fastening after the second year.

No summer pruning is done. Winter pruning consists of thinning out unnecessary laterals and weak spurs. As a rough guide it has been suggested that the new growth be cut back to eight buds each year until the desired extension is reached.

A second lateral may be trained along each wire if the planting distances are sufficient. More laterals than this leads to shading and weak wood. Strong-growing varieties will spread to 10 to 12 feet in the row. Weaker-growing or spur-type varieties will extend to 6 to 8 feet. In this connection it should be noted that varieties that are terminal-bearing and willowy in growth habits, such as Rome, are not so well adapted to the method as are the spar types, such as Golden Delicious and Stark-rimson. At all events, the bearing surface of the trees is kept close to the trellis to give room for orchard operations. The general appearance and management of the trees is strongly suggestive of the grape.

Trees in British Columbia have averaged 40 pounds of fruit per tree, with occasional individual tree yields of 100 pounds. With planting distances of 7 × 12 feet, this represents yields of 500 to 1,000 boxes of fruit per acre (Fig. 124).

Dwarf Bush in Hedgerow

In refined training there is a distinct difference between a tree trained as a spindle bush, a pyramid, a bush, and a free-growing form. As grown under American conditions, there is much less difference. The general procedure is to develop a tree with good framework that is compact but with well-spaced branches and spurs so that the tree is open to light and air. Trees are planted fairly close (4 feet) and are grown as small bushy trees that develop into a continuous or hedgerow habit. A single wire at 3 or 4 feet from the ground is sufficient support. Rows are 12 feet apart, or even 15 feet apart, to facilitate operations. The trees are grown as any small bush tree, with minimum of pruning. The significant features of the method are the relatively close planting distance in the row approaching a hedgerow, and the modest support of a single wire.

Yields

Reported yields of fruit from dwarf apples on the EM IX rootstock have been exceptional in many instances. For example, Golden Delicious trees in the state of Washington grown on trellises, planted 6 × 12 feet ('605 trees per acre) have produced some fruit the second and third growing seasons, and have yielded 1½ boxes per tree, or 907 boxes per acre the fourth year. This is higher production than standard size trees under comparable orchard conditions at fifteen years of age. Acre costs are said to be much less than for standard trees of the same age. The labor required to keep the trees on the wire trellis has not been a problem. Picking has been greatly accelerated.

In British Columbia, yields of 500 to 1,000 bushels have been secured from plantings 7 × 12 feet (518 trees per acre) at five years of age. In Oregon, with trees set 10 × 12 feet (363 trees per acre) on the EM IX rootstock, computed yields by Roberts for Golden Delicious for the first ten years of the orchard have been 5,082 boxes. Another planting of this

Fig. 124. Apple trees of a hedgerow planting in British Columbia, showing branches bowed and fastened to adjacent trees with clothespins. Trees 4 years old, on the EM IX rootstock. *Courtesy of J. H. Harris*

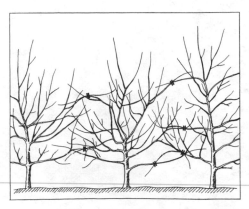

same variety at 8 × 15 feet (363 trees per acre) produced at the rate of 2,700 to 3,500 boxes of fruit at eight years of age. Estimates are that yields two to three times more than from standard trees can be secured during the first fifteen years of the orchard.

The plan has been to keep a clear 6-foot space between rows for the movement of vehicles and pickers. One procedure has been to seat pickers on each side of a low, self-propelled platform from which the pickers can conveniently harvest the fruit halfway through the tree from each side.

Finally, records from Ontario, Canada, covering a twenty-year period for eight varieties of apples on the EM IX rootstock trained on a four-wire trellis are given in the following table:

Cumulative Yields per Tree of Apples (20 years) on the EM IX Rootstock (Upshall). Trees Planted 10 × 12 feet (363 trees per acre), Trained to Four Wires at 2, 3, 4, and 5 feet from the Ground

Variety	Yield (Pounds)
Northern Spy	972
Delicious	858
Secor	814
McIntosh	804
Coxe's Orange	595
Jonathan	576
Melba	376
Early McIntosh	371

It must be remembered that these records are for the entire twenty-year period, including the first four years in the orchard when the trees were just getting started. In comparison with standard trees planted 35 × 40 (31 trees per acre), the cumulative yield of McIntosh/EM IX has been over twice that of standard trees; for Delicious, four times; and for Northern Spy, five times. At full bearing, average yields of 2 bushels per tree were not uncommon, or about 700 bushels per acre. The early-maturing varieties, Melba and Early McIntosh, were the lowest producers, largely for the reason that they were strongly biennial. It is suggested that a higher level of nutrition and blossom thinning might have increased the overall yield of these two varieties during the period.

Similarly, the yields of five varieties of pears are given in the following table:

Cumulative Yields per Tree of Pears (20 years) on the Quince A Rootstock (Upshall). Trees Planted 10 × 12 (363 trees per acre), Trained to Four Wires at 2, 3, 4 and 5 Feet from the Ground

Variety	Yield (Pounds)
Clapp	363
Vermont Beauty	340
Bartlett	247
Howell	215
Beurré d'Anjou	207

Here again it must be remembered that the yields are for the entire twenty-year period, and include production during the early years when the trees were just getting established. All the varieties performed well with the exception of the Clapp variety, which grew a little too vigorously. It is suggested that strong growers, like Clapp, might be worked on Quince C, which is more dwarfing, and using an interstock of Old Home or Beurré Hardy to improve congeniality.

In the final analysis, both the apple and the pear, grown on a wire trellis in Ontario, Canada, more than paid back the entire outlay for posts, wires, additional trees, and tying every year. The ease of harvest and the excellent quality of the fruit were special features.

Hedgerow Planting

In addition to the hedgerow system for very dwarf apple trees (EM IX) on wire trellises, which has been described in a previous chapter, dwarfed trees may also be grown in an unsupported hedgerow on semidwarf or semistandard rootstocks.

The EM II apple rootstock has been used for apple plantings of this kind; also, double-worked trees have been tried that are dwarfed by an intermediate stempiece of a dwarfing rootstock, such as Clark Dwarf (EM VIII) and EM VII. The rootstock is of vigorous seedling origin or of some vigorous clonal material like EM XVI. The chief consideration insofar as rootstock is concerned is to have a well-anchored, sturdy tree that will stand without support.

There is still insufficient practical experience to say just what planting distances should be employed, and what pruning and management details should be followed. However, plantings have been made with trees 10 to 15 feet apart, in rows 20 to 30 feet apart. The branches eventually touch within the row, presenting a hedgelike appearance. The trees are kept below 10 feet in height and are sheared in the direction of the row, on both sides.

Some growers have developed mechanical pruning devices patterned after those used for citrus in large commercial citrus plantings. A boom is erected with either sicklebar cutters, as on a grass mowing machine, or with a series of circular saws. The apparatus is operated from a power takeoff from a tractor. The boom or arm may be placed in a horizontal position at the height at which the trees are to be topped. Or it may be placed vertically and operated down the row as a hedge might be sheared. How effective this method may be has yet to be determined.

In this connection it is interesting to note that Thomas Rivers, English nurseryman and fruit grower, commented in 1864 that a friend near Paris grew pears on the quince in a hedgerow system. The trees were spaced 30 inches apart and clipped in July with garden shears much as a hedge might be clipped. In the winter months, any excessive spur development and weak, shading shoots were cut out.

In the great fruit area of northern Italy, near Ferrara, a system has been used with apple trees that suggests a hedgerow system. The trees are planted 10 to 15 feet apart in rows 20 to 30 feet apart. The rows run north and south so that light may reach both sides of the tree. The trees are kept in hand by cutting back and thinning out on the sides facing the open row but permitting branches to develop in the direction of the row so that they touch and run together. The result is a hedge or wall of apple trees that is about 4 feet in thickness and that is not allowed to reach beyond a height of 8 or 10 feet.

The growth of foliage is relatively thick, but there is sufficient light intensity to mature fruit of good size and color. This system is used especially for vigorous-growing varieties that seem to be unharmed by the heavy pruning that is of necessity employed.

22

SOIL MANAGEMENT, NUTRIENT REQUIREMENTS, AND FERTILIZER APPLICATION

GENERAL CULTURAL REQUIREMENTS

T H E culture and general management of dwarfed trees growing in the garden or in the orchard are not unlike those for standard orchard trees. Both kinds of trees require water, an adequate supply of nutrients, and proper attention to training and to general orchard care. Yet beyond this, dwarfed trees have certain cultural requirements that spell the difference between real success, mediocrity, and failure.

The dwarfed tree cannot bear neglect in its first few years. Standard-size fruit trees on seedling rootstocks, which are planted in an extensive orchard operation, are too often somewhat neglected the first year or two because of the press of other seasonal orchard operations. They may suffer from competition with weed growth. They may be subject to minor foliage injury from insects and diseases. They may fail to receive the water or the nutrients that would improve their growth. Or they may not be trained and pruned as carefully as they should be. Nevertheless, these trees will somehow continue to exist, and when the grower returns them to good management they will respond and move into production even though somewhat delayed.

This is not so with dwarfed fruit trees. And the more dwarfing the rootstock, the more exacting the trees seem to be. They virtually require garden care. Once they have been stunted through partial neglect or adverse conditions, they fail fully to recover. This is partly due to the early-bearing tendency of the very dwarf forms, which is accentuated and accelerated by conditions of neglect so that the tree comes into bearing as a stunted, scraggly individual. With standard trees, which may not be in full bearing until ten to twelve years of age, one year may not make so much difference. But with dwarfed trees, which may be in bearing at three to five years of age, a single year makes a great difference.

The introduction of dwarfing rootstocks into commercial fruit growing quickly separates the good grower and plantsman from the one who is not. The requirement is for more intensive care; better attention to soil, nutrients, and water; and protection from grass competition and from insects and diseases that may injure the foliage even to a minor degree.

The Importance of Water

While nutrients are important, the first requirement of a fruit tree is for water. The soil is the reservoir of water, which is held in the pore spaces between the soil particles.

The water-holding capacity of the soil is altered by its physical make-up. With large particle size, as with sandy soils, the pore spaces are large but relatively few, so that the ability to retain and supply water is limited. With smaller particle size, as with a clay loam, the pore spaces are smaller but numerous, and the capacity to hold water is increased. The addition of organic material increases the water-holding capacity of the soil.

Deep soils are preferred for fruit trees. This is in part because a deep soil provides a large reservoir of water. The soil need not be deep if there is a constant and adequate supply of water. Thus in the delta region of the Mississippi River below New Orleans, fruit trees are grown in soil with a water table at perhaps 18 inches. Yet because the water table is constant, the trees grow satisfactorily. They would not survive in 18 inches of soil under conditions of a fluctuating water table.

It is often said that dwarfed trees are shallow-rooted and that they will grow in shallow soil. This is not an accurate statement. It is certainly true that a small tree does not have so extensive a root system as a large tree. Roots have been traced for 57 feet for very large trees. Yet small trees will do best on deep soil with good water-holding capacity to supply the necessary water requirements steadily throughout the growing season. And practices that conserve moisture are the ones to adopt.

CLEAN CULTIVATION, MULCHING, AND WEED CONTROL

Although for many years clean cultivation was accepted as the preferred system of soil management for fruit trees, this supremacy no longer remains. The chief virtue of clean cultivation is that it reduces or eliminates weed competition for moisture. If there is no weed growth, or if weed growth is suppressed, there is no need for cultivation.

In fact, cultivation is not in itself a moisture-conserving practice. More water is lost by evaporation from soil that is frequently stirred than from soil that is not. In addition, feeder roots from young trees may be disturbed and destroyed. There is some evidence that mechanical hoeing close to the tree may favor the entry of damaging soil root-rot organisms, such as *Phytophthera*.

Chemical Weed Control

With the advent of chemical weed killers, competing grass cover can be eliminated close to the tree by this means. They should be used at the manufacturer's recommendation. Effective materials are simazine (2-chloro-4, 6-bis[ethylamino]-s-triazine), and amitrole (3-amino-1,2,4-triazole). An especially effective combination is 3 parts of simazine and 1 one part of amitrole.

Only a light spraying of the ground cover is needed—no drenching. With heavy applications there is the possibility of penetration into the soil and injury to tree roots. When applied to a limited 6 × 6-foot area around each tree, and with 100 trees to the acre, the gallonage suggested for an acre is extended to 11 acres of area treated. A single application, under favorable conditions, has been effective for two seasons.

Other materials, such as dalapon (2,2-dichloropropionic acid), maleic hydrazide (diethanoleamine salt of 1,2-dihydropyridazine-3, 6-dione) and diuron (3-[3,4-dichlorophenyl]-1,1-dimethyl-urea) have also been used. Dalapon is effective only on grasses, and does not give control much beyond the middle of the summer; but is preferred for this reason by orchardists who desire ground cover in late summer to slow down tree growth and improve fruit color. Maleic hydrazide is useful for merely suppressing the growth of the ground cover. Diuron is similar in action to simazine but is more soluble and has given some injury to trees in light soils.

Some caution should be observed when using chemical weed killers around dwarfed trees. The root system of young trees is relatively close to the soil surface. Further, when dwarfed trees are given the garden care that they deserve, the soil is usually in high fertility and is favorable to root development in the upper 6 or 8 inches. Accordingly, any application that may penetrate the soil even shallowly may come into contact with roots, and cause injury. The weed growth should never be drenched. There should be no runoff onto the ground. The ground cover should be touched only lightly. There is already some evidence that the fine-rooted Paradise-type EM I may be injured by herbicides, while the coarser-rooted Doucin-type EM II may not. It would not be surprising if EM VII and EM IX were also susceptible to injury because of a disposition to form root primordia freely on the trunks. It is always possible to make a second application if the first has not been sufficiently effective.

Another successful practice involves a strip of sod in the middle of the rows between the trees in one direction, and clean cultivation in a strip on either side of the trees. This method is especially suited to plantings in which the trees are closer together in the row than between rows, as when they are set 4, 6, or 8 feet apart in the row, and 12 feet apart between rows. In such a planting, a 6-foot strip of mown sod may be maintained in the center, and a 3-foot strip may be clean cultivated either side of the trees in the same direction. A chemical herbicide may be used to keep the strip clean close to the trees. Wider planting distances can be similarly adapted.

Before herbicides are used in the orchard, careful attention must be given to official regulations. Some chemicals are not permitted in bearing orchards, and others are restricted to certain crops.

Mulching

Mulching or a combination of mulch and sod has been employed with great success. Both are widely practiced in America, especially for apples and pears, but have been used only to a limited extent in Europe.

Mulching consists essentially in bringing organic material into the orchard and placing it as a cover. Many attempts have been made to grow the mulch in the orchard between the trees, but this is rarely satisfactory because of shading and competition with the trees. From the standpoint of conserving moisture, preventing erosion, and maintaining the fertility of the soil, it is ideal. It is especially adapted to the dwarfed orchard for reasons that will appear.

Numerous organic materials have been used effectively for mulching, including hay, straw, peanut hulls, buckwheat hulls, corncobs, sawdust, seaweed, and tobacco stems. In fact, several inert materials have also been used to advantage, such as granulated foam rubber, glass wool, plastic film, and coarse gravel.

Although the value of mulching has often been thought of in terms of the organic nutrients that are added, there is much to indicate that a major effect is upon the physical condition of the soil. Thus, summer soil temperature has been found to be reduced at least 5 to 10 degrees. Temperatures of clean-cultivated soils in Indiana, for example, have been found as high as 110 degrees at 4 to 6 inches. On a sandy soil a temperature of 130 degrees has been recorded. Roots of apple trees prefer temperatures of 65 to 80 degrees. In fact, tests by Nelson and Tukey with some of the Malling apple rootstocks would seem to indicate that high summer temperatures may be a limiting factor to their adoption by certain areas. The cooling effect, therefore, becomes additionally important with dwarfed trees.

Winter soil temperatures are higher under mulch than under clean cultivation. Snow is caught and held, so that still further protection is given to roots from penetrating winter cold.

Soil moisture is also conserved by mulching. Records by Tukey and

Schoff from Indiana have shown only about one-third as much moisture lost from soils under organic mulch as from soils which were clean cultivated. In a droughty year, the soil under organic mulch never reached the wilting point, whereas the clean-cultivated soil did. At the end of a five-year period, the soils under mulches averaged 70 per cent more water than did clean-cultivated soils. In comparison, soils under non-organic mulches contained only 5 per cent more moisture than did soil under clean cultivation. The larger amount of moisture in soil under organic mulches is explained by the increase in water-holding capacity of the organic material.

Of equal importance is the increased ability of soils under mulch to absorb moisture, especially water from quick rains or showers. On heavy clay, mulched soils can absorb moisture nine to eighteen times as fast as can clean-cultivated soils; and even on sandy loam the increase may be twice as great as for clean-cultivated soil. Obviously, aeration is also improved.

While there are differences in the effect of various organic mulching materials, the differences are less than might at first be expected. The important point is that mulching seems an ideal practice for dwarfed fruit trees. Some growers are so satisfied with its benefits that they would not attempt to grow dwarfed fruit trees without mulching with organic materials.

When an organic mulch is used, it must be remembered, additional nitrogen must be added to compensate for the nitrogen used by the organisms that flourish in the mulched soil. One grower in Michigan has had excellent results with a corncob mulch and one pound of ammonium nitrate per tree. Good results have been experienced when trees were planted directly in sod, with the addition of mulch around each tree applied immediately after the tree is planted. If the application of mulch is delayed, the young trees may suffer in midsummer from lack of moisture.

An especially effective way of using mulch in the young orchard is to place it in the fall of the year where each tree is to be planted the next spring. That is, the orchard may be laid out in the fall of the year, and a stake placed where each tree is to be planted. A half-bale of loose mulch is thrown over each stake. Then, the following spring, the mulch is pulled back from the stake, the tree is planted, and the mulch replaced around the tree. This is one way of ensuring that the mulch is in place sufficiently early in the spring.

Spoiled hay from adjacent farm lands has been widely employed for mulching. Some growers have found it worthwhile even to grow millet, rye, alfalfa, and similar crops especially for mulching material. The equivalent of one or two bales of hay are commonly used for each young tree. Additional amounts are added from time to time as material becomes available. It should be placed back 6 to 12 inches from the trunk of the tree. Under older trees, it should be spread principally under the drip of the branches, where greatest root concentration occurs. Since mulch will harbor mice, a mouse-eradication program must be maintained. There is no excuse for damage from mice now that efficient chemical poisons are available with which to control them.

NUTRIENTS AND FERTILIZERS

A fruit tree, whether dwarfed or standard, requires carbon, oxygen, hydrogen, nitrogen, phosphorus, potassium, calcium, magnesium, manganese, copper, chlorine, iron, zinc, sulfur, boron, and molybdenum. The first three (carbon, oxygen, and hydrogen) are supplied from water and from the air. Nitrogen, phosphorus, and potassium are the major mineral nutrients and are found in the soil in varying amounts, depending on the fertility of the soil. All the other materials may be considered together as minor elements. Sulfur, iron, calcium, and magnesium are present in the soil in relatively large amounts in relation to plant needs. Boron, copper, zinc, manganese, molybdenum, and chlorine are present in the soil in much smaller but usually adequate amounts.

Nutrient Needs of Dwarfed Trees

Because of the close spacing of dwarfed fruit trees, the competition for nutrients is accentuated. Further, the root systems of the very dwarfed forms are not so extensive as those of free-growing standard trees. In consequence, dwarfed fruit trees require more critical examination of their nutritional status and more attention to their fertilizer needs. This is a point that should be repeatedly emphasized.

Supplying Nutrients

Nutrients may be supplied to the trees either in organic form, such as barnyard manure, or in mineral form. The mineral forms may be used singly or mixed in combination with other minerals. The analysis of a fertilizer is designated on the manufacturer's label in per cent of the major elements, namely, nitrogen (N), phosphorus pentoxide (P_2O_5), and potassium oxide (K_2O). Thus a 10–5–10 fertilizer contains 10 per cent nitrogen, 5 per cent phosphorus pentoxide, and 10 per cent potassium oxide.

Fertilizers are never placed in the tree hole or close to the tree at planting. Not only is there danger of injury to the newly developing roots; the start a tree makes is also dependent upon the stored materials within it rather than upon an external supply of nutrients. Further, if the area has been properly fitted for planting both in terms of physical condition and fertility, there is really no need for applying fertilizer the first year. However, if it is deemed necessary to improve the growth of the trees a month or two after growth has started, an application of two to four ounces of nitrate of soda or the equivalent may be placed in a ring 6 to 12 inches from the trunk. A like amount of sulfate of potash may also be similarly placed. As a general principle, fertilizers are placed in a band out under the drip of the branches. If trees are in close rows, as in a trellis or hedgerow system, this would mean in a band along the row at a distance of 1 to 2 feet from the trees.

Nutrients may also be supplied as foliar sprays. Nitrogen may be supplied in this way in the form of urea. Both manganese and magnesium may be applied similarly as sulfates. Elements that are frequently tied up in the soil by chemical action, such as phosphorus, magnesium, copper, zinc, and iron, may also be applied to advantage as foliar sprays. Zinc has been also applied effectively to trees in a dormant condition.

Nitrogen

Nitrogen is the element that commonly produces the greatest response in a fruit tree. It is applied in various amounts in relation to the desired growth and performance of the tree. It is needed especially in early spring, when it has a pronounced effect on the setting of the fruit. Peaches and other stone fruits generally require more nitrogen than do apples and pears. Trees deficient in nitrogen have a yellowish cast to the leaves, the terminal growth is weak, leaves drop early in the fall, blossoms and fruits drop excessively, and the wood ripens early in late summer.

Nitrogen is available in organic form in manure and in urea (48 per cent nitrogen). There are various inorganic or mineral forms of nitrogen, such as ammonium sulfate (20.5 per cent nitrogen), ammonium phosphate (11.0 per cent), sodium nitrate (15.7 per cent), calcium nitrate (12.5 to 17.0 per cent), ammonium nitrate (35 per cent), and calcium cyanamid (22 per cent). Applications are made once a year. Ammoniacal forms are best applied in fall, since they become slowly available. Nitrate forms may be applied either in fall or in spring, but preferably in very early spring. It is important that the nitrogen be available for the beginning of new growth.

A common application for apple and pear trees is 50 pounds of actual nitrogen per acre or about 300 pounds of sodium nitrate or equivalent. Another rule-of-thumb guide is a quarter-pound of sodium nitrate, or its equivalent, for each year the tree is in the orchard. Thus, a four-year-old tree would require one pound. For small areas or for single trees, a figure of one ounce of sodium nitrate per square yard of soil surface is useful. Half again as much nitrogen should be used for the stone fruits. Foliar sprays of urea are also effective in supplying nitrogen. The material is used at the rate of 3 to 5 pounds per 100 gallons of water for apple trees, and 12 to 16 pounds for peach trees. Two or three applications may be made at fortnightly intervals, beginning at petal fall.

Potassium

Potassium plays an important part in protein synthesis and in photosynthesis. Deficiencies are not generally serious in America, but may be observed from time to time, especially on stone fruits. There is a crinkling and cupping of the leaf, and a marginal scorch. Certain of the EM apple rootstocks, especially EM V, are subject to potassium deficiency. Annual applications of 100 pounds of potassium sulfate or potassium chloride per acre, or 400 pounds made every third or fourth year, are sufficient to supply the needs for potassium to apples and pears in most situations. A figure of ¾ ounce of potassium sulfate per square yard of ground surface is useful for computing the needs in small areas. For the stone fruits, the amount of potash should be increased to half again as much, so as to balance the nitrogen. Foliar sprays of potassium sulfate at the rate of 10 pounds per 100 gallons of water will correct deficiencies promptly. They may be made at fortnightly intervals, beginning with full bloom, for three to four applications.

Phosphorus

Phosphorus is necessary to growth and to photosynthesis. It plays an important part in the translocation of carbohydrates. There is usually sufficient phosphorus in the soil, so that it need rarely be added. The greatest response from phosphorus may be seen in a cover crop or ground cover rather than in the trees. Deficiency symptoms are early-ripening fruit and narrow, oblong leaves. Foliar sprays of orthophosphoric acid may be used for immediate correction. Soil applications of acid phosphate, rock phosphate, and bone meal at the rate of 150 pounds of phosphorus pentoxide (approximately 500 pounds of acid phosphate) every fourth or fifth year are sufficient for both tree and ground cover whenever it is found that phosphorus is needed.

Magnesium

Magnesium is a constituent of chlorophyll, and is essential for photosynthesis. Deficiency symptoms are chlorotic areas between the leaf veins. Prompt correction is secured from foliar sprays of 10 to 15 pounds of magnesium sulfate (Epsom salts) in 100 gallons of water made at fortnightly intervals for three or four times beginning with petal fall. Heavy applications of potassium tend to reduce the supply of magnesium and induce deficiencies. Soil applications of Dolomitic limestone, which contains magnesium, are also a correction.

Manganese

Manganese deficiency is rarely seen. It is manifested by chlorotic areas between the leaf veins, checkered yellow or green in color, later becoming reddish blotches. Considerable injury may result to the tree in severe cases of deficiency. Foliar sprays of 2 to 3 pounds of manganese sulfate in 100 gallons of water applied at petal fall, and soil applications of 1 pound of manganese sulfate per tree, will correct the trouble.

Calcium

Calcium is required by growing roots and shoots. Symptoms of deficiencies are not easily detected, and deficiencies are seldom observed. Limestone applications will correct any disorder.

Boron

Boron is necessary to the oxidative system of the tree. Deficiency symptoms appear as small sunken and necrotic spots on the branches, and corky "drought spots" and cracking of the fruit. It is mostly observed on apples and pears. It may be corrected by soil application of ½ pound of borax per tree, as well as by a single application of borax at petal fall at the rate of 2 or 3 pounds to 100 gallons of water.

Zinc

It is thought that zinc bears some relation to auxin. Where zinc is deficient, shoots do not elongate properly, typical rosetting of terminal growth occurs, and fruits drop. It is more common in the western part of America than in the East. It is found on both stone fruits and on apples and pears. It may be corrected by spray applications of 25 to 40 pounds of zinc sulfate per 100 gallons of water made to the dormant tree, or by ½ to 1 pound of zinc oxide in 100 gallons of water applied to the foliage at petal fall.

Copper

Copper deficiency is less common even than zinc deficiency, which it very much resembles. Corrective treatments are soil applications of 1 to 5 pounds of copper sulfate per tree, or foliar sprays of ½ pound of the same material in 100 gallons of water made toward the end of June, and 40 pounds in 100 gallons of water applied during the dormant season.

Iron

Iron is needed for chlorophyll development. Deficiency symptoms are evident as a yellowing or chlorosis. The trouble is associated with lime and with other materials that tend to lock up the iron and render it unavailable to the plant. Iron deficiency is seldom seen in America on deciduous fruit trees, but it is found on citrus trees. It is more common in Europe. It may be corrected by soil applications of iron chelate, ethylene diaminetetraacetic acid (EDTA). One ounce per tree is sufficient for one or two years. Successful treatment is also made with ½ to 1 pound of iron chelate in 100 gallons of water applied to the foliage two or three times at fortnightly intervals, beginning with petal fall.

Foliar Analyses

The needs of a fruit tree for nutrients are usually gauged by tree growth and other symptoms enumerated in the preceding paragraphs. But this approach is aimed principally at correcting a condition that has already developed. To anticipate nutrient needs, chemical analysis of the foliage has been found helpful. In fact, rapid spectrographic determinations may be made in a few hours during the growing season, and corrective measures can be taken promptly.

The method depends upon comparing the analysis of the leaves in question with analysis of similar trees under ideal nutrient conditions. These are used as the standard of comparison. In general, the different classes of fruits, such as apples, pears, peaches, plums, and cherries, show considerable consistency in analysis. Thus, in Michigan, the standards determined by Kenworthy for apple, cherry, and peach trees of good performance are roughly as follows, in per cent of the element in the leaves:

	Apple	Cherry	Peach
N	2.3	2.9	3.2
K	1.6	1.5	1.5
P	0.2	–	–
Mg	0.4	0.8	0.8
Mn	.015	–	–
Ca	1.25	2.0	2.0
B	.004	–	–

In England, similar comparisons by Bould for healthy apple trees are as follows, given in parts per million in the leaves:

Molybdenum	0.3	Phosphorus	2,000
Copper	7	Magnesium	2,500
Zinc	25	Calcium	10,000
Boron	30	Potassium	15,000
Manganese	40	Nitrogen	20,000
Iron	100		

Differences Between Varieties and Rootstocks

While in general, as has been said, the similarities are striking in comparisons of this kind, nevertheless there are differences between varieties (cultivars) within a group. Further, the rootstock has an effect. Thus, in 1924, using Bramley's Seedling and Worcester Pearmain apples, Hatton and Grubb reported that there was twice as much leaf scorch (potassium deficiency) with EM V as the rootstock as compared to EM I as the rootstock. And there was still less scorch with EM X as the rootstock. Comparing eight of the EM rootstocks, the relationships were as follows:

Much scorch	EM V
Considerable scorch	EM VII
Slight scorch	EM II
Very slight scorch	EM I
Little or no scorch	EM IX, X, XIII, XVI

Again, the nutritional requirements of one variety on one rootstock may be different than that same variety on another rootstock. Bramley's Seedling/EM V, according to Hoblyn, has suffered more from potassium deficiency than Bramley's Seedling/EM I; but Coxe's Orange and Beauty of Bath have shown the reverse on the same rootstocks.

In other observations of Worcester Pearmain on various EM apple rootstocks, the following relationships from leaf analyses have been determined by Brown for nitrogen, potassium, and phosphorus:

N	VII (high)	>	XIII	>	II	>	I	>	V (low)
K	XIII (high)	>	I	>	VII	>	II	>	V (low)
P	V (high)	>	I	>	II	>	VII	>	XIII (low)

Studies in America by Kenworthy for the McIntosh variety of apple on several EM rootstocks have shown the following relationships:

N	no differences									
K	XII (high)	>	XIII	>	VII	>	IV	>	V	>	II (low)
P	no differences									
Ca	XII (high)	>	XIII	>	II	>	VII	>	V	>	IV (low)
Mg	XII (high)	>	XIII	>	II	>	VII	>	IV	>	V (low)
Mn	no differences									
Fe	II (high)	>	XII	>	XIII	>	VII	>	IV	>	V (low)
Cu	XII (high)	>	XIII	>	VII	>	IV	>	V	>	II (low)
B	no differences									
Zn	no differences									

The rootstocks with high analysis of a given element would be less likely to evidence a deficiency of that element. Thus, EM II and IV are recorded as relatively low in potassium and would be most likely to show potassium deficiency on soils of low potassium content. Conversely, EM I, with high potassium content, will tolerate a soil with lower potassium content.

From these observations it was concluded that differences in fertilizer requirements are to be expected from trees growing on different rootstocks; however, the differences are not so great as to cause the rejection of any of the rootstocks studied. With proper use of diagnostic methods, any nutritional disorder can be quickly identified and corrected. Further, with an understanding of the needs of the different rootstocks, potential difficulties can be anticipated.

IN SUMMARY

The principal and most common fertilizer need of fruit trees is for nitrogen. Potassium is next in frequency, and is increased by applications of nitrogen. Magnesium follows as the next most common nutrient required, which is again accentuated by applications of potassium. Phosphorus is seldom needed, excepting as it may be used in growing a cover crop between the trees. Other elements are minor and can be taken care of by an occasional "shotgun" application of minor elements or by attention to each one as symptoms of need appear.

A general recommendation for fertilizer applications annually per acre to fruit trees is as follows:

Apple and pear

50 pounds of actual nitrogen, supplied in 300 pounds of nitrate of soda or equivalent under clean cultivation;

50 pounds of potash (P_2O_5), supplied in 100 pounds of muriate of potash or potassium sulfate; and

half again these amounts if the trees are in sod.

Peach, plum, cherry, and apricot

75 pounds of actual nitrogen, supplied in 450 pounds of nitrate of soda or equivalent; and

75 pounds of potash (P_2O_5), supplied in 150 pounds of muriate of potash or potassium sulfate.

There are differences between the nutritional needs of different varieties (cultivars) and of different rootstocks. These differences are not so great that a radical change in fertilizer practices is necessary for dwarfed trees as compared to free-growing standard trees. Nevertheless these differences, small though they may be, are present, and should not be entirely ignored.

In the case of nitrogen, there are no appreciable differences between rootstocks, excepting that trees with more leaves and greater growth require more nitrogen.

In the case of potassium, both the EM II and the EM V apple rootstocks are likely to show deficiencies earlier than other EM rootstocks. Attention should be given to providing EM II and V with a higher level of potassium in the soil. EM I will tolerate a lower level.

In the case of magnesium, EM I and VII show deficiencies earlier than other EM rootstocks, so that soils on which they are growing should contain a higher level of this element.

As regards manganese, EM II is likely to show deficiency symptoms where EM XIII does not. In Illinois, manganese toxicity killed all of a number of Starking/II apple trees, whereas 67 per cent of Starking/I were killed, and none of Starking/XIII were killed.

23

POLLINATION, FRUIT SET,
AND THINNING

POLLINATION, fruit set, and thinning are no different in principle with dwarfed trees than with vigorous standard trees; but the problems differ in degree. And the techniques and methods used in meeting these problems operate more easily and more favorably with dwarfed trees, for reasons that will become apparent.

POLLINATION AND FRUIT SET

For fruit to develop from a flower, as has been discussed in an earlier chapter, the flower must be pollinated and fertilized. To be sure, there are fruits which will form without the requirement of fertilization, but these are the exception among deciduous tree fruits. They are not to be relied upon for good commercial crops. Generally, the higher the seed content, the better the set, the shape, and the development of the fruit. And the higher seed content is the result of practices and conditions favorable to fertilization.

Planning and Planting to Insure Cross-Pollination

In planning the orchard, pollenizers must be provided for. The ideal is to place a pollenizer as every third tree in every third row. This is at the rate of one in nine, and places a pollenizer adjacent to every tree. The difficulty with this procedure comes in harvesting scattered trees.

The more common practice is to plant solid rows of varieties, but to select and plant two or more varieties in rows side by side so that pollination needs are met. The ideal would be to alternate single rows of the variety. Or, better yet, if three varieties are planted, they may be similarly set in successive parallel rows, or in two rows of each in succession. But in commercial plantings there is usually one variety which is preferred over another and which constitutes the bulk of the planting. To meet this situation, four rows of the main variety may be alternated with a single row of the pollenizing variety.

With dwarfed trees, which are set closer together than vigorous standard trees, some growers have planted as many as seven rows of the preferred variety and then a single row of the pollenizer. While it may seem at first glance that the shorter distances for bee flight may thus insure cross-pollination, the fact is that bees work intensively in close areas, so that it is the amount and kind of bloom in the area which is important rather than the distance. In this respect, then, a block which is dominated largely with a single variety does not set fruit so well as one that is not. In fact, long before the principles of cross-pollination were understood it was a common statement that the best yielding blocks of fruit were those which were of mixed plantings.

Another method of insuring cross-pollination is by grafting a pollenizing variety into the permanent trees. The difficulty is that one variety tends to overgrow or outgrow the other. Also the harvest becomes confused. The method usually has no place in dwarfed plantings, where the better solution is to plant another tree or two of a pollenizing variety. In fact, the possibility of insuring pollenizers in this way is one of the features of dwarfed fruit trees.

Still another method is the use of bouquets of pollenizers placed in barrels or pails of water throughout the orchard. The blooms must be kept fresh.

Artificial Pollination

Pollination may also be done artificially. A simple method is to cut a branch from a pollinating variety when the flowers of both varieties are in full bloom. The branch or bouquet may then be brushed gently over the flowers of the tree which is to be pollinated. This transfers the pollen artificially to the blossoms which it is desired to cause to set as fruit. This is a good method in a small planting to insure cross-pollination. Obviously, dwarfed trees because of their small size are well suited to the method.

A more common and more extensive method of hand pollination is with pollen collected by hand, ripened, and applied at the appropriate time by hand or some other artificial means. This procedure is again quite satisfactory for small plantings, and it is also used in extensive commercial operations. It is ideal for dwarfed trees.

To collect the pollen, blossoms are gathered when they are in the balloon stage, just before opening and before the stamens have dehisced. The stamens are scraped out of the flower into a shallow jar. This can be done with a knife or scalpel with each individual flower, or the flowers can be rubbed across coarse screen through which the stamens will fall. The pollen is placed at 68 to 70 degrees F. for 48 hours in a dry place for "ripening," and is then ready for use. It can be stored at 34 degrees F. in a dry refrigerator.

It is applied when blossoms are receptive. This can be determined by a glistening moistness or sticky exudate that appears on the stigmas of the pistils. The pollen may be applied with the finger tip or with a small brush. All that is necessary is to touch the center of a flower lightly.

Not all flowers need be pollinated—only those which it is desired to set as fruit. In this way considerable fruit thinning may be avoided. In fact, if the gardener who grows dwarfed fruit trees has the time and the interest he can resort entirely to hand pollination, thus relieving him of concern about bees and bee flight, and provision for pollinating varieties. He can space the fruit as he prefers and avoid much labor in thinning.

Self-Fruitful, Self-Unfruitful, and Incompatible Varieties

Some varieties of fruit are self-fruitful and will set fruit with their own pollen. Other varieties are self-unfruitful and will not set fruit with their own pollen. They require pollen from another variety. Still further, some varieties will not set fruit even with pollen from another variety; the pollen must be from a compatible variety. Such varieties are termed incompatible.

While this may seem complex, it is not a difficult problem. The incompatibilities which exist with major varieties of fruits are recognized and understood. This is especially so for the major varieties of fruits found in commercial plantings, where routine recommendations take care of the situation. On the other hand, the amateur fruit grower who desires to plant odd and unusual varieties may find himself dealing with more complicated problems of incompatibility, and he should pay more than passing attention to them. The sweet cherry and the plum pose the greatest problems. Happily here, too, the relationships are understood and may be provided for.

The apple is self-unfruitful or only partially self-fruitful. Some few fruits may form without cross-fertilization on trees of Yellow Transparent, Winter Banana, Oldenburg, McIntosh, Jonathan, Wealthy, Baldwin, R. I. Greening, Golden Delicious, King, Grimes, Yellow Newtown, Rome, Early McIntosh, and Milton. But a full crop is not to be expected. It is only seldom if ever that fruits will form from self-pollination on trees of Northern Spy, Cortland, Gravenstein, Delicious, Starking, Richard, Twenty Ounce and Wagener. These observations are for New York State and may differ to some degree in other areas under different climatic conditions, but the general relationship persists.

There are no incompatibilities between major varieties of apple, although some produce little or no visible pollen and so are not effective cross-pollenizers. Among these are Baldwin, Bramley, R. I. Greening, Gravenstein, King, Stark, and Winesap.

The pear is also self-unfruitful or only partially self-fruitful. A few fruits may form without cross-fertilization on trees of Beurré d'Anjou, Beurré Bosc, Doyenné du Comice, Seckel, Flemish Beauty, Duchess d'Angoulême, Dana Hovey, Beurré Gifford, P. Barry, and Winter Nelis. Seckel is more highly parthenocarpic than any of the other varieties mentioned. In some areas even the Bartlett variety may set full crops of parthenocarpic fruit. But again this is the exception and cannot be relied upon for full crops.

There are a few incompatibilities between varieties of pears. Belle Lucrative, Laxton's Superb, Louise Bonne de Jersey, Seckel, and Bartlett will not cross-fertilize each other. Neither will Beurré d' Amaulis and Conference cross-fertilize each other. Among these, the major incompatibility to be called to attention is between Bartlett and Seckel.

The peach is self-fruitful with few exceptions. The J. H. Hale and June Elberta do not have good pollen and will not set fruit by themselves. Others generally considered self-unfruitful are Candoka, Halberta, and Chinese Cling.

The sour cherry is self-fruitful insofar as the major varieties are concerned, including Montmorency, Early Richmond, and English Morello. No cross-pollination is necessary.

The sweet cherry is the reverse of the sour cherry and is self-unfruitful, with the possible exceptions of Napoleon, Black Tartarian, Elton, Coe, and Kirtland. These varieties are, however, much improved by cross-fertilization. There are many incompatibilities in the group, but the only ones of major importance in America are between Bing, Lambert, and Napoleon (Royal Anne) which will not cross-fertilize one another. In Europe, where many more varieties of sweet cherry are grown than in America, an intensive study has been made of incompatibilities. Varieties have been placed in twelve groups; the varieties within each group will not cross-fertilize one another. There are, however, several varieties which are called "universal donors" which satisfactorily cross-fertilize all varieties with which they have been tried. These include Belle d'Orleans, Early Purple Guigne, Noir de Guben, and Florence. All of this information is valuable to amateur growers who may wish to plant some of the interesting, little known varieties of cherry.

The Japanese and American plums are generally self-unfruitful, although some are listed as being partially self-fruitful. Among these are Beauty, Climax, Elephant Heart, and Wickson.

The European and Damson plums are self-fruitful in the case of some varieties but not in the case of others. Among varieties commonly listed in America as self-fruitful are Italian Prune, Reine Claude, Yellow Egg, German prune, Stanley, Sannois, Agen, Sugar, Diamond, and French Damson. In all instances, however, cropping is improved by cross-fertilization.

There are incompatibilities among varieties of plums, but not as many as there are among sweet cherries. Here again the problem is not important with the varieties grown commercially in America. In Europe, where many more varieties of plum are grown, there is more attention paid to the problem but it is far less severe than with the sweet cherry. There are three small groups of plums given in European literature within which one variety will not cross-fertilize another, but the problem is not major.

The apricot is self-fruitful.

Some Factors Affecting Pollination, Fertilization, and Fruit Set

There are a number of factors that affect pollination, fertilization, and fruit set. An obvious but important factor is weather at blossoming time. During wet weather the activity of bees is reduced if not entirely eliminated. Also, both low and high temperatures have an effect on bee activity. Thus, bee flight is noticeably reduced at 60 degrees F. or below; and pollination, fertilization, and fruit set are accordingly reduced.

Pollen germination is slowed by temperatures of 51 degrees F. or below. At 60 to 80 degrees F., the pollen tube usually will grow down the style and enter the ovarian cavity within 48 hours, or sometimes within 24 hours. These conditions are generally considered favorable for good fertilization and fruit set. However, hot, dry conditions at this time may cause the style to dry out, reducing fruit set.

The nutritional condition of the tree at the time of full bloom and during the early development of the fruit is also very important. Many developing fruits may drop from the tree during the first month following full bloom and greatly reduce the potential crop. Speaking in general terms, fruit drop occurs in three distinct waves, ending with the so-called "June drop" in late June. These series of drops are related to hormones that are produced in the embryo and the endosperm of the seed. The "June drop" is the largest of the three.

Fruit drop may amount to 50 to 75 or even 90 per cent of the total number of blossoms on a tree. Fortunately, only a small percentage of blossoms need be set as fruit in order to ensure a crop. The dropping of young blossoms and fruit is frequently mistakenly ascribed to poor bee flight and poor pollination. Yet, examination of thousands of sour cherry fruits, for example, in the first two waves of drops has revealed that pollination was successful in approximately 95 per cent of the cases and that fertilization occurred in at least 29 per cent. Almost all of the fruits which dropped in the third wave, or "June drop," were found to have been fertilized. In other words, neither lack of pollination nor failure of the pollen tube to traverse the style and enter the ovarian cavity was found responsible for poor set of fruit. Further, when there is a heavy bloom, the percentage of drop is greater than with a light bloom. This indicates that fruits compete with each other for materials essential to their adherence and development, including water and nutrients.

Additional evidence pointing to the importance of nutritional factors in determining fruit set is shown by studies involving artificial shading of the trees. Heavy shading will reduce set, and it will also reduce the sugar content and the dry-matter content of the leaves. During the first two weeks after pollination, heavy shading will cause heavy embryo abortion and fruit drop. Severe defoliation at blossoming time or soon after also greatly reduces set. On the other hand, supplemental light has been shown to increase set, in which artificial light has been provided from sundown until midnight for a period after bloom.

A substantial increase in the size of spur-borne leaves, and to some extent in the size of shoot-borne leaves, can be induced by additions of nitrogen. The number of leaves per spur does not seem to be influenced appreciable by nitrogen supply, but the size does. This relationship has significance since leaf size influences the photosynthetic potential of a tree.

From these facts, one should place increasing importance upon maintaining trees in good nutritional balance and vigor, with strong healthy foliage for photosynthesis. This applies especially to the time of full bloom and for about 30 days after. It also applies to trees during the preceding growing season, when blossoms buds are being initiated and developed.

THINNING OF BLOSSOMS AND FRUIT

A fruit tree commonly produces many more blossoms than should be permitted to set as fruit. Dwarfed trees are heavy producers of flowers, often approaching a "snowball bloom." Ten per cent of the flowers of apples and pears are sufficient for a full crop. With cherries, peaches, and plums the necessary percentage crop may be two or three times this amount or 20 to 30 per cent. If all precautions are taken to ensure a good set of fruit, it is easy to see that something must be done in some seasons to reduce the possible overload. Although the natural competition between fruits may accomplish this, there is no prediction as to what degree of reduction may occur, and the results may be quite unsatisfactory. Fortunately there are practices that will more accurately control the number of fruits on a tree. And the earlier in the season they can be performed, the better is the control.

As has already been noted, trees should not be allowed to fruit too heavily at too early an age. This will depend upon the rootstock and the variety. If a type variety is permitted to overload, the leader may be bent severely, and a low, scraggly tree may develop. Some growers in Michigan have felt that apple varieties of this type on the EM II and VII should not be allowed to fruit until the fourth year in the orchard. It has seemed important to get the trees through the sixth or seventh year without bending and overloading. In the case of very dwarf apple trees, as on the EM IX rootstock, it is considered wise to remove all blossoms the first year in the orchard. With trees supported on trellises, there is no problem of developing a central leader; in fact, this is one of the features of the trellised tree.

The effect of blossoming and fruiting early in the life of the tree is shown by studies by Preston in England, in which trees were deblossomed for the first seven years after planting and compared with trees not deblossomed. Surprising as it may seem, the *first* crop from the trees that had been deblossomed for seven years was almost equal to the *accumulated* crops of the trees that were not deblossomed. On the other hand, there was a tendency for the deblossomed trees to become biennial following the first heavy crop; also, the economic importance of early fruiting in the establishment of an orchard must be borne in mind. The results do, however, emphasize the significance of blossom removal.

There is still another point about blossom thinning that should be emphasized. When apple and pear trees bloom full and set a heavy crop of fruit, they may become biennial bearers; that is, they may bloom and fruit one year and they may fail to bloom and fruit the next year. This habit continues year after year, producing a crop every other year. Dwarfed apple and pear trees can easily become biennial, particularly varieties that naturally possess this tendency. When grown upon a very dwarfing rootstock, as the EM IX, for apples, or the Quince C for pears, this alternate fruiting habit is accentuated. It becomes important, therefore, to prevent too heavy a set of apples and pears, and this must be controlled very early in the season, as by blossom thinning or thinning the young fruits very shortly after bloom. Thinning done in midsummer is too late to affect next year's crop materially, since the time is then past when blossom buds for next year's crop are initiated.

First of all, pruning during the dormant season may be used as a thinning operation. With dwarfed fruit trees, which are relatively small and easily accessible for pruning, this can be efficiently done. Bearing in mind the fruiting habit of the kind and the variety of fruit, some of any potentially excess fruit buds may be cut off. For example, in the case of apples and pears that bear fruit in spurs, some of the spurs may be cut off or otherwise reduced. In the case of the peach which fruits on one-year wood, potentially excess buds can be reduced by cutting back some of these shoots. It is best to delay the operation until all danger of winter injury to buds has passed, and after examination of the buds has been made to determine how many are alive.

The next opportunity to thin is at blossom time. If the bloom is potentially heavy some of it can be literally knocked off with a rubber-tipped pole. This has proved an efficient and practical method for peaches, plums, apricots, and cherries. It can also be used for apples and pears, but is less satisfactory because of the spur-bearing habit of these fruits. Even though a considerable number of the blossoms may be removed in this way, the crop need not necessarily be reduced, and the marketability is improved. The blossoms that remain will set as fruit in high proportion.

Chemical Thinning

Chemical thinning of blossoms and fruit is very effective for apples and pears, but less certain with the stone fruits. The chemicals most commonly used for apples are classified as plant regulators, such as naphthaleneacetic acid (NAA) and naphthaleneacetamide (NAd). They are available under various trade names.

Varieties differ in their response to these chemicals, and may be divided into three groups: (1) easy to thin, (2) intermediate, and (3) difficult to thin.

Where NAA is used, it is applied 5 to 7 days after petal fall at the following concentrations:

1. *Easy to thin varieties:* such as McIntosh, Delicious, Jonathan, Northern Spy, and R. I. Greening: 4 grams of actual NAA to 100 gallons of water (11 parts per million).

2. *Intermediate varieties:* such as Grimes, Oldenburg, Fameuse, Hubbardston, Wagener: 6 grams of actual NAA to 100 gallons of water (15 parts per million).

3. *Hard to thin varieties:* such as Yellow Transparent, Wealthy,

Golden Delicious, Rome, and Baldwin: 8 grams of actual NAA to 100 gallons of water (20 parts per million).

If the first application does not thin sufficiently, the concentration should be increased by 2 to 5 parts per million and repeated 7 to 10 days later.

Where naphthaleneacetamide (NAd) is used, it is applied at 60 parts per million at petal fall. Applications made *after* petal fall have resulted in no thinning and have actually caused an increase in set and an overload of small-sized fruit. This material is used principally with varieties that mature earlier than McIntosh, and for varieties likely to be injured by NAA. These include Yellow Transparent, Oldenburg, Early McIntosh, Wealthy, and Northern Spy. It may also be used on other varieties, but in the case of Delicious it has sometimes led to an increased crop of small-sized fruit, as previously mentioned.

Fruits that have been affected by the chemical will not grow, whereas fruits that have not been affected will continue to grow. The differences in size are apparent in 7 to 10 days after application, so that there is opportunity to estimate the degree of thinning that has been accomplished. A second spray of NAA may then be applied if more thinning seems desirable.

The effectiveness of chemical thinning agents is affected by various environmental conditions that must be taken into consideration. Adjustments must be made to meet these conditions. For example, trees not in good vigor are thinned more easily than vigorous ones. Weather that is unfavorable to pollination increases thinning action. NAA gives best results under fast-drying conditions, with temperatures between 70 and 75 degrees F. NAd prefers slow drying conditions, and is often applied in the evening.

Each grower must work out the concentrations that best suit his trees. Too high a concentration of NAA can remove all the fruit and can also injure the foliage. When conditions are such that overthinning seems a possibility, NAd is a safer material to use.

Part Six

ORCHARD PERFORMANCE OF DWARFED FRUIT TREES

24

ORCHARD PERFORMANCE
OF DWARFED APPLE TREES,
WITH SPECIAL REFERENCE
TO AMERICAN CONDITIONS

SOIL AND CLIMATIC PREFERENCES

T H E Malling apple rootstocks as a class have been employed successfully in the relatively heavy soils of England and North Europe. This region is also characterized by ample rainfall and by an equable climate. The extremes of winter cold and summer heat commonly found in parts of America are seldom, if ever, experienced.

It is not surprising, therefore, to observe that apple trees on these rootstocks have become more popular in those areas in America where there is some similarity to these conditions. Thus, the greater success has come from Nova Scotia, Massachusetts, western New York, the Niagara Peninsula of Canada, and Michigan in the East, and from the Pacific Northwest. The performance of Malling rootstocks as a class has been less impressive in the areas south of this latitude, where high temperatures prevail in summer, as in parts of the Virginias, southern Indiana, southern Illinois, Missouri, and Kansas.

On the other hand, for trees on strong-growing American rootstocks, but employing an intermediate stempiece, such as the Clark Dwarf, the more favorable reports have commonly come from regions where trees on the Malling rootstocks do not perform so well. Thus, many successful plantings of Clark Dwarf have been made in parts of the Virginias, southern Ohio, southern Indiana, southern Illinois, Missouri, Kansas, and the Great Plains areas. It is in these same regions that trees on the Malling rootstocks have been less successful.

Hardiness

In the principal apple-producing areas of North America, the desirable Malling and Malling-Merton rootstocks have proved sufficiently hardy to winter cold, although some differences in this regard do exist among them. Similarly, the interstocks commonly used for dwarfing the apple in the major commercial fruit areas have proved sufficiently hardy to winter cold. It must be remembered, however, that these fruit regions are characteristically among the more temperate and favored areas for fruit growing in North America.

Outside these favored regions there are large geographical areas that possess only limited horticultural advantages, and where conditions reach the minimum for success with dwarfed fruit trees. Even in what might be called nonhorticultural areas, there are spots where snow cover is sufficient to prevent the soil from freezing to any great depth, thereby permitting trees to survive that otherwise would not. The Georgian Bay region of Canada and the Traverse Bay region of Michigan are examples. The real record of a stock-scion combination, therefore, is its actual performance under field conditions. Nevertheless, there is much helpful suggestion to be derived from studies of the actual low-temperature tolerance of rootstock material itself, as well as from observation in various locations, fragmentary though they may be.

Observations have been made on winter hardiness of extensive populations of Malling apple rootstocks in the nursery in western New York. The rootstock material was in stool beds, and the tops were exposed to air temperatures of −25 to −30 degrees F. No serious injury was observed to the wood of EM I, II, III, IV, V, VII, IX, XII, XIII, and XVI. In comparison, trees of Baldwin and R. I. Greening in adjacent commercial orchards were severely damaged by the same temperatures.

Further observations led to three groupings; namely, (1) relatively hardy —EM I, II, III, IV, V, VII XII, and XVI; (2) less hardy—EM IX and XIII; and (3) still less hardy—EM VI, VIII, X, and XV.

In Kansas, during the severe winter of 1947, all Malling apple rootstocks were killed down to the ground line. Similar reports have been made from the Great Plains areas as a whole.

Artificial freezing studies at —15 to —20 degrees F. with ten of the Malling apple rootstocks, grouped them into three classes; namely, (1) very hardy—EM III; (2) hardy—EM IV, VII, XIII, and XVI; and (3) moderately hardy—EM I, II, V, IX, and XII.

Under orchard conditions the rootstock is seldom exposed to prevailing winter air temperatures. This is fortunate, since temperatures of 12 to 14 degrees F. are likely to cause severe root injury, if not complete killing. Observations made variously between 1860 and 1875 in the Niagara Peninsula of Canada and in western New York indicated no greater injury to trees on dwarfing rootstocks than to those on seedling rootstocks. Air temperatures of —20 degrees F. were recorded.

Attempts have been made to induce hardiness in the rootstock by working a hardy scion variety upon it. For example, with the hardy Dolgo crab apple used as the scion and with the bud union below the ground line no injury was observed to several of the Malling apple rootstocks during five winters in North Dakota. Yet adjacent apple seedlings of hardy varieties of apple exhibited considerable winter injury. The explanation seems to lie not so much in the transference of hardiness to the Malling rootstocks as in the fact that the scion variety was hardy enough to withstand the low air temperatures to which the stock-scion combination was exposed.

Similarly, tests have been made to determine the possible effect of Malling rootstocks on the hardiness of the scion variety worked upon them. The EM IX, II, and XIII rootstocks have been used. Detailed, critical examination has indicated no effect upon the scion variety.

In Massachusetts, the Malling apple rootstocks have been found as hardy as, or hardier than, French Crab seedlings, with the possible exception of EM IV. But the EM IV induces heavy fruiting in the scion variety, so that any seeming lack of hardiness may have been due to heavy cropping. In fact, the heavy fruiting of trees on some of the dwarfing rootstocks confuses the interpretation of winter hardiness. It is often not so much the relative hardiness of the rootstock as the condition of the tree brought about by the rootstock and the environment. During the severe winter of 1933–1934 in America, it was observed that trees that had borne very heavy crops or that were in poor vigor were injured, whereas trees in good vigor, which had not borne excessively, were not seriously hurt.

In addition, cold injury has been reported to the roots of ten- and eleven-year-old fruit trees on EM I, XII, XIII, and XVI; but the roots were exposed from hoeing around the trees for mouse control.

As indicated from the above discussion, winter hardiness must be judged in relation to factors other than temperature alone. Further, the region of the union in any stock-scion combination is generally one of late maturity. Trees that are budded high in the nursery and that are planted with the union several inches above the ground are subject to injury because of the exposure of the union to low air temperatures. While there is danger of scion-rooting if the union is not kept at or slightly above the ground line, yet the danger from injury to the union by low winter temperatures is equally important and is often lost sight of.

Double-worked trees propagated on hardy rootstocks and with an interstem of dwarfing rootstocks are similarly subject to damage at the unions. Thus, in tests with the Clark Dwarf (EM VIII) used as an interstock, severe cold (−25 degrees F.) has resulted in killing of tissues both at the union of scion and interstock and at the union of interstock and rootstock. The scion, the intermediate stempiece, and the rootstock were not severely injured, but the injury to the two points of union resulted in death to the entire tree. This is also the explanation of survival of trees on the EM IX rootstock in South Dakota and in Minnesota under conditions where trees on the supposedly hardier Clark Dwarf interstock (EM VIII) were killed. The EM IX union was below the snow line, and protected, whereas the unions of the double-worked trees were subject to the low air temperatures.

There is some evidence that trees on the EM I rootstock are subject to collar rot associated with the *Phytophthera* organism. But it is not understood whether the organism is primary or secondary. It may be that the same stock-scion combinations with EM I are susceptible to winter injury at the crown.

Soil and Moisture Preferences

It is generally accepted that apple trees on the EM IX, VII, I, XIII, and XVI rootstocks are well adapted to relatively heavy soils. The EM XIII is not drought-resistant; instead it will tolerate a wet, heavy soil. EM IV has done poorly on droughty soils of low fertility, but performs well with adequate moisture on soils of good nutrient level. By contrast, trees on the EM II will succeed on gravelly and sandy soils and seem especially adapted to deep, well-drained soils. MM 104 behaves somewhat similarly.

From tests by Tukey and Brase with ten Malling rootstocks (EM I, II, IV, VII, IX, XII, XIII, and XVI) in soils of high, medium, and low moisture content, EM VII and I succeeded at all three levels, with the latter performing especially well at high moisture level. MM 109 likewise seems tolerant of soils with high moisture.

On the other hand, EM IV, II, and XII appeared intolerant of low moisture. Intermediate between these two groupings were EM IX, XIII, and XVI, with the latter performing especially well at the higher moisture level.

These findings correlate well with field observations in which EM VII, I, and XIII have appeared useful as "general purpose stocks," adapted to a range of soil moisture, and tolerant of high moisture levels. The seeming ability of EM VII to perform well at different levels of soil moisture is of particular interest in light of the wide acceptance and adaptability of this rootstock for orchard planting in America. EM IX and XVI seem to prefer adequate moisture. EM XII does not enjoy droughty conditions. EM II is not generally suited to a wide range of soil moisture and does not accept droughty conditions very well (Fig. 125).

Summer Soil Temperatures

Summer soil temperatures have a significant bearing on the performance of trees on various Malling rootstocks. In studies by Nelson and Tukey with roots of EM I, II, VII, IX, and XVI growing at 44, 55, 66, and 77 degrees F., the EM IX appeared the most tolerant of low

Fig. 125. Clonal rootstocks are as distinctive as are scion varieties, and retain their characteristics even when budded to the same scion variety (Baldwin): (Lower left) Baldwin/EM XII; not long, downward-spreading root system, very few fine roots; makes very vigorous or standard-sized tree. (Upper center) Baldwin/EM I; note strong, downward-spreading root system, not so extensive as EM XII, many fine roots; makes early-fruiting, well-anchored, semistandard to semidwarf tree. (Lower right) Baldwin/EM XIII; note heavy, extensive, horizontal root system with numerous fine roots; makes semistandard tree, well-anchored, adapted to wet soil.

temperatures and the most intolerant of high temperatures. Even at 44 degrees F., this rootstock regenerated a few new roots. It performed best at 55 degrees, poorly at 66 degrees, and poorest at 77 degrees.

EM II likewise preferred the lower temperatures and did not do well at the highest temperature. In contrast, the EM VII rootstock performed well over a wide range of temperatures, followed closely by EM XVI. And seedling rootstocks failed to regenerate roots at low temperatures, but were prolific at the high temperatures. Recent tests with MM 106 indicate that it, too, is adapted to a wide range of soil temperatures.

The unsatisfactory performance of several of the Malling rootstocks in southern areas of America, where soil temperatures may on occasion reach 90 to 100 degrees F., may be associated with these findings. Likewise, the rather satisfactory performance of EM VII over considerable areas of the United States and Canada may be associated with its adaptability to a wide range of soil temperatures as suggested by the studies. The fact that EM XVI appeared tolerant of relatively high soil temperatures suggests its further trial in southern regions. And the fact that seedling rootstocks generated roots profusely even at relatively high soil temperatures tends to explain why trees on seedling rootstocks, with dwarfing interstems, are successful in southern areas.

SIZE, SHAPE, UNIFORMITY, AND OTHER TREE CHARACTERS

Root Systems, Anchorage, and Support

Trees on the EM IV have been reported frequently to have a high top/root ratio and a tendency to lean or blow over. Similarly the EM IX does not have an extensive root system, and the EM VII is none too well anchored while the tree is young. The EM I has, in several instances, been stated to have produced trees that were poorly anchored or leaned, but for the most part this condition has been with certain scion combinations such as Delicious and Turley. Trees on EM II have also usually had sufficient anchorage without support, although some varieties, such as Delicious, have been reported as tending to lean. Trees on EM XII, XIII, and XVI have required no support. The MM rootstocks are reported to be better anchored than their counterparts in the EM series. MM 106 is better anchored than EM IX; MM 104 than EM IV; and MM 109 and 111 than EM II. Yet trees on the MM rootstocks have blown over in heavy winds in Iowa.

Trees with dwarfing interstems, such as Clark Dwarf, EM IX, EM VII, and K 41, are fairly well anchored. They do not commonly require support if the tops are kept in balance and if the trees are not permitted to fruit too early.

Whether a tree requires support or staking is as much or more dependent upon the scion/rootstock combination as upon the rootstock alone. Thus, varieties that are "heady" and strong-growing, such as Northern Spy, are more likely to require support, regardless of rootstock, than are less heady varieties, such as Jonathan.

Combinations with a high top/root ratio usually require support. A ratio of 5 to 2 has been reported for EM IX, 3 to 5 for EM II, and 3 to 9 for EM XIII. EM IV characteristically has a high top/root ratio, which is accentuated by characteristic early, heavy fruiting.

Further, the scion/rootstock combination is altered by the system of pruning employed. For example, trees on the EM IX rootstock that are headed in and grown compactly as pyramids are less in need of support than are trees left to grow unchecked or trained as open-center trees.

Finally, too early fruiting of the more dwarf forms, such as EM IX, tends to make the tree top-heavy and subject to leaning. If the trees are kept from fruiting for two or three years until they have become established, they are less likely to need support.

Observations on the Use of Interstocks

In order to circumvent the poor anchorage of some of the dwarfing Malling rootstocks that produce trees of desirable size, nine-inch stempieces of EM IX, II, and XIII were used in Canada in a test of Bramley Seedling. French Crab seedlings provided the root system. There seemed little difference between trees involving the EM II and the EM XIII interstocks, and trees with EM IX interstems were half the size of the other two.

In Oregon, considerable success has been experienced by using the EM XVI as the rootstock, with a twelve-inch stempiece of Clark Dwarf (EM VIII) or EM IX between the rootstock and the scion variety. The resulting trees have been well anchored and have not required support, yet the trees have been as small as those worked directly on the EM IX rootstock.

In Michigan, both the EM VII and K 41 have shown promise as interstocks for several varieties of apple, including McIntosh, Jonathan, Delicious, and Golden Delicious. Trees came into bearing earlier than on standard seedling rootstocks, and they were so strongly anchored that no support was required. In New York State three-inch stempieces of EM IX interposed between French Crab seedling roots and

McIntosh and Delicious scion tops were compared with trees on EM IX rootstocks. The interpiece of EM IX did not reduce the height of the trees as much as did EM IX used as the rootstock. In fact, the reduction in size was only one-tenth to one-third. In performance, the trees more nearly resembled those on French Crab roots than those on EM IX roots. They were sturdy and well anchored, requiring no special support.

The conclusion was reached that by the use of an interstock of dwarfing material, trees could be produced that were somewhat smaller than standard-sized trees, but were not at all comparable in size and performance to trees propagated directly on rootstocks of the same interstem material. The question was raised as to why it was necessary to resort to an interstem when similar performance could be secured by propagating directly upon the appropriate clonal rootstock. The best argument for the use of an interstem seems to be in adapting a tree to a soil condition to which some of the clonal dwarfing rootstocks are apparently not well adapted, as in the case of soils with very high summer temperatures or very low winter temperatures. Also trees with appropriate interstems may be sturdier and more desirable under some conditions for this reason.

The use of interstocks in America received a serious setback when it was found that the Virginia Crab rootstock that was commonly used in combination with Clark Dwarf (EM VIII) was infected with the stem-pitting virus. As a result of this peculiar and unfortunate experience, many growers associate double-worked trees with poor performance, whereas it was the virus and not the double-working in itself that was responsible for failure. Nevertheless, in view of the latent viruses found in fruit trees, it is obvious that trees composed of wood from three sources (scion/interstem/and rootstock) have a greater chance of incurring difficulty than do trees composed of wood from two sources (scion/rootstock).

Scion-Rooting

In tests of six varieties of apples on the EM IX and the EM II rootstocks in West Virginia, there was no scion-rooting found on EM II. There was some on EM IX. In Canada, trees on EM I scion-rooted more freely than those on the French Crab seedlings, EM II and EM

XVI. Vigorous, vegetative trees have been found less prone to scion-rooting, regardless of rootstock. That the nature of the scion variety is also important is shown by tests in West Virginia with Golden Delicious, York, Gallia, Jonathan, Staymared, and Starking on EM IX. Of these, Golden Delicious/EM IX was most prone to root. In all these tests, the unions were placed just at the ground line or slightly above.

Insect, Disease, and Rodent Resistance

EM V has been reported susceptible to borer injury in Pennsylvania. EM I, IX, X, and III were susceptible to the fire-blight organism (*Erwinia amylovorus*); and EM II, XII, and XIII were found resistant. EM VII has appeared subject to crown gall (*Bacterium tumefaciens*) in Michigan, EM IX is generally conceded to be especially attractive to rabbits and mice. The roots of EM IX have been reported as inordinately susceptible to woolly aphis (*Eriosoma lanigerum*) in Pennsylvania, and EM I, IV, VII, and IX have shown more injury from this insect in Michigan than have EM II, XII, XIII, and XVI, yet in New Zealand the EM XVI is considered susceptible. All the MM series are reported resistant to woolly aphis. EM I and Merton 789 are reported as subject to collar rot (*Phytophthera cactorum*) in some regions, but not universally. Resistant clones are EM IX, VII, II, and XXV, and MM 104, 106, and 111.

In southern Illinois observations have been made on the susceptibility to apple measles of nine varieties of apples on EM I, II, VII, IX. In general, stions involving EM IX as the rootstock were free of the disorder. On the other hand, stions involving EM II were susceptible, with EM I and VII intermediate. The Starking variety is commonly considered very susceptible to apple measles in southern Illinois. However, Starking/EM VII and /EM IX showed no injury from this trouble, whereas all trees of Starking/II were killed. Blackjon and Rome on EM I and VII were all killed, whereas none were injured on EM IX.

Effect of the Rootstock upon Tree Shape

The shape of the tree may be altered by the rootstock. Trees on the EM IV are characteristically low and spreading. In Indiana, trees on

EM I and XII have been lower and wider than trees on EM XIII. In Michigan, Delicious/EM II develops a wider angle of branching than does Delicious/EM VII; and since the angle of branching of Delicious is typically sharp, the EM II is sometimes preferred for this variety over EM VII. In Pennsylvania, a wider angle of branching has been observed for young trees on rootstocks EM I, EM II, and EM XV. Again, the effect of the scion is evident. In Canada, trees of McIntosh on EM I and EM II seemed as broad as on French Crab seedlings, whereas R. I. Greening was reduced in both height and width. As should be expected, the precocity of fruiting and the weight of crop on the different stock/scion combinations have a marked bearing on the height, spread, and general conformity of the scion top.

Size Relationships of Various Stock-Scion Combinations

The relative size of trees resulting from several of the Malling and Malling-Merton rootstocks is shown in the accompanying chart prepared from records secured in Michigan (Fig. 126). The smallest trees occurred with EM IX, and increased in size progressively through EM 26, MM 106, EM VII, EM IV, MM 104, EM V, EM I, EM II, MM 111, EM XIII, MM 109, MM XXV, EM XVI, EM XII, and French Crab seedlings. Observations in Massachusetts agree generally with these size relationships as follows: (*a*) VIII, IX; (*b*) III; (*c*) I, IV, V, VI, VII; (*d*) XIII; (*e*) X, XV, and (*f*) XVI, XII. For comparative purposes, the EM IX may be considered to produce a tree about 6 feet high; EM VII, 8 to 10 feet high; EM II, 10 feet high; and EM XVI, 12 to 14 feet high. Size relationships in England are shown in Fig. 127.

These size relationships are fairly constant within a variety. But the differences between scion varieties on the same rootstock are in many cases as great as differences between the same variety on different rootstocks. In West Virginia it has been shown that York and Gallia are much more dwarfed on EM I than are Starking and Staymared. These size relationships are also altered by climate, soil, and season.

Vigorous, strong-growing varieties will develop into larger trees on

Fig. 126. **Size relationship in Michigan of trees on several Malling and Malling-Merton clonal apple rootstocks.**

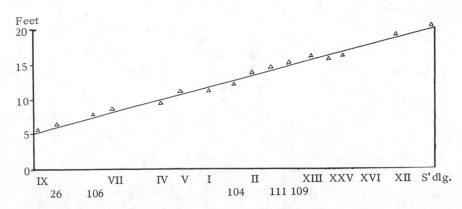

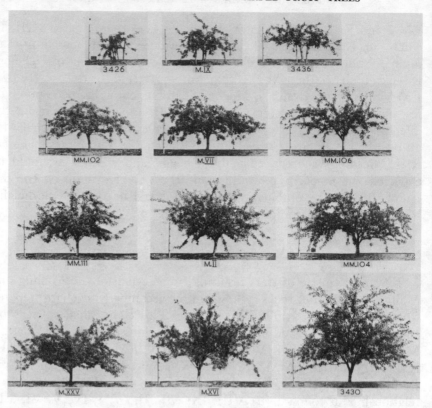

Fig. 127. Size relationships in England of trees on several Malling and Malling-Merton clonal apple rootstocks. *Courtesy East Malling Research Station*

the same rootstock than will weaker-growing trees. It is possible, at least theoretically, to develop trees of the same size by matching strong scions with very dwarfing rootstocks, and weak scions with slightly dwarfing rootstocks. Thus, the vigorous Northern Spy/EM IX may become as large as the moderately vigorous McIntosh/I and the smaller-growing Cortland/XIII. Practically, this presents something of a problem in that a top/root ratio prevails with a strong scion variety on a very dwarfing rootstock, and the trees require special bracing.

Environmental factors alter the expression of the stock-scion combination. Pruning is a typical example, as has already been commented upon. The heavy pruning of young trees tends to obscure the effect of the rootstock and make the differences smaller; also, conditions that favor exceptionally vigorous growth seem to obscure size differences. Some conception of the variations in size relationships due to environment and culture is evident in the following series of observations. EM I, commonly thought of as semidwarfing, was found in New Hampshire to produce a tree larger in height, spread, and trunk circumference than was produced by EM III, V, X, and XVI. In Canada, five-year-old trees of Northern Spy were as large on EM I as on EM XVI, although at the end of a thirteen-year period the Spy/EM I were smaller than Spy/EM XVI. In Indiana, EM I produced trees larger than EM XIII and of about the same vigor as EM XII. EM I is considered a semidwarfing rootstock; yet at Vineland, Canada, trees on EM II were larger than those on EM I, while in Nova Scotia the reverse of this condition prevailed. In West Virginia, Jonathan/EM II was very different in size from Jonathan/EM XVI at six years of age; yet Melba/EM II was as large as Melba/EM XVI at thirteen years of age.

Uniformity of Trees on Clonal Rootstocks

It has been commonly accepted that clonal rootstocks will produce trees of greater uniformity than will seedling rootstocks. In New Hampshire, for example, McIntosh/EM XVI has been exceptionally uniform in size and performance. Similarly, in New York State where a wide range of scion varieties and of rootstocks have been studied, results have emphasized the feature of uniformity (Fig. 128).

On the other hand, other reports are contradictory. In Canada, trees on EM I and II were found to be less uniform than the same varieties on French Crab seedlings; and trees on EM I were more variable than those on EM II. In Massachusetts, tests with EM I, XII, XIII, XV, and XVI showed that these rootstocks had done nothing to improve the uniformity of tree size. Reports of similar nature have been received from other parts of America.

However, it must be remembered that apple trees in the major commercial orchards of America are for some reason remarkably uniform, in spite of the fact that the rootstocks used are of seedling origin. This is in contrast to trees on seedling rootstocks in Europe, where a noticeable

lack of uniformity obtains. The explanation has been made that the growing and selecting of trees for orchard planting in America is very rigorous. During the several steps in rootstock production and in nursery management, it has been estimated that no more than one potential tree in ten ever reaches the orchard.

While this may be the explanation, it is also true that soil and other environmental factors play a very large part in the performance of one tree in comparison to that of another, adjacent tree. This has been shown by careful experimentation. Trees have been measured at planting time, and records have then been made of the growth each season for several years. It has been found that the trees that were largest at planting did not necessarily remain the largest trees. Rather, the trees that made the best growth the first two years in the orchard, regardless of planting size, became the largest trees, and remained so.

Accordingly it would seem that trees on clonal rootstocks carry the potential for greater uniformity, but that this uniformity may be over-balanced by the interplay of local environmental factors.

Fig. 128. (Opposite) Clonal rootstocks improve the uniformity of trees in size, conformation, time of coming into bearing, and general performance: (Upper) Semidwarf apple trees (Baldwin/EM I) in fifth growing season in western New York. (Lower) Semistandard apple trees (Macoun/XIII) in fifth growing season in western New York.

25

APPLE YIELDS, FRUIT QUALITY, AND ORCHARD COSTS

COMPARATIVE YIELDS OF DWARFED APPLE TREES

IT IS a common statement that dwarfed fruit trees will outbear standard-sized fruit trees. It is often not clear, however, whether this refers to the early bearing years of the tree, to yield in relation to tree size, or to acre yield. Each is a separate topic.

Precocity

Generally speaking, dwarfed trees bear earlier in their lives than do standard-sized trees, and the most dwarfing are the most precocious. The EM IX rootstock, which has the greatest dwarfing effect, produces flower buds at a very early age, even in the nursery, with some varieties.

Sixteen varieties of apples on EM IX rootstocks were planted in the orchard as two-year-old trees in New York State, including Baldwin, Coxe's Orange, Early McIntosh, Gallia, Grimes, Jonathan, McIntosh, Northern Spy, Red Spy, R. I. Greening, Stark, Tompkins King, Turley, Wagener, Winesap, and Wolf River. All but Northern Spy and Red Spy carried some blossoms the first year set in the orchard, and Baldwin and Stark carried some fruit. The second year, all varieties blossomed, and all but Northern Spy, Tompkins King, and Winesap bore fruit. The third year, all blossomed and all set fruit, the amount ranging from 16.7 and 13.6 pounds per tree for Stark and Early McIntosh, respectively, to 0.8 and 1.2 pounds per tree for Red Spy and Tompkins King, with an average of 8 to 10 pounds per tree. EM 26 is similar to EM IX in early fruiting, but no actual figures of performance are available in America.

EM VII has been found in general to be superior in inducing early bearing; even the typically late-bearing Northern Spy produced fruit the fifth year in New York State when worked on this rootstock. In Michigan, McIntosh/EM VII produced 7 pounds of fruit the fourth season and 60 pounds the fifth. MM 106 is reported to be intermediate in behavior between EM IX and VII.

EM I has also proved precocious, especially for the McIntosh variety, to which it seems well suited. In Michigan, it has produced 2 pounds of fruit per tree the third growing season, and has fruited heavily with 78 pounds the fifth growing season. Other records for the McIntosh variety in Michigan report EM II to be a year or two later in fruiting than EM I, carrying 2 to 3 pounds of fruit the third and fourth years and 28 pounds the fifth. EM IV, although not so early in bearing as the preceding, has fruited relatively heavily the fifth season with a production of 44 pounds per tree. MM 104 has behaved similarly to EM IV as regards early fruiting. EM V has been a year later than EM II, which it resembles, and has produced smaller initial crops.

Earliness of fruiting is related to the characteristic of the scion variety. Those varieties that have been found to fruit early on seedling rootstocks are early fruiting on dwarfing rootstocks. For example, Gallia, Jonathan, Cortland, and Golden Delicious bear earlier than do McIntosh and Delicious, and these, in turn, fruit earlier than does Northern Spy.

Yield in Relation to Tree Size

There is a relationship between early fruiting and small tree size. It has been proposed that since fruiting is a dwarfing process, it is the bearing of fruit in itself that is responsible for the small size. It is as though young trees were to grow at somewhat similar rates during their early years, and as though growth were arrested in each, successively, as each rootstock induced fruiting.

That the matter is more complicated than this is shown by the performance of various stock-scion combinations. Thus, some varieties on the EM XIII rootstock may not grow much larger than on the EM I rootstock, yet the latter induces much earlier and heavier fruiting. Similarly, the EM IV is a heavy yielder for its size. In other words, although there is a trend for small trees to be early croppers, this is true only in a general sense, and fails under critical examination. Each stock-scion combination must be considered by itself.

Efficiency

Dwarfed trees produce more fruit with a lower leaf/fruit ratio. Standard-sized trees may require 35 to 40 leaves for the development of a good apple, whereas trees on the EM IX rootstock may require only 25. But this seemingly greater efficiency of the leaves on dwarfed trees may be associated with the open nature of the trees and the reduced shading of one branch by another in comparison to standard-sized trees.

An interesting comparison has been made in England between the weight of the tree and the weight of the fruit, comparing the dwarf EM IX and the very vigorous EM XVI. At the end of 27 years, trees on EM XVI had produced only five times the weight of the tree in fruit, whereas trees on EM IX had produced fourteen times their weight in fruit.

Total Fruit Yields

Although growth-controlled trees on dwarfing rootstocks come into bearing somewhat earlier than do standard-sized trees on seedling rootstocks, the more important consideration is that dwarfed trees are smaller and that more can be planted on an acre. The evidence is that there is a level of fruit production possible on an acre of land when it is covered by a bearing surface of fruit trees. And this coverage can be achieved with many small units or with a few larger ones. Only when the somewhat earlier fruiting of the dwarfed tree is multiplied by the larger number of trees per acre does the dwarfed tree become really significant.

Figures by Roberts from Oregon show this relationship very well, computed for Golden Delicious trees on six different rootstocks over a ten-year period. The total yield per individual tree during this period was over three times as great for seedling rootstocks, for example, as for EM IX rootstocks. However, with 363 trees per acre of Golden Delicious/IX and only 48 of Golden Delicious/seedling, the calculated yields of 50-pound boxes of fruit were 5,082 per acre for the former and only 2,112 for the latter:

Rootstock	Trees per Acre	10-Year Total Yield of 50-Pound Boxes (after A. N. Roberts)	
		Per Tree	Per Acre
EM IX	363	13	5,082
EM VII	134	28	3,752
EM II	70	42	2,940
Seedling	48	44	1,488

Computations from records of production of Coxe's Orange Pippin in England for trees on several rootstocks at different planting distances also show the importance of the number of trees to the acre. The figures given in the accompanying table are for trees between 11 and 15 years of age. This is not cumulative yield, but yield for a single year for trees of the ages noted. The Coxe's Orange Pippin, of course, is not a heavy cropper, although recognized for the high dessert quality of its fruit. This fact should be kept in mind when comparing other yields per acre given in this chapter. Nevertheless the relationship between the number of trees per acre and the yield per acre is comparable.

Rootstock	Planting Distance (feet)	Trees per Acre	Average Annual Yield per Acre (bushels)
EM IX	10 × 12	363	454
EM 26	13 × 15	223	334
EM 26	10 × 15	290	435
EM II and 26	17 *	150	337
EM II and IX	12 *	300	506
EM II and IX	24 × 12 *	150	319
MM 104 and 111	17 *	150	524

* Providing for thinning at about 15 years of age so as to give a final planting of 24 feet square, but which could, of course, materially affect yields in subsequent years because of the reduced number of trees per acre (after A. N. Preston, East Malling Research Station).

In Canada, comparison has been made by Upshall of cumulative production between three varieties on the EM IX rootstock and the same varieties on seedling rootstocks. At the end of the twenty-first year, the production in pounds of fruit per tree favored the trees on seedling rootstock:

Variety	Pounds per Tree on Seedling Rootstocks (21-Year Total)	Pounds per Tree on EM IX Rootstocks (21-Year Total)
McIntosh	4,864	851
Delicious	2,692	962
Northern Spy	2,699	999

However, when computed on the acre basis, with 363 trees per acre for the dwarfed trees and 31 per acre for the standard-sized trees, the yields were two to four times in favor of the dwarfed trees:

Variety	Bushels per Acre on Seedling Rootstocks (21-Year Total)	Bushels per Acre on EM IX Rootstocks (21-Year Total)
McIntosh	3,351	6,855
Delicious	1,854	7,760
Northern Spy	1,859	8,058

Production records in Michigan for the McIntosh variety on six of the Malling rootstocks show the same trend. Although the yields in pounds of fruit per tree were greater for the smaller trees in the early years in the orchard, the larger ones were catching up by the tenth to thirteenth years:

Rootstock	4th Year	7th Year	13th Year	13-Year Total
EM I	79	121	859	2,727
EM II	28	144	531	1,974
EM IV	44	96	720	2,383
EM V	25	93	472	1,973
EM VII	60	198	490	2,007
EM XII	1	84	558	1,519
EM XIII	4	198	895	2,283

But computed on the acre basis, with a larger number of the smaller trees per acre, the yields are much higher for the smaller trees.

A clear picture of rootstock effect on cumulative yield for individual trees is shown in the following table, based on results at the East Malling Research Station with apple rootstocks EM I, II, IV, IX, and XVI at five-year intervals from ten to thirty-five years in the orchard:

Years	Lowest to Highest in Yields (Left to Right) (after A. N. Preston)				
10	XVI	II	I	IV	IX
15	XVI	I	IX	II	IV
20	I	IX	XVI	II	IV
25	IX	I	IV	XVI	II
30	IX	I	IV	XVI	II
35	IX	I	IV	II	XVI

It is at once observed that the trees with the lowest cumulative yield at ten years of age (EM XVI) are the highest after thirty-five years in the orchard. Conversely, the trees with the largest cumulative yield at ten years of age (EM IX) are the lowest after thirty-five years. The others are intermediate. It is interesting, also, to observe for each five-year period the fairly steady and regular shift to the left or to the right, as the case may be, for each rootstock.

Cumulative yields of Coxe's Orange Pippin at the same station for six-year and eleven-year periods are of additional consideration, as follows:

Rootstock	Planting Distance (feet)	Trees per Acre	Cumulative Bushels per Acre	
			At 6 Years	At 11 Years
EM IX	12	302	543	1,313
EM VII	15	193	699	1,987
MM 106	15	193	699	1,785
EM IV	17	150	645	2,066
EM II	17	150	416	1,646
MM 111	17	150	585	2,242
MM 104	17	150	675	2,666
MM 109	24	75	210	984
EM XXV	24	75	262	1,125
EM XVI	24	75	180	660

Yield records from Holland for Golden Delicious and Jonathan on the EM IV and the EM IX rootstocks for an eight-year period are given below. Of interest are the early fruiting and the tendency toward alternate bearing:

| Year After Planting | Yields in Bushels per Acre (689 Trees per Acre) | | | |
| | Golden Delicious | | Jonathan | |
	EM IV	EM IX	EM IV	EM IX
1st	0.0	0.0	0.0	0.0
2nd	6.6	33.4	0.0	0.0
3rd	222.7	371.8	232.0	219.0
4th	311.7	249.4	441.1	220.0
5th	396.4	418.6	120.4	62.3
6th	1,561.0	946.1	835.0	394.0
7th	739.2	550.0	904.4	298.4
8th	1,314.5	825.0	757.1	247.0

Representative acre-yields for four varieties and four rootstocks in Michigan are given in the accompanying table. The trees were planted 15 by 20 feet in 1945. In 1952 the trees were beginning to touch, and half of them were removed, leaving a planting distance of 20 by 30 feet. There was one complete crop failure in 1955 due to spring frost.

Golden Delicious/EM IX in British Columbia, planted 8 × 12 feet, produced 300 boxes of fruit per acre the third season. Spartan/II planted 22 × 22, were producing 1,000 boxes to the acre from the tenth through the twelfth year. Golden Delicious/Clark Dwarf planted 20 × 20 were averaging 1,296 boxes per acre the eighth year.

In Michigan, both EM VII and K 41 have induced early bearing and high yields when employed as intermediate stempieces between the scion variety and seedling rootstocks. For example, Jonathan yielded at the rate of 968 bushels of fruit per acre the sixth year, planted 6 × 15 feet, or 484 trees to the acre. Golden Delicious under similar conditions yielded at the rate of 1,355 bushels per acre.

Golden Delicious/EM IX in Oregon, planted 8 × 15, attained a production of 1,000 boxes per acre the sixth year, 1,600 boxes the eighth year, 1,700 boxes the tenth year, and 2,400 boxes the twelfth year. A tendency prevailed for strong biennial fruiting under these conditions. Delicious/EM IX, under similar conditions, produced approximately half these amounts.

ANNUAL AVERAGE YIELDS PER ACRE OF AN ORCHARD IN MICHIGAN

(after K. F. Carlson)

Variety/Rootstock	144 Trees per Acre (15 × 20)									72 Trees per Acre (20 × 30)					
	1947	1948	1949	1950	1951	1952	1953	1954	1955*	1956	1957	1958	1959	1960	1961
						Bushels per Acre									
McIntosh/EM I	9	9	274	228	461	432	648	122	0	1,224	598	1,548	792	1,706	1,440
McIntosh/EM II	12	9	101	352	244	522	522	243	0	766	432	956	468	1,332	1,008
McIntosh/EM VII	0	26	215	392	381	718	513	95	0	656	608	883	720	1,145	936
McIntosh/EM XIII	0	0	13	123	381	718	639	104	0	737	405	1,612	604	1,649	1,224
Cortland/EM II	36	35	220	430	244	763	600	94	0	794	648	875	1,296	929	1,152
Cortland/EM VII	6	35	111	271	201	727	414	52	0	513	389	423	792	583	1,008
Cortland/EM XIII	0	1	18	150	223	540	459	112	0	494	300	792	1,080	677	1,008
Golden Delicious/EM XIII	0	0	3	207	569	573	580	112	0	940	680	1,199	788	1,378	1,296
Northern Spy/EM VII	0	0	40	54	248	299	169	167	0	623	389	842	403	1,008	1,008

* Crop destroyed by spring frost.

In another planting of Golden Delicious/IX the yields in bushels per acre for the first four years were as shown:

Planting Distance	1st Year	2nd Year	3rd Year	4th Year
4 × 15	25	75	500	800
6 × 15	20	50	400	600
8 × 15	10	40	250	600

Under the same conditions, Starking/IX yielded one-half to one-third as much.

Production figures in New York State have been recorded as 64 bushels per acre of Delicious/VII at five years of age, compared to no fruit on standard-sized trees of the same variety. By the fifteenth year, the yields were 450 bushels per acre for the semidwarf trees and 431 for the standard-sized trees. For the ten-year period between the fifth and fifteenth year, production on the semidwarf trees was approximately 30 per cent greater.

Most of the high production figures per acre that have been noted are for relatively close plantings of semidwarf or very dwarf fruit trees. Lower yields are recorded for wider spacings. This is in agreement with what has already been said; namely, that the principal gain in high early yields from dwarfed fruit trees is in the large number of trees per area of land. As standard trees become larger, they will outyield smaller trees. The closer spacing that is possible with dwarfed trees makes the difference.

On the EM IV and MM 104 in England yields of approximately 700 bushels have been observed for plantings 15 × 15. Older plantings on EM II have averaged approximately 500 bushels per acre for over thirty years. As is to be expected, depending on the management, the site, and other environmental factors, some orchards have averaged only 300 bushels per acre. At Summerland, B.C., outstanding differences in yields have been noted between orchards, depending upon their ratings as good orchards or inferior orchards. For example, a good, well-cared-for orchard of Delicious/EM VII yielded 392 boxes per acre, whereas a poor orchard of the same variety and rootstock bore only 36 boxes. Similarly, Spartan/EM I bore 761 boxes per acre in the ninth year in a good orchard and only 130 boxes per acre in a fair orchard. In other words, high yields with dwarfed fruit trees are as much or more dependent on factors associated with good fruit growing as they are with standard-sized fruit trees.

Other yield data will be found in the discussion of special methods of training in subsequent chapters.

FRUIT MATURITY, COLOR, SIZE AND STORAGE QUALITY

Fruit Maturity

Certain rootstocks have a pronounced effect upon the time of maturity of the fruit, as well as upon the color and storage quality. Generally speaking, the more dwarfing rootstocks tend to cause the fruit to reach maturity earlier. The EM IX apple rootstock, for example, may cause the fruit to mature a week to ten days early. However, this earlier date is not due entirely to a shortening of the time interval between full bloom and maturity; trees on the EM IX commonly blossom four days to a week early in spring.

Similarly, pears on quince roots mature several days earlier than on pear roots. It is common knowledge that varieties of pear that require a long growing season may be grown successfully in regions of shorter growing seasons when they are worked upon the quince. Sour cherries mature two to three days earlier on mahaleb roots than on mazzard roots. Peaches and plums also mature earlier on rootstocks that tend to dwarf the tree.

Color

The rootstock also affects the color of the fruit. Apples on the EM IX rootstock and pears on quince roots commonly develop higher color than on vigorous rootstocks. Since the development of anthocyanin pigment is associated with sugar accumulation, it would seem reasonable to expect that dwarfing rootstocks would have this effect. However, it must also be remembered that fruit on dwarfed trees is better exposed to sunlight than is fruit on more vigorous trees. Not only may the fruit be somewhat shaded by heavy foliage on vigorous trees, but also the restrictive pruning of dwarfed trees still further exposes the fruit to sunlight.

Finish and Russeting

It has been observed that the rootstock may affect the finish of the fruit and the amount of russeting, although here again the effect may be indirect rather than direct. Thus, Jonathan apples in Michigan have been more russeted on EM IX, I, II, and VII than on the more vigorous EM XIII and XVI. This russeting, however, was associated with the vigor of the tree rather than with any direct effect of the rootstock. Similarly, more russeting was found on trees in low vigor suffering from a deficiency of magnesium.

Fruit Size and Uniformity

As for size of fruit, there is a tendency for fruit to be larger on the more dwarfing rootstocks. This characteristic is especially noticeable on young trees, but is perhaps more apparent than real. That is, the first fruits borne on a tree are usually large for the variety. This is true for large, vigorous trees as well as for dwarfed trees. But in the case of the latter, the first few fruits amount to considerable tonnage to the acre, and the appearance is noticeable.

As the trees become older, this difference in fruit size is not so apparent. In fact, as with any tree, unless dwarfed trees are kept in good vigor and are prevented from overloading, fruit size can be small.

One of the most striking features of fruit on dwarfed trees is the uniformity of size. This is especially noticeable on the more dwarfing rootstocks, such as the EM IX for the apple and the quince for the pear. There may be variation in fruit size from tree to tree, but the fruit on a given tree is remarkably uniform in size. Thus, with Jonathan apple trees on the EM IX rootstock it has been observed that a given tree may carry fruit that ranges from $2\frac{1}{2}$ to $2\frac{3}{4}$ inches in diameter, depending largely upon the number of fruits on the tree. An adjacent tree may, however, range from $2\frac{3}{4}$ to 3 inches. And a third tree may bear fruits ranging from 3 to $3\frac{1}{4}$ inches. This is in contrast to fruit from trees on vigorous seedling rootstocks where fruit may range from 2 inches to 3¼ inches on the same tree. If the fruits are thinned to the appropriate number that a dwarfed tree can properly develop, the uniformity of the crop is striking.

Storage Quality of Fruit

In England it has been observed that Worcester Pearmain apples develop a greater amount of internal browning on the EM IX rootstock than on EM I, II, V, and XII; and Bramley's Seedling develops more on EM I than on EM V. In Switzerland, the storage life of apples was shortened when grown on EM IX.

One wonders, however, how many of these differences may be due directly to the rootstock and how many to other, indirect factors. For instance, it is difficult at best to determine the proper stage of fruit development for harvest. It would seem that in view of the typical earlier maturity of fruit on the EM IX rootstock, the fruit might be somewhat past the proper stage when harvested. At least this possibility should not be overlooked, and full attention should be given not to delay the harvest.

In this connection, the writer has observed a greater degree of "Jonathan spot" on Jonathan apples on the EM IX rootstock. But here the trouble was definitely associated with delayed harvest due to failure to observe that the fruits on this rootstock had matured before the typical date for Jonathan harvest on standard seedling rootstocks.

On the other hand, tests in Michigan have shown no differences in storage qualities of McIntosh apples grown on EM I, II, IV, V, VII, XIII, and XII.

RELATIVE COSTS OF ESTABLISHING DWARFED APPLE ORCHARDS

A comparison of the cost of bringing an orchard of standard-sized fruit trees into bearing in comparison with an orchard of semidwarf trees has been made in southwestern Michigan. The figures were compiled by J. H. Mandigo of Paw Paw, Michigan, from a group of fifteen experienced commercial growers in the area, whose average holdings were between 150 and 200 acres each. Actual figures were secured where possible; others were estimated carefully and used to complete the study.

Computations were based on semidwarf orchards producing sufficient fruit by the sixth growing season to support the operation. For the standard-sized trees, the age at which operations were supported by fruiting was taken as the eighth year.

COST OF ESTABLISHING AN APPLE ORCHARD ON STANDARD ROOTSTOCKS IN SOUTHWESTERN MICHIGAN
(after J. H. Mandigo)

Preplanting year

Materials cost: cover crop seed, $6.50; fertilizer, $17.00—$23.50

Labor: bulldozing, weed and stone removal, $20.00; plowing and fitting, $14.00; fertilizing and seeding, $2.50; other (lime), $9.00—$45.50. $ 69.00

Planting year

Materials cost: trees (54 @ $1.00), $54.00; cover crop fertilizer, $9.00; seed, $6.50; sprays (6 @ $1.66), $10.00; 54 tree guards, $11.88; mouse bait, $3.00—$94.38.

Labor: planting, $20.00; power hoeing (3 @ $2.50), $7.50; hand hoeing (2 @ $3.00), $6.00; placing guards, $2.70; spraying, $10.00; fertilizing cover, $2.50; cultivation, $8.00; other (training and management), $15.00 —$71.70. $ 166.08

Second through 7th years

Materials cost: fertilizer, $7.00/A. × 6 years, $42.00; sprays, $30.00 × 4 (first four years), $120.00; mouse bait (materials and labor), $3.00 × 6, $18.00; cover crop $2.50 × 6 years), $15.00; miscellaneous, $5.00 × 6 years, $30.00—$225.00.

Labor: fertilizing, $3.00 × 6 years, $18.00; tree hoeing (power plus hand), $7.50 × 6 years, $45.00; spraying, $15.00 × 6 years, $90.00; pruning, $7.25 × 6 years, $43.50; other cultural practices, 20¢ per tree, $10.80 × 6 years, $64.80; other ($10.00 × 6 years for management), $60.00— $321.30 $ 546.30

Use of Equipment (based on 50 acres of orchard)
Investment in equipment (tractors, sprayers, and other tillage tools), $7,500., depreciated at 12½ per cent per year—$7,500 × .125 ÷ 50 A. = $18.75 × 8 years $ 150.00

Use of land (valued @ $250.00 per A.) at 6% interest—$15.00 × 8 years—
$120.00; taxes on land, $5.00/A. × 8 years, $40.00; 6% interest on tree
investment of $54.00/A. × 8 years—$25.92 $ 185.92

Grand total $1,117.30

COST OF ESTABLISHING AN APPLE ORCHARD ON SEMIDWARF ROOTSTOCKS IN SOUTHWESTERN MICHIGAN
(after J. H. Mandigo)

Preplanting year

Materials and labor cost the same as for standard rootstocks $ 69.00

Planting year

Materials cost: 90 trees @ $1.50 each, $135.00; fertilizer for cover crop,
$9.00; seed, $6.50; six sprays @ $1.66, $10.00; 90 guards @ 22¢, $19.80;
mouse bait, $3.00—$183.30.

Labor: planting, $30.00; power hoeing (3 @ $3.25/A.), $9.75; hand hoeing
(2 @ $4.50), $9.00; cultivation, $8.00; training and management,
$17.00—$90.75 $ 274.05

Second through 5th year

Materials cost: fertilizer, $7.00 × 4 years, $28.00; spray materials, $35.00
× 4 years, $140.00; mouse bait, $ 3.00 × 4 years, $12.00; cover crops,
$2.50 × 4 years, $10.00; miscellaneous expenses, $7.50 × 4 years, $30.00
—$220.00

Labor: fertilizing, $3.00 × 4 years, $12.00; tree hoeing (machine plus hand),
$10.00 × 4 years, $40.00; spraying, $16.00 × 4 years, $64.00; pruning,
$10.00 × 4 years, $40.00; other cultural practices 20¢ per tree, $72.00;
other, $10.00 per acre × 4 years for management, $40.00—$268.00 $ 488.00

Use of Equipment (based on 50 acres)
Investment in equipment (tractors, sprayers, and other tillage tools),
$7,500.00 depreciated at 12½ per cent per year—$7,500 × .125 ÷ 50
acres = $18.75 per year × 5 years $ 93.75

Use of land (based on evaluation of $250.00 per acre)
Interest at 6% for 5 years, $75.00; taxes on land @ $5.00 per acre, $25.00;
interest on tree investment, $135.00 = $8.10 per year × 5 years—$40.50 $ 140.50

Grand total $1,065.30

The conclusion from the study was that the actual cost of bringing
a semidwarf orchard into production with 90 trees to the acre was

actually less than for bringing an orchard of standard-sized trees of 54 trees per acre into production. With closer spacing of dwarfed trees and wider spacing of standard trees, the differences would be even greater and still further in favor of the dwarfed planting.

When the earlier cropping of the semidwarf plantings is added to these figures (two years), the advantages of the smaller trees are seen to be substantial.

Comparable figures from cost studies in England, but including land, buildings, and all operating expenses, have been made for close plantings of 257 trees per acre in comparison to wider spacing of 68 trees per acre. At the end of nine years it was determined that total costs were $1,716 per acre for the close planting and $1,058 for the wider planting. Income from fruit was $1,520 per acre for the close planting and $324 for the wider planting. The deficit to this point was, therefore, $196 per acre for the close planting and $734 for the wide planting. The higher rate of expenditure was accordingly instrumental in increasing the speed of repaying the capital investment. It was concluded that close planting serves a means of getting started more quickly in business.

Costs for closely planted trees on trellises on the EM IX rootstock in the state of Washington have been given as $1,125.00 for the trees, and $224.10 for the trellis, or a total of $1,349.10 per acre, not including labor. To offset this high cost, a row of 150 Jonathan trees planted 4 by 12 feet yielded 1,089 pounds, or at the rate of 131 boxes per acre the second year. Lodi produced 200 pounds, or at the rate of 24 boxes per acre. Delicious produced 509 pounds, or at the rate of 60 boxes per acre. Golden Delicious produced 3,000 pounds, or at the rate of 360 boxes per acre, also the second year.

Part Seven

DWARFED FRUIT TREES
AS SPECIAL INTERESTS:

IN POTS, UNDER GLASS, AS BONSAI,

AND AS ORNAMENTALS

26

TREES TRAINED TO ARTISTIC AND NOVEL FORMS

ONE of the common sights in Europe is the dwarfed fruit tree trained to special forms and employed variously as borders along walks, as wall coverings, and simply as interesting individual specimens (Fig. 129). The intricacies of pattern and design seem limited only by the imagination and ability of the workman. Trees may be seen growing in flat form covering the entire side of a two-story house, leaving only the windows and the doors uncovered. They may be found shaped like a tabletop, a Grecian urn, a goblet, or even as a basket-like structure 4 or 5 feet in diameter and supported by a trunk 8 feet tall.

Again, for varieties of fruit which require a long growing season for proper maturity, trees may be seen trained on walls that provide protection, warm temperatures, and a generally favorable situation. And extensive commercial acreages may be found of fruit trees growing on simple low trellises or trained symmetrically on substantial tall supports.

In America, interest in fruit trees has been concerned with commercial fruit production. Although settlers in colonial America succeeded to a limited extent in patterning trained fruit trees after the style of their homelands, yet America has been so busy with other things that it has had little time for the attention that such trees demand. But there is no limit to the satisfaction that comes from working with small fruit trees trained to special forms. Gardening with fruit trees is the aristocratic form of gardening. The enthusiasm and fascination to be derived from daily contact with individual trees in bud, flower, shoot, and fruit are difficult to describe.

It is perhaps doubtful that success may be expected as easily and to the same degree in America as in Europe. The climate of America is rigorous, with extreme of summer heat and winter cold unlike the more equable climate of Europe. Nevertheless, there are enough remnants of earlier attempts to justify the feeling that some degree of success and satisfaction may be derived from training trees to some of the forms described in this chapter. In any event it is of interest to see what has been and can be done with dwarfed trees trained to special forms.

CLASSIFICATION OF TREES AS TO FORM AND SHAPE

Fruit trees, as regards general form and shape, may be grouped into (a) natural shapes, and (b) artificial shapes. The natural shape is assumed by a tree that is left to grow as it will, untouched by pruning and training techniques. This is the form one sees along roadsides or in open fields. An apple tree is easily identifiable by its characteristic shape, as is a pear tree, a cherry tree, or a peach tree. In fact, cultivated varieties within these classes naturally assume a shape that is also characteristic and identifiable, as a McIntosh apple tree, a Seckel pear tree, or Windsor sweet cherry tree.

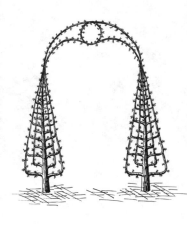

In contrast to the natural form, man has variously pruned and shaped fruit trees to artificial forms adapted to his particular needs and desires (Fig. 130). The artificial forms, in turn, have been divided by French horticulturists (Du Breuil and Rivière) into (a) nonpalisade or non-supported forms (*formes non palissées*), and (b) palisade or supported form (*formes palissées*).

Fig. 129. (Above, left) Verrier palmettes used as a wall covering, consisting of a high-stemmed double palmette and two six-armed palmettes. (After Lucas)

Fig. 130. (Below, left) Archway formed from two winged pyramids with intertwined terminal arms. (After Lucas)

The nonpalisade or nonsupported forms are those that can be trained and grown successfully without the need for firm, permanent support, as from strong lattices, stakes, trellis fences, palings, and palisades. They include the vase or bush trees, pyramids, and columns. They are essentially three-dimensional.

The palisade or supported forms are those that cannot be trained and grown without being firmly braced and supported, as by fastening to walls, or to strong lattices, stakes, trellis fences, palings, and palisades, which may or may not be themselves attached to walls. They must be more than merely temporarily staked or braced; they must require support throughout their existence. They are not only *formes palisées*; they are *formes soutenues* (supported forms).

In order to be properly supported against walls and palisades, the forms are for the most part similarly flat or two-dimensional. This is in contrast to the three-dimensional bush, pyramid, and columnar non-palisade forms. Both the palmette or palmlike forms and the cordons or single-stem forms are included as palisade forms. In addition, some miscellaneous forms are often included that are so intricate in design that they must be permanently braced and staked even though they are not two-dimensional forms.

American horticulturists (Waugh) have divided artificial forms into three classes rather than two, based on the form itself rather than the requirement of support. Thus, they speak of: (*a*) forms of three dimensions, as vase, bush, pyramid, and column; (*b*) forms of two dimensions, as Verrier palmette, fan forms, and various espaliers; (*c*) forms trained to a single stem (one dimension), as cordons.

In England (Whitehead) it is not uncommon to find a classification based on the principal forms used there: (*a*) pyramids and bushes; (*b*) espaliers, forms trained against a wall or supports, including both the true espalier and the contre-espalier; (*c*) cordons, single straight stems, but also including single and double U-shaped, and triple cordons.

There is, indeed, merit in simplified classifications of this kind for different parts of the world based upon the forms most commonly found successful there. Yet it would seem best in the interests of clarity and uniformity to accept a well-approved and complete international classification. The system employed in French horticulture is of this type and is used in this book.

NONSUPPORTED OR NONPALISADE FORMS
(*Formes non palissées*)

Nonsupported or nonpalisade forms (*formes non palissées*) are free-standing forms that do not require permanent support. They are for the most part three-dimensional, represented by the vase-shape or bush, the pyramid, the column, and their variations.

The Bush, or Vase, is the common low-headed, open-center-type tree. The Germans call it *Buschbaum* or *Kugelbaum*. It is essentially a low-headed, dwarf tree grown in the open-center or V-shape style. Good examples are seen with peach trees, which assume this open-center shape almost naturally, or with very little help. Plums, cherries, apricots, pears, and apples are also grown in this form. Very vigorous, upright-growing varieties are more difficult to handle than are weaker, spreading varieties.

Pyramidal forms are numerous in their variation. Essentially, as the name implies, they are forms in which the greatest tree width is at the base and which taper regularly upward toward the top. They may be low-headed dwarf forms, strong-growing tall forms, narrowly columnar, broadly cone shape, and forms with scaffold branches artfully arranged in tiers or whorls, in ascending spirals. Voltaire (1694–1778) is said to have been among the first to favor the pyramidal form and to have grown them in his garden.

The Dwarf Pyramid is a very successful form, both for apples and pears, but especially for the latter. It has reached considerable acceptance in England both for garden planting and for orchard planting. It is a short, low-headed form with the largest branches at the base, and with progressively shorter branches in succession upward toward the top. This gives a pyramidal effect, as the name implies. Planting may be done as close as 3 feet by 6 feet in the garden, although in commercial plantings the distances are commonly somewhat greater in order to accommodate orchard equipment.

When properly handled, such trees do not require staking. Some growers employ a single wire the length of the row at 30 inches from the ground to steady the trees. Because of the value of this form, it is discussed more in detail in the chapter dealing with commercial fruit growing.

The Spindelbusch, or spindle bush, as grown in the Rhineland of Germany, is similar to the English dwarf pyramid, but the trees are larger and planted farther apart. Less summer pruning is done than with the dwarf pyramid. Instead, branches are trained horizontally by tying them down to the main stem, or by bending and fastening them with metal clips. A full discussion of this form in commercial orcharding is given in Chapter 22.

The Winged Pyramid (*pyramide ailée, Flügel pyramide, and Arm-leucht pyramide*) is a strong-growing pyramid form with sturdy, un-branched scaffolds beginning about one foot from the ground and spaced upward at about 1-foot intervals (Fig. 131). It must be made clear that these branches are not just scaffold branches in the ordinary sense of the word. They are scaffolds that are trained stiff like cordons, with no branching. They are trained at an angle of 30 to 45 degrees from the horizontal.

Further, the scaffold branches may be developed in tiers, usually five tiers, but sometimes three tiers for weak-growing varieties. A special non-tier form can also be grown, and the scaffolds may be arranged alternately in the tier arrangement.

When the scaffolds are arranged one above the other in five tiers, there are five open spaces, alleys, or wedges between the tiers of fruiting branches. This permits the entry of light and air and also favors pruning and harvesting. Sometimes the end of each scaffold is bent upward and inarched into the scaffold immediately above. The result is a sort of five-ribbed tree.

The width of the tree is proportioned to the height. Thus, trees 12 to 14 feet tall are usually grown with lower scaffold branches about 4 feet in length. For trees 14 to 16 feet tall, the lower scaffolds are about 5 feet in length.

The form is confined almost exclusively to the pear, and is best for strong-growing varieties on good soil. Trees on quince roots are planted 10 to 12 feet apart, and trees on pear roots are set 12 to 15 feet apart.

The Cone is also a variation of the pyramid. In fact, sometimes the

terms "cone" and "pyramid" are used interchangeably. Yet there is a difference. In the true cone, the tree is top-shape, being about as wide as it is tall. Branches are trained upward and outward at about 35 degrees from the horizontal for the lower branches. The pear is best suited to this form, followed by the apple. Plums, cherries, and peaches are not suited. Best results are secured with vigorous varieties in good soil.

The Winged Cone (*cône ailée*) is a cone-shaped tree with sturdy, unbranched scaffolds, grown in the wing form described under the winged pyramid.

The *Column* or *Spindle* (*fuseau* or *Saulenpyramide*) is a relatively tall, narrow form, as the name implies. The main stem may reach to a height of about 10 feet. The diameter of the tree at the base is about 2 feet, while toward the top it is about 1 foot.

This form is successful both for pears and for apples, using Quince A for the pear and EM II for the apple. Trees are set 6 feet apart for vigorous varieties and 5 feet apart for weaker ones.

M. Lorette is credited with having encouraged this form in his planting at Wagnonville, France. It is also found in Lorraine, Flanders, Normandy, and Belgium. Trees grown in the so-called "pillar system" for apples in England are not unlike this shape.

The *Guenouille* is a form often confused with the pyramid. It differs from the pyramid in that the widest part of the tree is the middle section. The tree tapers from the middle downward to the base and also upward to the tip. It is rarely seen, and seems not generally known outside France.

SUPPORTED OR PALISADE FORMS
(*Formes palissées*)

The palisade or supported forms require permanent support during the time of their existence. These supports, as has already been noted, are frequently walls, or they are substantial lattices, special palisades, or fencelike palings or stakes. Rows of trees grown in the palisade form,

Fig. 131. (Opposite) Winged pyramids, horizontal and vertical cordons, and Verrier palmettes in a European garden.

usually in gardens, with scaffold branches attached directly to a wall or indirectly to a lattice or trellis that is itself attached to the wall, are properly called an espalier or *arbres en espalier.*

In historical review, it should be noted that the wall served both as a protection and as a support. The warmth of the sun was increased so as to favor fruit maturity; and reflected sunlight from the wall led to improved fruit color. Besides, a protective screen could be dropped from the wall over the trees and fruit to protect against adverse weather.

Subsequently, lattice frameworks or trellises were erected parallel to the walls but at some distance from them. Rows of palisade trees that were fastened to such supports away from the wall were called contre-espaliers or *arbres en contre-espaliers,* meaning "away from the wall." Contre-espaliers were often used along a walkway paralleling a true espalier on a wall. While not given the protection from cold winds and spring frost nor the warmth and reflected light of the sun that a wall provided, yet the contre-espalier was less likely to be shaded, and aeration was better. It was therefore deemed superior in some situations, and was so used.

Gradually, the term "contre-espalier" is being lost, and is being replaced in common usage by the single word "espalier" to designate both espaliers and contre-espaliers. Further, the meaning of the word has changed to refer not so much to the method of training and arranging a palisade of trees in rows (*en espalier*). Rather, it is being used to designate two-dimensional forms of trees in general. Thus a specimen of Verrier palmette is often loosely called an espalier. It is true that it is an espalier form and that when trained against a wall it is properly an espalier; but it would be better to call it precisely a Verrier palmette, a four-arm Verrier palmette, or whatever the form may be. It is a pity that the terminology concerning palisade trees is losing its preciseness and specificity. The reader is encouraged to use the terms in their proper senses.

Origin of the Word "Espalier"

According to Gibault the term "espalier" is seemingly derived from two possible sources. Some linguists feel that it has come from the Italian *spalliera,* which refers to a support for the shoulder or back, as the back of a chair. This in turn is derived from *spalla*, shoulder, in the sense of supporting or propping. The same meaning is expressed in military science in the word *épaulement*, which signifies a rampart of earth designed to contain the enemy or provide support against the enemy.

Other linguists, however, suggest that a contribution to the word "espalier" has come from old French *pau, aspau*, which means stake or prop. In other words, an espalier becomes a living hedgerow of fruit trees or other materials supported by stakes, or by *espaux*. Both etymologies are equally acceptable. Certainly, to shoulder or support is related in function to a prop or a stake.

Early Mention of Espaliers

The first use of the term "espalier" appears in the French language in the sixteenth century in the intermediate form *espaulière*. The French naturalist Pierre Bélon, of Mons, in 1558 mentioned hedgerows on both sides of garden walkways. The Italians, he said, called these hedgerows *Spalieres*, with "we say supports or espaliers." At about the same time, Charles Estienne, in speaking about the cultivation of orange trees, says ". . . make for them an espalier (*espaulière*) and side edgings of laurels, planted in double row. If there are no laurel, the same can be done with cypress."

Thus the word *espaulière* was restricted to mean a palisade, a living hedge of leafy trees, twisted and shaped in the form of a wall and designed for the decoration of gardens or to serve as a shelter for delicate plants.

While these are the first references in literature to the term "espalier," it is certainly true that the practice of growing trees in this form was known considerably before the sixteenth century. In fact, Georges Gibault calls attention to the portrait of a woman hanging in the Louvre, belonging to the French-Flemish school of painting dating from the fifteenth century. The picture is named *Une donatrice*, and shows her in the middle of the garden, which is bordered by a notched wall against which grapevines are growing. The vines are supported by a system of palisading that consists of four bands of white cloth, regularly spaced, extending horizontally and fastened with nails, not to the wall but to slight green-painted stakes.

M. Gibault points out that the Flemish painters of that time reproduced objects accurately and in great detail. Thus in the painting mentioned, even the heads of the nails are faithfully reproduced so that "a photograph would not be more exact."

From this observation that grape espaliers were well established with formal rules by the fifteenth century in the gardens of northern France, he concludes that this practice has been known and occasionally employed since the Middle Ages in northern France and perhaps also in Belgium, Holland, and Germany. And although the beginning of espaliers is usually associated with the reign of Louis XII or Louis XIV, it seems likely that the use of espaliers at that time was merely an elaboration of the methods perfected by preceding generations of horticulturists.

As for information on the actual establishment and growing of plants beside garden walls, one of the first references occurs in *Horti Germaniae* (German Gardens) by the Swiss naturalist Conrad Gesner, which appeared in 1561. He made the observation that the heat of the sun is increased by reflection of its rays against a wall. He then recommended that certain plants be grown beside a wall, such as the fig, the currant, and jasmine. However, this is not a true espalier in the literal sense, since the plants were not supported on the wall.

A little later, Olivier de Serres (1539–1619) described a sheltered hedge of "freui" trees supported by a series of stakes and horizontal strips. He had already recognized that trees that are too thick produce little fruit. From this observation, he advised growing fruit trees in narrow hedges. But here again, he was not speaking strictly of espaliers, but rather of contre-espaliers, since the espalier is supported on a wall, whereas the hedgerows that he described were supported on a trellis or frame independent of a wall.

Rapid Rise of Interest in Espaliers

Claude Mollet, head gardener of the Gardens of Fontainebleau for Henry III and Henry IV, was the first writer to discuss espaliers as such. He is often credited with being the individual who introduced the espalier-trained fruit tree. Writing at the beginning of the seventeenth century, although his book (*Théâtre des plans et jardinages*) was not published until 1652, he described espaliers in a manner that indicates that they were not novelties at that time. He advised the supporting of apricots and peaches on a frame against a wall facing the southern sun.

In 1638 there appeared *Traité du jardinage* by Jacques Boyceau de la Barauderie, caretaker of the gardens of Louis XIII, which contained a complete chapter on espaliers. He was the successor to Claude Mollet. He described the spreading of branches in the shape of a fan or an open hand on a trellised wall. He discussed also espaliers of peach, apricot, "gid," almond, plum, and pear. For protection from frost at blossom time he covered them with a cloth shelter.

By the middle of the seventeenth century the espalier had become "the principal ornamentation of gardens." Numerous churchmen and magistrates became great enthusiasts of this form of fruit culture. About this time, a number of books appeared that treated the subject fully. Among these was *Jardinier français* by Nicolas de Bonnefons, which appeared in 1651. A year later (1652) Le Gendre, Curate of Henouville, published the first edition of *La Maniere de cultiver les arbres fruitiers*. He called attention to the fact that trees cultivated in espalier were so close together that they looked like rows of spines. From descriptions it would seem that they were cultivated like ornamental hedges. In 1653 there appeared *Instruction pour les jardins fruitiers* by Triguel, prior of Saint Mark's. Another churchman was the author of *Jardinier royal* which was published in 1661.

It has been said the publication of these four works marks an important period in horticulture, with "the establishment of the first espaliers worthy of the name and the beginning of pruning, which is the direct consequence of cultivation of espaliers."

Two other important names in French horticulture who published books discussing trained fruit trees are La Quintinye (1690) and Duhamel du Monceau (1768). To these, Auguste Rivière adds a long series of famous French horticultural names: Roger Schabol (1767), Le Berriais (1768), Thouin (1780), Butré (1793), Bosc (1806), Du Petit Thouars (1809), Noisette (1812), and Mozart (1815). To this list must be added the more recent names of J.-B. Lhomme, L. Verrier, Jamin, Du Breuil, Cossonnet, Alexis Lepère, Rose Charmeux, Ch. Baltet, Chevalier aîné, Hardy, Forney, Marc and Gabriel Luizet, Salomon, Barret, Cuissard, Guindon, L. Lhérault, Pinguet, P. Passy, O. Opoix,

A. Nomblot, Chasset, and others. These names are repeated here as a record of those who made substantial contributions to horticultural art. They should not be forgotten even though the interest they represented has waned.

CORDONS
(*Cordons, Quirlandbäume*)

A *Cordon* (*Cordon, Quirlandbäume*) in its simplest form is a stem or "string," which is the literal translation of the word *cordon* (Fig. 132). It may be thought of as a single-dimensional tree, possessing neither breadth nor thickness, but only height, and with fruiting spurs close to and along the single stem. Simple cordons may be vertical, oblique, or horizontal, depending upon the angle to which the stem is trained. Sometimes the stems are doubled into a narrow U form at about a foot from the ground. This is called the double cordon, and may similarly be trained in vertical, horizontal, or oblique position. The horizontal cordon may be developed with two arms in opposite directions, both in the horizontal position; and the vertical cordon may be trained in serpentine or reverse S form (*cordon ondulé*).

The cordon system of cultivation was highly developed in France by Du Breuil during the middle of the nineteenth century. Sir Henry Scudmore Stanhope is said to have planted cordons between 1861 and 1865 in Herefordshire, England.

Pears and apples are well suited to training in cordon forms, using Quince A or C as a rootstock for the pear and EM IX for the apple. They may be set 2 to 3 feet apart in rows 6 feet apart. EM VII and EM II are sometimes used; if so, the distance between trees should be increased to 3 or 4 feet. While these distances may seem relatively close, it must be remembered that close planting is one means of restricting vegetative growth and keeping the trees in hand.

Tip-bearing varieties of apple, like the Worcester Pearmain, Cortland, and Gallia, can be grown successfully as cordons, but the spur type, such as Golden Delicious, Coxe's Orange Pippin, and Starkrimson are better. Among pears, the compact varieties, such as Conference, are to be preferred to upright and very vigorous sorts, such as Beurré Superfin. Varieties that tend to develop bare wood as cordons may be put on EM

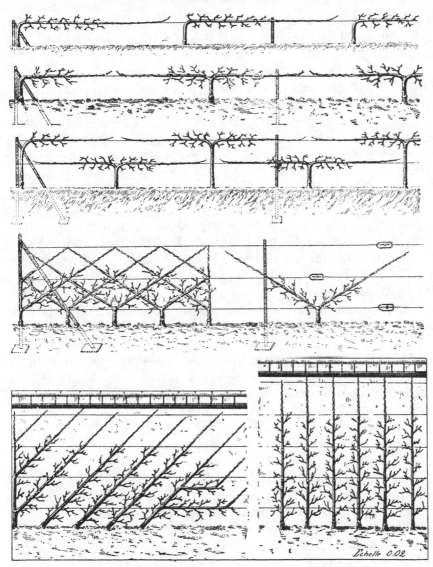

Fig. 132. (Upper) Horizontal cordons with one arm; horizontal cordon with two arms; horizontal cordons arranged in two tiers. (Center) Palmettes with two oblique arms, arranged in systems variously termed Croisillon (crossarm), Losange (diamond), and Belgian fence. (Lower left) Oblique cordons. (Lower right) Vertical cordons.

II and VII, which tend to encourage the development of vegetative buds. These will develop into lateral shoots that can be pruned to form fruiting laterals.

Peaches are also used successfully in cordon form, especially against a house or wall where they are protected and where they serve as pleasing ornamentals. Cherries and plums are not adapted to cordon form.

The cordon is the simplest and most satisfactory form for the amateur gardener. It is especially useful where a collection of different varieties is desired. Placed along a garden walk, or on both sides of the walk, at intervals of 1 or 2 feet, they make a most attractive, as well as most useful, espalier.

Since some of the cordon forms, especially the oblique cordon, are used successfully in commercial fruit production, their general culture and performance are discussed more in detail in the chapter dealing with commercial fruit production. Planting distances, methods of pruning, and other specific procedures are discussed under major headings in other parts of the book.

The principal forms of cordons may be enumerated as follows, together with a few specific comments:

The Vertical Cordon is especially suited to the pear on Quince A and C. It is also useful for the apple and sometimes the peach, but is rarely successful for plums and cherries. It is grown to a height of 10 to 12 feet, with the first fruiting wood developed at about 12 inches from the ground for the pear and the apple and about 15 inches for the peach. Spurs are always parallel to the wall or support, and not back toward the wall or out from it. This form is very useful as an ornamental espalier against a house (Fig. 133).

The Vertical Double Cordon or U Form is developed by training two arms vertically in U shape about 15 inches apart. This is really a palmette with two vertical branches, and properly belongs with that grouping.

The Oblique Cordon is one of the best forms for both the amateur and the commercial grower. It is frequently trained as a contre-espalier

Fig. 133. (Opposite) (Upper left) Vertical cordon (grapes) arranged in two tiers. (Upper right) Vertical cordons (gooseberries). (Lower left) Horizontal cordons (grape) arranged in five tiers, known as Thomery form. (Lower right) Vertical cordon (grape).

on a wire trellis. The main stem is grown erect to a height of 12 to 15 inches before being bent at that point to the oblique position of 45 degrees. Properly speaking, merely planting a tree in a slant or oblique position does not make it an oblique cordon, but for all practical purposes the results are the same. If trees are planted slantwise in this way,

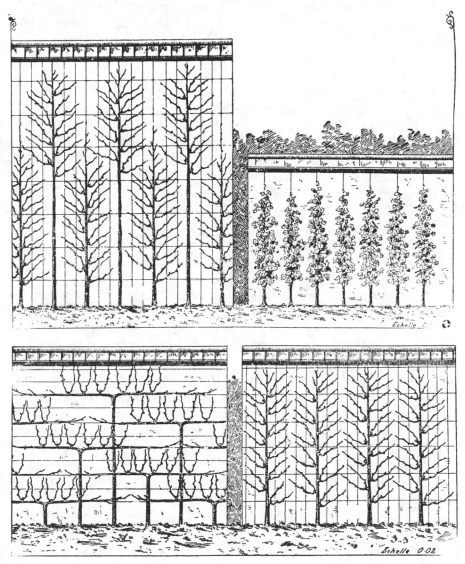

care must be taken to place it so that the budded scion is on the top side. The terminal should not be cut. Instead, if the cordon grows beyond the height of the trellis, which is usually 6 to 8 feet, the cordon is lowered to a greater angle with the vertical. Or the tip may be trained horizontally along the top wire when it reaches that height.

The Oblique Double Cordon is variously treated. In some forms, it is developed in a U form similar to the vertical double cordon but trained at the 45-degree angle. In another form, the vertical portion of the stem is permitted to grow erect beyond the first obliquely trained arm, and a second arm is developed parallel to the first, and above it. These forms are difficult to grow because the upper cordon grows more rapidly than the lower. It is not suited to amateur garden needs.

In still another form, the second arm is developed at a 45-degree angle in the opposite direction from the first so as to develop a V form, which is more properly called a palmette. A series of these V-form cordons placed close together so that the branches overlap and cross each other is called a Belgian fence. This form is the basis of the Delbard system of fruit production, which is described in the section dealing with trellised trees for commercial fruit production.

The Horizontal Cordon with One Arm, trained at 15 to 18 inches from the ground, is useful when grown on a wire along a garden walk. The plants are set at intervals of 4, 6, or 8 feet, depending upon the vigor of the variety, just so that the tip of one reaches the next cordon so as to form a continuous chain. The apple succeeds better in this form than does the pear, but both are very successful. The latter is said to be often unfruitful. Apples on the EM VII and II are reported as more fruitful than on the EM IX.

The Horizontal Cordon with Two Arms consists of two arms trained in opposite directions, usually along a wire, as for the one-arm form.

The Horizontal Double Cordon is similar to the vertical double cordon, or U form, except that the two arms are grown horizontally and parallel. It may be developed either in the one-arm or the two-arm form, mentioned above.

The Horizontal Cordon with Two or More Tiers of Branches is, as the name implies, a cordon in which the central axis is continued vertically and from which two or more tiers of branches are developed one above the other. An example is the Cordon Ferraguti of the French, which is really a palmette form and belongs in that class. This form is not well recommended because of the fact that the lower branches are often weak.

The Vertical Serpentine Cordon (*ondulé*) is a vertical cordon developed into a series of reverse S's to give a serpentine or undulating effect. The method is used with somewhat vigorous varieties where it is desired that the height of the cordon be not over 6 or 7 feet, the idea being that a greater length of stem is compressed into this height. Further, the bending hastens fruiting and checks vegetative growth. Planted about 2 feet apart, they make a most interesting espalier.

The Spiral Cordon is merely a cordon trained spirally around a column. Two or three cordons may be planted at the base of a column, and grown in parallel spirals or in crisscross fashion.

The Arched Cordon (*Arcure*) is formed from a series of one-year-old trees without side shoots, which are planted about 3 feet apart and slanted in the direction of the row. When the young tree, which is grown as a cordon, has reached a height of about 45 inches, it is arched over and fastened near the base of the next adjacent tree. The top of the resulting arch is about 20 inches from the ground level. A single vertical shoot is now developed from each arched tree near the top of the arch or slightly beyond. When this shoot has reached a height of about 45 inches, it is arched over similarly, but in the reverse direction, and fastened to its neighbor. This procedure is continued so as to form a series of arches one above the other, developed to the height that seems appropriate to the vigor of the trees (Fig. 134).

This is the Lapage system, which has reached commercial proportions in the vicinity of Angers and which is discussed further in the section dealing with commercial fruit culture.

PALMETTES (PALMETTES)

The term "palmette" refers to the shape of a palm leaf, an open fan, or an open hand with spread fingers (Figs. 135, 136, and 137). Originally the term was restricted to trees grown only in this form, as espaliers against walls, or as contre-espaliers supported on trellises or lattices. However, this terminology has now been broadened to include all flat or one-dimensional forms except the single-stem forms, or cordons.

The palmette is the elegant form for fruit trees. But the palmette has developed not alone because of its pleasing appearance. There are very practical reasons why it has succeeded and continued throughout several centuries. By virtue of the bending of the branches, the growth is readily controlled and balanced. And the many forms that have been devised are associated not alone with the whims of the originators; they have developed because they were well adapted to the length of the growing season, the general climate and soil of a region, the vigor of a given variety, and other factors. That some of these forms are still grown commercially in large acreage attests to their value. It is obvious that they are also very effective in gardens and as ornamentals.

The palmette is formed by developing lateral scaffold branches from the erect main stem or trunk, beginning at about 12 inches from the surface of the ground. These laterals are in pairs and are equally balanced, in opposite directions. Other scaffolds are in turn variously developed but are kept about 12 inches apart. These figures apply principally to apples and pears. For peaches, the height of first branching is a little higher, namely, about 15 inches, and the scaffolds are kept at about 15 inches apart.

Palmettes may be divided into four main groups, (1) horizontal palmettes, (2) oblique palmettes, (3) candelabra palmettes, including the Verrier palmettes, and (4) fan palmettes.

Fig. 134. (Upper left) Simple palmette (peach) with two tiers of branches. (Upper right) Simple palmette (peach) with one tier of branches. (Lower, —left to right) Double-U form: Palmette Verrier with seven branches; Palmette Verrier with six branches; Palmette Verrier with five branches; Palmette Verrier with four branches; Palmette Verrier with three branches; Double-U form; U form (pear); U-form (peach).

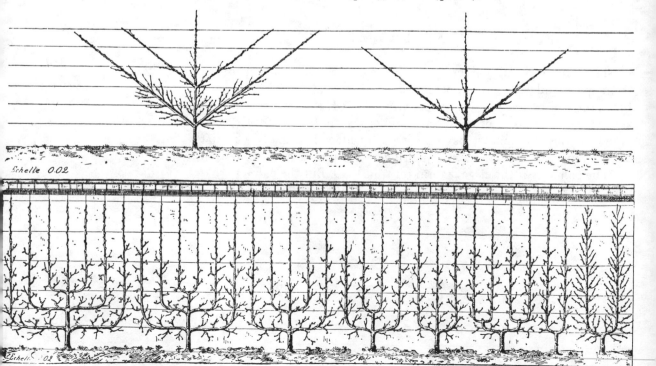

Horizontal Palmettes

The Palmette with Horizontal Branches consists of a single central axis from which tiers of equally spaced horizontal scaffold branches are developed in pairs (Fig. 135). There may be as many tiers as desired, depending upon the fertility of the soil and the vigor of the variety employed. A common number of tiers is six or eight. The scaffolds may be equal in length, or they may decrease in length toward the top, so as to give a pyramidal effect. When this is done, a horizontal palmette may be alternated with an oblique palmette in the Cossonnet system, as discussed in subsequent paragraphs.

The horizontal palmette is used mostly for low walls. It is not one of the most desirable palmettes. The lower scaffold branches frequently become weak as the tree becomes older, so that the value is lost at just the time it is most desired.

A modification of the horizontal palmette was exhibited in 1780, attributed originally to Legendre. It is essentially a palmette with six horizontal arms arising in both directions from the central axis. On each of these horizontal arms, three small arched shoots are developed (Fig. 136).

Oblique Palmettes

The Palmette with Oblique Branches is similar to the palmette with horizontal branches except that the branches are trained upward at vari-

Fig. 135. (Upper, left to right) Palmette with two tiers of branches; Palmette with three tiers of branches; Palmette with one tier of branches. (Lower, left to right) Contre-espalier formed of oblique and horizontal palmettes (Cossonnet).

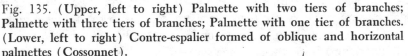

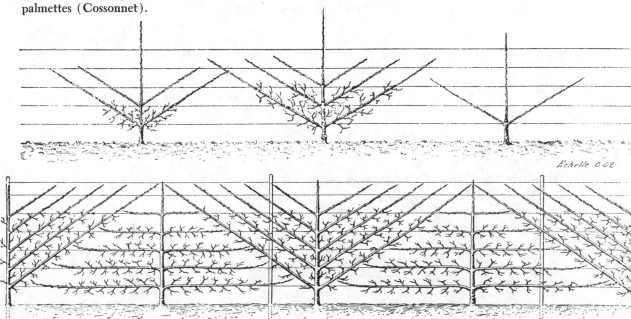

ous desired angles (Fig. 137). This form does not make either a good wall cover or a good contre-espalier, since an awkward bare space is left below the lowest scaffolds. Because of this fault, a system of training was devised by M. A. Cossonnet of Longpont, France, about 1849, consisting of alternate horizontal palmettes and oblique palmettes. Known as the Cossonnet system, the horizontal palmettes were grown in an inverted-V form, with the length of scaffolds decreasing from the lowest to the highest. Into this V, the oblique palmettes fit nicely when planted alternately with the horizontal palmettes, to provide a solid wall covering.

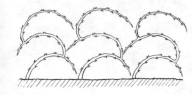

The Palmette with Two Oblique Branches is, of course, a form of the oblique palmette often called a simple fan palmette. It is sufficiently different from the large oblique palmettes mentioned above, to warrant special discussion. It consists of palmettes formed with two, and only two, oblique scaffolds originating at about 12 inches from the ground level, arranged variously at 35-, 40-, or 45-degree angles from the horizontal. The trees are planted at distances of about 3 feet apart so that the branches of adjacent trees overlap in crossarm fashion. The lower angle of branching is combined with wider spacing so as to accommodate to the vigor of the variety. The lower the angle, the more the vigor of the tree is checked.

Because of this arrangement and the general appearance, this system is also called the Croisillon (crossarm) system, as well as the Losange (diamond) system. This is also the popular Belgian fence.

The system is seldom used other than for contre-espaliers, mostly with apples and pears. Trees are maintained at a maximum height of about five feet, and are most attractive along a walkway. Trees bear early and heavily but are said to diminish in productivity as they become older. For many years this form was very popular in the Lyons region of France. It is again being revived under the leadership of Georges Delbard, Paris, and is in consequence sometimes called the Delbard system.

Fig. 136. (Above, left) Arched (*Arcure*) form. The Lepage method of growing fruit trees (*L'arcure Lepage*) described in Chapter 21 is a development of this form.

Fig. 137. (Below) Palmette with arched branches. (After M. A. Du Breuil, *Culture des arbes et arbisseaux*, Paris, 1876)

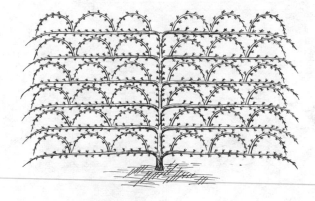

Because of its importance it is described more in detail in the section dealing with commercial fruit production.

Candelabra Palmettes

The candelabra palmette differs from the oblique and horizontal palmettes in the direction of the scaffold branches. In the candelabra form the scaffolds are first trained horizontally and are then trained vertically for most of their length.

The Palmette with Two Vertical Branches is the simplest of the vertical palmettes. It is sometimes called a "double vertical cordon" or a "U form." It properly belongs, however, to the palmette group, and not to the cordons, which are single-stem forms. In developing this form, two horizontal scaffold branches are developed to the right and to the left from the main stem or trunk at about 12 inches from the surface of the ground. The main stem is not continued beyond this point. The two horizontal branches are then turned vertically at about 6 inches from the main stem, so that the two vertical branches are about 12 inches apart. This is a good distance for scaffold branches for the pear and the apple. For the peach, the distances are nearly doubled, that is, 10 to 12 inches from the main stem for each of the vertical scaffolds and 20 to 24 inches apart from each other.

The palmette with two vertical branches is not difficult to develop. It is preferred by many growers to the single vertical cordon. Fruiting is early, growth is well contained, and a row makes an excellent ornamental planting.

The Double Palmette is formed from the palmette with two vertical branches, just described. Outward from each of the two main scaffolds, various modifications may be developed of horizontal, oblique, and serpentine scaffolds. The form is seldom seen.

The Verrier Palmettes

These forms (Fig. 135), because of their importance, may be elaborated upon at this point before entering into an enumeration of the several shapes. They are considered the ultimate in grace, elegance, simplicity, and manageability. They are composed of vertical stem or axis from which scaffold branches arise in even numbers, regularly spaced, and in the same vertical plane. The continuity of the central axis is the feature that clearly identifies them.

The scaffold branches or arms are first trained horizontally or in a slightly oblique direction. They are subsequently turned vertically near their tips when the desired length has been attained. The arms that constitute the lowest level are longer or wider than those in each succeeding tier. The greatest length of the scaffolds is in the vertical direction, which may vary from 4 to 5 feet to 10 or 12 feet.

Because the width of each palmette is thus clearly defined, planting distances can be accurately gauged. The number of tiers of arms can be adjusted to the vigor of the variety, the soil, and the general growing conditions. With horizontal or oblique palmettes and similar forms, the length to which the branches may grow cannot be foretold with certainty, and the plantings become irregular.

The form was named for Louis Verrier, chief gardener of the Regional School of Agriculture at Saulieu, France, about 1849. He was not the originator of the design, which had been known for some time, but he was its most ardent advocate. His pupils, several of whom became influential horticulturists, attached his name to this form. In fact, they attached his name to all candelabra shapes.

The Palmette with Three Vertical Branches is also appropriately called a "trident palmette" and a "candelabrum with three branches." Really, it is the simplest of the "Verrier palmettes." It differs from the palmette with two vertical arms by virtue of the fact that the central stem is permitted to continue, thus giving three vertical branches. The two outside vertical scaffolds are, of course, spaced farther from each other so that all three scaffolds will be about 12 inches apart for the pear and apple, and about 15 inches for the peach. This form has the advantage of small size, but it is somewhat difficult to balance because of the strength that is forced into the continuous central axis.

The Palmette with Four Vertical Branches takes three principal forms, as follows:

a. True Candelabra Palmette, Gril de Saint-Laurent (Gridiron of St. Lawrence). This form recalls a four-tined fork (Fig. 138). From the usual short main stem or trunk, two horizontal arms arise; and from these, four vertical arms are formed, equally spaced (Fig. 138). The central axis is not continued. The number of vertical branches need not be confined to four. It may be increased to six, eight, or any number, depending upon the vigor of the tree. The form is difficult to maintain for the reason that the two innermost vertical scaffolds tend to outgrow the others. Large forms with up to ten vertical branches have been used successfully for the peach.

b. Double-U Candelabra Palmette. This form is composed of two simple and equal U's developed side by side in the same plane and at the same level, from a connecting horizontal axis. Two horizontal arms are developed in opposite directions from the main stem. Their tips are turned upward when they are about 18 inches long. From each of these two vertical shoots, a U is developed, so arranged that the four vertical branches are about 12 inches apart. This is a very lovely and very effective form. Because of the way the scaffolds are bent and balanced, the trees are easy to maintain.

c. Verrier Palmette with Four Vertical Branches. This form consists of two U's of unequal size, superimposed one above the other in the same plane and on a common central axis. This is in contrast to the Double-U Palmette described in the preceding paragraph, in which the U's are of equal size and are at the same level as well as in the same plane. It is formed from two horizontal arms that are turned vertically near their ends and about 18 inches from the central axis, to form the outer U. The central vertical axis of the tree is continued for 12 inches beyond the point from which the two lower horizontal arms arose. Here, two parallel horizontal arms are developed, which are turned vertically at their tips when they have reached a length of about 6 inches, so as to form a second U within the first, with all four vertical branches about 12 inches apart. The central axis is terminated at this point. The height of the tree is kept to about 6 or 8 feet for weaker-growing varieties, but may reach 10 feet with stronger sorts.

The Palmette with Five Vertical Branches (*Verrier Palmette with Five Branches*) is perhaps the most satisfactory of all the palmette forms, and the most widely grown. It is used successfully in commercial fruit growing on the Continent of Europe, especially with the pear. It is similar to the Verrier Palmette with Four Vertical Branches, described

Fig. 138. **Candelabra with vertical arms.** (After Lucas)

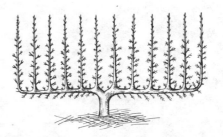

in the preceding paragraph, except that the central axis is permitted to grow beyond the point at which the second U arises. It is this extension of the vertical central axis that adds the fifth vertical branch. Obviously, the spacing of the vertical branches must be such as to adjust to this additional one, which means that the lower U must be formed by turning up the tips of the horizontal arms when they have reached 24 inches. And the second U must be formed by turning up the horizontal arms when they reach a length of 12 inches.

This form covers a wall rapidly and well. It is easy to grow and to manage. The central branch does not bear much fruit. It is inclined to outgrow the others and must be kept in hand, but this can be done with no difficulty.

Palmettes with Six (or More) Vertical Branches (*Verrier Palmettes with Six* [*or More*] *Vertical Branches*) may be developed with vigorous varieties under optimum growing conditions. Specimens with seven and with nine vertical branches are to be seen, but they are not common.

The Half-Circle Verrier Palmette (*Hemicycle Palmette*) is a modification of the true Verrier palmette in which the angles at which the scaffolds are bent are not right angles. Instead, the arms are bent in graceful curves or arcs. Although perhaps more easily developed than the preferred right-angle form, the half-circle type does not cover the wall at the base.

The Candelabra Palmette with Converging Branches is used for the peach (Fig. 139). Two lateral arms are trained horizontally to the desired length and then turned upward at right angles and trained vertically to the desired height. Scaffold branches are developed upward and inward from both the horizontal and the vertical framework so that the oblique scaffolds come together in a series of V's. This form is used for large wall coverings, and single specimens may be 10 to 12 feet long and 8 feet high.

The Candelabra Palmette with Crossed Branches is another large form, also used for peaches. Two horizontal arms are developed to the desired length and turned upward to form two vertical arms that outline the specimen. Inside this rectangular framework, scaffold branches are developed at 45-degree angles from both the horizontal and the vertical

Fig. 139. **Candelabra with oblique arms.** (After Lucas)

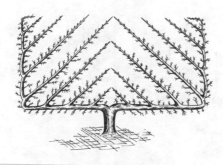

scaffolds. They are trained in such a way that they cross each other to form a series of diamonds.

Fan Palmettes (Éventails, Fächerspaliere)

The fan palmettes are composed of scaffold branches radiating upward and outward from a short (12- or 15-inch) main stem, like the rays of the sun. Because of their resemblance to peacock tails, they are called *Éventails* in French horticulture. In Germany, they are termed *Fächerspaliere*.

They are especially suited to the peach, and are usually found on walls as true espaliers, where they receive the warmth of the sun and protection from frost and wind. They are grown to considerable size as individual specimens, often reaching a width of 15, 20, or even 30 feet, although grown not much higher than 8 or 10 feet. They are among the oldest of artificial forms, having been grown by La Quintinye in the Gardens of Versailles during the seventeenth century. Such large, elaborate specimens are seldom seen today, but they are of interest in showing the detail and the intricacies that have been developed at one time or another.

The Simple Fan is successful with the plum, the cherry, the apricot, and the peach, but especially the latter. Using the peach as the example, begin with two main scaffold branches at about 2 feet from the ground, trained to the right and to the left at about 45 degrees from the horizontal. Two evenly spaced laterals are permitted to develop in oblique direction from the upper sides of both main scaffolds, and a single lateral is trained in almost horizontal direction from the undersides. From the upper sides of the laterals, in turn, are developed two upward-growing secondary laterals, while from the lower side, single laterals are produced. These branches constitute the scaffold system of the tree. Along the scaffold branches, shoots are permitted to form at about 6-inch intervals. These are the fruit-bearing shoots. They are summer pruned and pinched so as to keep them in hand. A renewal system is necessary so as to keep new shoot growth developing, since the peach bears on one-year wood.

The same general effect can be accomplished in other ways. For example, one system involves the development of a central leader and two oblique side arms. All three of these main scaffolds are induced to develop a central shoot and two strong laterals at about 20 inches from the main stem or trunk. This makes a nine-arm form. Each of these arms may again be divided in similar fashion as the fan develops.

In another system, two lateral arms are developed in nearly horizontal direction, and the central stem is continued. From near the crotches, another set of two oblique arms is trained at about 30 degrees with the horizontal. The central stem is continued. Again from near the crotches of the second pair of branches another pair is trained at about 60 degrees with the horizontal. This provides a nicely balanced tree with a central stem and three evenly spaced arms radiating outward on each side. Subsequently, these seven main scaffolds may be divided as desired as the tree extends outward and upward.

The Rectangular Fan (*Lepèresche Carrée-Spalier, Éventail carrée*) is a form used for covering low walls, and is grown nicely to a width of about 20 feet and a height of about 7 or 8 feet. It begins with two main scaffold branches at about 2 feet from the ground, trained to the right and to the left at about 45 degrees with the horizontal. As the tree develops, these branches may be brought down to about 35 or 40 degrees. From the underside of each of these two main branches, three arms are spaced at about equal distance in nearly horizontal but somewhat upward slanting direction. From the upper side of each main arm three branches are similarly developed but in a nearly vertical but somewhat outwardly slanting direction.

The Dumoutier Fan (*Éventail Dumoutrie*) is developed in rectangular form with two nearly horizontal lower arms, two nearly vertical central arms, and two oblique arms between the horizontal arms and vertical arms on both sides. Each of these arms is again divided once and sometimes twice so as completely to fill the rectangle and give good coverage to the wall.

The Montreuil Fan (*Éventail carré de Montreuil*) is developed from two main scaffold branches inclined at a 45-degree angle. From each of these oblique arms, four arms are trained in nearly horizontal position and four arms in nearly vertical position. This form differs from the common rectangular fan in having four scaffolds instead of three on both the upper and lower sides of the two main scaffolds. Further, these lateral scaffolds are closer to the horizontal and the vertical, respectively.

The Converging Fan (*Éventail à branches convergentes*) is similar to the Montreuil fan excepting that the arms that are trained in nearly vertical position in that form are not so developed. Instead they are trained at a 90-degree angle with the main oblique arms so that they converge at right angles to each other.

The Dalbret Fan (*Éventail de Dalbret*) is formed from two oblique main branches. From these, four nearly horizontal branches radiate outward. Two other scaffold branches are trained vertically, and from these there are developed oblique branches that are parallel to the main oblique branches.

The Wayward Palmette, Devil's Palmette (*Palmette à la diable*) is a free-growing fan shape in which the scaffold branches, instead of being carefully and geometrically designed, are merely fastened to the support much as they happen to develop. Little or no pruning is done. Instead, as each shoot arises it is tucked in and fastened much as one might handle a climbing rose. Attention is given, however, to spreading and bending and spacing so that the branches are well distributed and the tree becomes not unsatisfactory for the amateur gardener. It is a simple system and not without considerable satisfaction. Unlike most fan shapes, this form is used principally for apples and pears. It has been grown by commercial fruit growers in parts of France and is discussed more fully in the section dealing with commercial fruit production.

THE GOBLET FORMS
(*Formes en gobelets, Becherbäume, Keffelbäume*)

The goblet forms are made in the shape that the name implies. They are hollow cylinders (Fig. 140). Such plants are exposed to free circulation of air and light. The fruit is not easily disturbed by wind, and is easily harvested. That trees should be grown in these complicated forms for fruit seems scarcely credible, but such is the case, although admittedly only on a very small scale and then mostly for the sake of interest. They are suited best to the apple and the pear.

Fig. 140. **Vase or goblet with vertical branches. (After Du Breuil)**

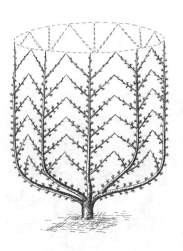

The goblet forms are divided by French horticulturists into large goblet (*grands vases, grands gobelets*) and small vases (*petits vases, petits gobelets*). The large ones are about 3 feet in diameter and 4 to 5 feet high. The small ones are about 2 feet in diameter and about 3 feet high. Both are developed on trunks at about a foot from the ground level.

The Simple Large Goblet is formed from three, four, or five scaffold branches that diverge from the main stem or trunk at about 12 inches from the ground line. They are trained in a horizontal or in an oblique direction until the desired diameter for the vase is attained, at about 3 feet. They are then turned vertically to a height of about 4½ to 5 feet. The scaffolds are trained on lath strips and are grown in cordon form.

For larger forms, the horizontal scaffolds may be divided again to provide six, eight, or ten spokes from which a similar number of vertical scaffolds arise. In even larger forms, these horizontal scaffolds may again be divided to form twelve, sixteen, or twenty spokes and vertical scaffolds.

The Simple Small Goblet is arranged similarly to the large vase described in the preceding paragraph, but with a diameter of about 2 feet and a height of about 3 feet.

The Goblet with Branches is developed as a simple goblet. From the vertical scaffold branches, short lateral branches are trained obliquely in pairs. They are arranged in tiers about 12 to 15 inches apart.

The Goblet with Crossarms is begun as a simple goblet with divided horizontal or oblique scaffolds. When the desired diameter of goblet is reached by the linear growth of these branches, they are trained diagonally in opposite directions at an angle of about 30 degrees with the horizontal and continued in a spiraling system upward. In this manner the scaffold branches cross and recross to give a most unusual and graceful effect.

The Double Goblet is merely one goblet within another. This is accomplished by first developing a simple goblet with vertical scaffolds. The central axis of the tree is continued above the point at which the first whorl of scaffolds diverge, and a second whorl is trained from it so as to form a smaller goblet within the larger and outer one.

The Goblet with Reversed Branches was developed by M. Maître, amateur gardener at Chatillon-sur-Seine (Fig. 141). The central stem

Fig. 141. **Vase or goblet with reversed vertical branches. (After Du Breuil)**

of the tree is grown to a height of 8 feet. Some twenty branches are then developed at this point and trained outward like the spokes of a wheel for a distance of about 18 inches. The branches are then turned downward to form an upside-down goblet. Further, at about 2 feet from the ground, the tips of the branches are reversed so that they grow upward to form what might be called the rim of the goblet.

The Plate Form was described from Odense, Denmark, about 1878. Five horizontal scaffold branches are developed spokelike from the central stem at about 12 inches from the ground level. Each of these may be divided again into two. A circular support is necessary, like the edge of a plate, to which the horizontal branches are fastened at their tips. The diameter of the circle is about 3 feet (Fig. 142).

Additional palisade or supported forms (formes palissées) include several of the shapes that have already been described under the nonpalisade forms. However, when these shapes are so included it is because they are grown in such a manner that they require firm bracing or supporting. Thus, the winged pyramid exists in a palisade as well as a nonpalisade form. In the palisade form it is held in place and trained

Fig. 142. **Horizontal and vertical cordons and high table or wheel form (Right) and Belgian fence (Left) in European garden.**

with the aid of strong guy wires by which each wing is firmly supported. Or each wing may be shaped by the aid of a sturdy wooden strip, to which each arm of the wing is attached. In the nonpalisade form it is developed without these aids. To be sure, the palisade form is more exact and is developed in more strict detail, but the general pattern is the same.

Similarly, other shapes that require strong bracing and support for any period of time, even though later able to stand alone, are classed as palisades. The distinction is, however, somewhat academic, and is mentioned here only to clarify the point.

HIGH–STEM OR HIGH–HEAD FORMS
(*Hautes tiges*)

Many of the palisade forms that have been described may be grown on long stempieces rather than on the conventional 12-inch stem commonly used (Fig. 143). They are employed mostly as palmettes on the sides of houses and barns, but they are also formed in other positions. For example, in Belgium there is a park in which tall-stem horizontal palmettes are grown overhead as contre-espaliers along a walk. Yet goblets, baskets, and similar forms may also be grown on high stems.

In high-head forms, the choice of variety is for the strong, vigorous kinds.

HOW FRUIT TREES ARE TRAINED TO SPECIAL FORMS

It is obvious from even a cursory glance at some of the artificial forms of fruit trees pictured and described in the preceding pages that the training operation is not something to be worked at haphazardly. It requires constant attention, together with some skill, but the former is a necessity. A tree is a living, growing plant. It is trained only by watching the growth regularly and giving the necessary guidance to the growth as it develops. This can only be done if the gardener is there to work with it at the proper time. If this is considered an arduous task, or if the gardener has

Fig. 143. Various vase and goblet forms, some with high stems, being trained in a French nursery.

other duties that prevent him being a regular, almost daily, observer, he had better not undertake the project.

On the other hand, if he has the time, the training of fruit trees to special forms is a most delightful and fascinating experience. It is ideal for those who for physical reasons cannot enjoy heavy work, or for those who have reached retirement age and wish an absorbing hobby. Each tree becomes a person, with its own characteristics of perversity and of tractability. It lives and grows from year to year, and is not lost as are herbaceous plants. Each day there is a new leaf unfolding or a shoot that requires attention in some way or other. And almost all of the work can be done in a standing position.

For anyone who does not wish to wait the time required by the formative years, there are a few nurserymen who can provide trees that have been already either partially or fully trained. The trees are quite naturally expensive, difficult to transport, and they do not come in a great variety of forms. If one is located near a nursery which specializes in such trees, he is fortunate.

Rootstock and Variety

Only trees that have been budded or grafted onto suitable dwarfing rootstocks will succeed as trained trees. It is necessary that the vigor of the tree be controlled and kept in hand. This is where the dwarfing rootstock functions. These have been discussed in detail in the section dealing with rootstocks. The following comments are, for convenience, in summary form.

For the apple, the EM IX is the most dwarfing. It is used for the more vigorous growing varieties, on fertile, well-drained soil. The EM 26 and MM 106 are less dwarfing and are suitable for less vigorous varieties and less fertile soils. The EM VII and II are also useful where still greater vigor is desired.

For the pear, the Quince EM A is best. Quince EM C is very dwarfing, and is suitable only for the more vigorous varieties. The quince root is a hazard where winter cold may be severe. It is tender to winter cold. And there is nothing so pitiful as to train a pear to a special form for eight or ten years and then have the entire tree destroyed by a severe cold spell during some winter that kills the quince root.

There are no reliable severely dwarfing rootstocks for the stone fruits comparable to those for the apple and the pear. The Common Mussel and St. Julien A are useful for the peach and nectarine, and the Brompton is used where there is plenty of room for the tree to grow.

For the plum and the apricot, the St. Julien A is the best available. Common Mussel can also be used, but is less dwarfing. For the cherry, mahaleb is used.

As for varieties, the slower-growing ones are the ones best suited to elaborate forms of training. Coxe's Orange Pippin, Jonathan, Golden Delicious, and Starkrimson are good examples for the apple. Conference is a good example among pears. Rapid-growing trees, such as the peach, nectarine, cherry, and Japanese plums, are better managed in somewhat simple form, such as the fan.

Supports

In Europe, trees are grown extensively on brick and stone walls. In fact, brick walls are sometimes built expressly for the purpose of accommodating fruit trees. The walls may run north and south or east and west. Both sides are used, even when one side faces north. Currants and gooseberries are grown on the north walls. West walls are for pears and plums. South and east walls are for apples, peaches, and cherries. Such walls may be 10 to 14 feet tall, and they may be built with a coping or projecting overhang of 10 to 18 inches from which curtains may be strung on wires and drawn for protection from hot sun, frost, or other inclement weather.

Brick walls would be expected to be less advantageous in America than in Europe. Such walls provide the warmth necessary to mature peaches and pears in cold, cloudy sections of Europe. While it might seem that there is no place in America for elaborate protection of this kind, it has been used. I. B. Lucas, of Ontario, Canada, has grown peaches successfully in an otherwise unfavorable climate. He has used portable wooden frames or screens, which he has placed vertically against the trees in winter, and so protected them from snow and excessive cold.

Where walls are not available, trellises of wire and wood are used. Five or six wires are needed for a good trellis, placed 12 to 14 inches apart, with the lowest wire about the same distance from the ground.

Woven wire fencing may also be used. Poultry netting is cheap and convenient, but it lacks strength and durability.

It is obvious that the trellis must be firmly anchored. If wooden posts are used they should be set in concrete. Iron posts are good. They should be driven so that at least 5 or 5½ feet are aboveground.

The trellis needs to be comparatively high for larger forms. Some trees can be contained at a height of 6 to 8 feet, but others are likely to require support to 10 feet. The entire height is not needed the first year, but vertical cordons of apples will reach a height of 10 to 12 feet in five years, and peaches and Japanese plums will reach this height in three or four years.

There are many different methods for training fruit trees to special forms. They are described in horticultural literature in great detail. But there is a general basic pattern. Following are general, simplified descriptions for the training of such representative forms as horizontal and oblique palmettes, the U-form palmette, and the Verrier palmette.

Training the Horizontal Cordon

The horizontal cordon may be made from one-year-old nursery trees. Straight-stemmed, unbranched "whips" are what are wanted. They should still retain lateral buds in the region 12 to 18 inches above the bud union. Sometimes in training in the nursery, the lower buds are rubbed off.

A single wire is stretched horizontally along the row and in the direction in which the arms are to be trained. It is placed at 15 inches from the ground line. The tree is cut back to the height of the wire and to a bud that points in the direction in which one arm is to be developed.

For a one-arm cordon, all other buds are rubbed off; and the remaining bud is encouraged into a shoot growth that is tied in position along the wire with raffia or soft string or tape. A useful procedure is to fasten a lath or bamboo along the wire, to which the shoot in turn is fastened.

For a two-arm cordon, two buds are selected and retained, from which to develop the two arms, running in opposite directions.

The training may also be done in the nursery with budlings; that is, with the young shoot forced from the bud that was inserted in the root-stock the previous summer. The training begins when the shoot reaches a little above the wire. It is then bent and twisted at the same time to form a sharp right angle (Fig. 144). The shoot is tied to a horizontal lath or bamboo that has been fastened to the wire. This is the one-arm horizontal cordon (Fig. 145).

For the two-arm horizontal cordon, the tip of the new horizontal shoot is pinched, and all buds are removed back to a bud at the bend. The bud is forced into growth to form the second arm running in the opposite direction.

Also, it is sometimes possible to bend and train a small one-year whip into the one-arm horizontal cordon the same year it is planted, thus saving a year. Instead of cutting the leader back to a bud at the height of the wire, the top is bent in an arc and fastened to the wire at one or two points along its horizontal position. The last foot or so of the shoot should be left unfastened so that the extention of the arm is encouraged. All buds below the wire should be removed, as well as those on the underside of the horizontal section. Of course, the angle produced by this method is not a sharp right angle, but with a little care and attention the arc can be reduced considerably as the cordon is developed.

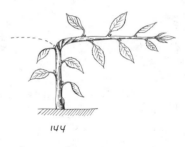

144

Training the U-Form Palmette

The U-form palmette is begun in exactly the same manner as the horizontal cordon. However, the support must be increased by a second wire that is strung parallel to the first but at about 30 inches from the ground line. Also, a lath or a bamboo must be fastened vertically, connecting the lower and the upper wires at about 6 inches either side of the main stem. The vertical branches are to be fastened to these wood supports. When the horizontal shoots on both sides have reached a length of about 10 inches, they are lightly twisted and bent into the vertical position. If one shoot makes more growth than the other, the more vigorous should be pinched back. It is best to cover bent and twisted sec-

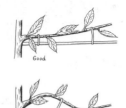

Good

Bad

Fig. 144. (Above, right) Developing a two-arm horizontal cordon by gently twisting and bending the shoot at about 15 inches from the ground, when it has reached a height of about 2 feet and is still soft. The bending tends to force a bud in the opposite direction to form the opposite arm (dotted line). (After Lucas)

Fig. 145. (Below, right) Proper method of tying oblique cordons. (After S. B. Whitehead)

tions with grafting wax or paraffin; bruises and slight breaks are often unavoidable.

Training the Oblique and Horizontal Palmette

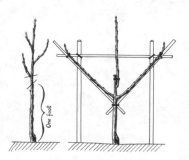

The start is again made with one-year-old whips or with a budling in the nursery row. Two wires are strung horizontally along the row in the direction the arms are to be trained. One is at 15 inches from the ground, and the other is at 60 inches.

The one-year whip is cut back so that three well-developed buds remain at the low wire. These are the buds that are to begin the main scaffolds of the tree. They should be in such position that the uppermost bud, which is to form a central leader, is to the front of the tree as it is faced. The other two should be to the right and to the left, insofar as possible. The buds along the stem below the wire should be rubbed off (Fig. 146).

As soon as the new growth is 10 to 12 inches long, actual training begins. The uppermost shoot that has come from the central bud is tied to a vertical lath, which is in turn tied to the two wires. The two lateral shoots are then tied to laths fastened to the wires at an angle of 45 degrees. The growth of the left- and right-hand shoots are thus trained along these 45-degree angle laths by tying the new growth in place as it develops (Fig. 147).

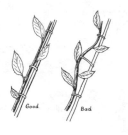

If the growth of the shoots is uneven, the longer one may be checked by pinching until the others have reached the same size. The next season the same general procedure is followed. The leader is cut back to 12 or 14 inches from the crotch of the last season's lateral shoots, while the lateral shoots are cut back to about the same length, namely, 12 to 14 inches. A second tier of oblique branches is now formed from the buds that will start from near the tip of the central shoot, at about 12 to 14 inches above the first tier. These branches are similarly tied to horizontal laths or bamboos. Additional tiers may be made as desired.

For the palmette with horizontal branches, exactly the same procedure is followed except that the arms are trained in the horizontal rather than in the oblique position.

Training the Verrier Palmette

The Verrier palmette begins with a uniformly balanced two-arm oblique or horizontal palmette, with the central leader still intact. Such a tree may be developed as described in previous paragraphs or may be purchased from a nursery that specializes in trained trees. The previously described two-wire trellis is also required.

The lateral branches are brought to the horizontal position and tied to laths or bamboos that have been fastened to the lower horizontal wire.

The central leader is cut back to 12 or 14 inches as measured from the first set of arms. Two buds are selected and retained near the tip, and all other buds on the leader are removed. These two buds will develop into shoots that will be trained horizontally for about 12 to 14 inches and then turned vertically to form a U, exactly as in the U-shaped palmette. Laths and bamboos must be placed so as to support the arms both horizontally and vertically (Fig. 148).

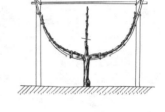

Branches of the first, or lower, horizontal tier must be extended about 24 inches to the left and to the right before being turned upward into the vertical position. This, then, spaces the five vertical arms 12 to 14 inches apart, including the central branch. This is the form known as the Verrier palmette with five vertical branches. Additional tiers may be made as desired, but the plan must be made in advance so that each vertical will be 12 to 14 inches from its neighbor.

Fig. 146. (Opposite, above left) Developing a simple palmette in the nursery: (Left) Method of pruning to force buds; (Right) Simple palmette with one tier of branches. (After Lucas)

Fig. 147. (Opposite, below left) Proper method of tying horizontal cordons. (After S. B. Whitehead)

Fig. 148. (Above, right) The beginning of a Verrier palmette in the nursery. (After Lucas)

27

FRUIT TREES IN POTS
AND UNDER GLASS

Historical Background

THE growing of fruit trees in pots is a very old practice. The Romans employed orange trees in tubs and large containers for decorative purposes. During the thirteenth century (1249), both flowering roses and bearing fruit trees are noted in horticultural literature as having been displayed at a winter festival at Cologne. The Italian author Ferrari, who in 1646 published the first book devoted entirely to citrus, depicted citrus trees in pots. Orangeries, or buildings for growing citrus, spread throughout Europe during the Middle Ages and the Renaissance, as a consequence of the Crusades (Fig. 149).

It was during the reign of Louis XIV of France (1643–1715), however, that potted fruit trees attained greatest prominence. La Quintinye, head gardener at Versailles, was very successful with figs in pots, as well as with oranges and other fruits. Not only were fruits grown in containers; they were also forced into early fruiting by being brought into warm houses. Figs are noted as having been fruited in June, and cherries in May. The "Orangerie" at the Palace of Versailles is still maintained as a feature, with the orange trees grown in large boxes and moved into position outdoors in spring.

During the early nineteenth century, the growing of fruit trees in pots and tubs, both indoors and outdoors, became a common practice in England and on the Continent. Thomas Rivers published a book in

498

1858 dealing with the culture of fruit trees in pots under glass, which stirred much interest both in Europe and in America. It is delightfully written and is still an authoritative treatise on the subject as well as a classic in horticultural literature.

DECIDUOUS FRUIT TREES

Uses of Potted Trees

In England, in particular, where the hazards of late spring frost are especially acute, the potted fruit tree reached considerable importance. There they have been employed not only for their value as ornamentals and for their interest to amateurs and plant lovers, but also for profitable commercial hothouse production of choice dessert fruits for market.

From the few trials that have been made in America with such trees, there seems no apparent practical reason why they cannot be and should not be equally enjoyed here, as well. This is especially so for the Pacific Coast and for southern regions, largely with citrus and related forms.

However, the native abundance of fruits in America and the relatively low prices at which they are sold in the markets no doubt precludes commercial production in this way. But for amateurs, and especially those who are interested in novel and interesting fruits, the potted dwarf tree holds possibilities that have hardly been touched, and badly need exploring.

Such plants, for example, may be used effectively in a small garden. They may be moved about and arranged as desired, as along a walk or in some special location. Specimens that have lost their usefulness or desir-

Fig. 149. **Orangerie at the Palace of Versailles.**

ability may be easily replaced with another potted plant, and a balance and satisfaction may be thus realized that permanently set trees cannot provide.

They may be used for their decorative features in conservatory or indoors on a table, as with a peach tree or cherry tree in full bloom. When the fruit is ripe, guests may be permitted to gather the fruit from the trees. Cherries, plums, and peaches are especially suited to this use (Fig. 150).

Further, the potted dwarf tree that has been started early and forced under glass may mature its fruit early in the season or develop an esteemed long-season variety, as the Doyenné du Comice pear, in a region where it would otherwise fail. Potted trees have also proved very successful in research laboratories for studies of pollination, insect and disease protection, nutrition and physiology. The fruit-breeding program can also be accelerated by this means.

The growing of trees in pots does, however, call for attention to particular details of potting, watering, pruning, fertilizer application, a wise choice of variety and rootstock, and some sort of greenhouse protection, preferably with some heat, especially if it is desired to force the trees.

Rootstocks for Deciduous Fruits

While the rootstock is a factor in successful culture of deciduous fruit trees in pots, it must be emphasized that it is the method of training and growing that imparts the greater dwarfing influence. The common rootstocks used in commercial orcharding are satisfactory for potted trees of the stone fruits, such as peach seedlings for the peach and nectarine, apricot seedlings for the apricot, Myrobalan seedlings for the plum, and mahaleb seedlings for both the sweet and sour cherry. Although the mazzard seedling may be used for the cherries, the mahaleb is a somewhat dwarfing rootstock, and assists just that much in controlling the size of the tree. Similarly, St. Julien A is somewhat dwarfing for the peach and the plum, and may be used to advantage if available.

Apples and pears are suitable for potted plants only when worked on dwarfing rootstocks (Fig. 151). For the apple, the EM IX (Jaunne de Metz) produces a very small and precocious tree in pots, considered by

Fig. 150. **Dwarf Mirabelle plum. (After Lucas)**

some to be too small and difficult. Where small size and early fruiting are desired, as in experimental work and in plant breeding, it has its place. A still better rootstock for this purpose, however, is EM 3431, which produces the most dwarf tree yet known.

EM I, II, VII, and IV are also suitable, producing trees less small and less precocious. Of these, EM II and VII are preferred, and EM 3461 is also esteemed. For the pear the Angers Quince A is best. Since not all pear varieties do well on the quince, only those varieties are commonly potted that are compatible with the quince (Chapter 11). If a special variety is desired that is not overly successful on the quince root, it is necessary to interpose an intermediate stock or body stock between the quince rootstock and the desired scion variety, the intermediate stock being compatible with both. Both the Beurré Hardy and the Old Home pears are successfully employed for this purpose.

Securing the Trees

Fruit trees are grown in two general forms for pot culture, depending in part upon the class and variety of fruit, since some lend themselves better to one form than to another. These forms are begun with the nursery tree, and in Europe are trained and developed as "bush" and "pyramid."

It is best if trees can be secured from the nursery already in pots or containers. If they are two or three years old in the container, and well trained, so much the better. The increased cost is well worth the difference. It is at present difficult if not impossible to secure many different kinds of deciduous fruit trees in pots. Citrus trees are the notable exception; they are grown and trained in containers in a most satisfactory manner by special nurseries on the Pacific Coast of America, and are described separately in this chapter. For most deciduous trees it is necessary to secure one-year-old trees and to head them and shape them to the desired form. Two-year-old trees as grown in the nursery are not satisfactory. They are usually headed at 30 inches, and all the buds below that height are rubbed off, making them difficult to use where a low-headed tree is desired.

For the bush type the trees are headed low, at 9 to 12 or perhaps 14

Fig. 151. **Dwarf pyramid of pear.** (After Lucas)

inches; for the half-standard they are headed at 18 to 24 to 30 inches; while for the compact pyramid they are headed at 30 to 36 inches. Peaches, apricots, nectarines, plums, and cherries are adapted to bush growth, although they may be grown as pyramids. Cherries are difficult to confine to any particular shape, but are good nevertheless. Apples and pears are best for pyramids.

Pots and Containers

The size pot used should be as small as possible to accommodate the tree, both for practical reasons of handling and for keeping the growth of the tree in hand. Ten-inch pots are satisfactory for the first year or two, though if only one size is wanted, the 12-inch size is probably better suited to general use and will be sufficiently large for three or four years. In fact, small peach trees have been fruited successfully in 8-inch and 9-inch pots. The 14-inch size is large enough for eight to ten years. Pots larger than the 16- or 18-inch size are not practical; wooden boxes are better. The smallest-sized pot possible is the one to use, always.

By repotting in the same size pot and with proper pruning it is possible to maintain the tree in a small pot for a surprising number of years. Peaches and nectarines have been grown for twenty or twenty-five years; cherries for thirty years; apricots, pears, and apples for twenty years; and plums for twelve to fifteen years.

Soil Mixtures and Potting

The soil used should be fairly heavy so as to retain moisture, but not clayey or of such nature as to interfere with drainage. Light soils are unsatisfacory. A mixture of ⅔ loam and ⅓ well-decayed and composted manure is good. With large containers, where weight is a factor, a mixture of ⅓ sand and ⅔ peat has been employed.

The suggestions made by Thomas Rivers in 1858 are still good: ". . . two-thirds turfy loam and one-third decomposed manure, to which some road or pit sand may be added. The loam should not be sifted; if it contains a large proportion of lumps as big as an egg, so much the better. If you examine an 11-inch pot, you will find it eight inches across at the bottom, and the aperture from one inch to one and a half in diameter; then place four or five large pieces of broken pot or tile across, so that they rest on the inside ledge left by the hammer, leaving interstices for the free emission of roots; on these place some of the most lumpy of your compost; then your tree, not too deeply, but so that the upper part of the roots is a little below the rim of the pot; if it has a ball of earth, loosen it; fill up with compost; ram the earth down firmly as you fill, with a stout blunt-pointed stick; place it on the border where it is to grow during the summer; give it two or three gallons of water, and a top-dressing of some manure to lie loosely on the surface, and the operation is finished."

Potting may be done either in late fall (end of September) when the trees are still in leaf, or any time during November or December, using dormant trees. A successful practice at the New York State Agricultural Experiment Station has been to pot dormant trees in late November or early December when they can be secured from the nursery, and to place the potted trees in a nursery cellar or cool storage until moved into the greenhouse in mid-January, February, or March (Fig. 152).

Because of the room and extra handling there is, however, no point to potting trees in the fall that are not to be forced until March. They will hold just as well if healed-in outdoors, or stored in a protected cellar as dormant nursery trees, and potted a few weeks before placing in the greenhouse.

English recommendations call for the potting of early peaches and nectarines first, apricots, cherries, and pears next, and pears and apples last, though this seems more a matter related to the desired season of fruiting, or the required rest period of the tree, than to any other factors of importance to tree behavior.

Nutrients

Manure or compost are the preferred fertilizers for potted trees. Concentrated commercial fertilizers may prove injurious if applied too strong or too frequently. Manure and compost are gentle fertilizers that are slowly and steadily available and that supply all the nutrients needed for

Fig. 152. **Dwarf fruit trees in pots plunged in sand or peat moss.** (After Lucas)

good tree growth and fruiting. Surface applications during the growing season are made to stone fruits at about the time the pits begin to harden, and to apples and pears when the fruits are the size of walnuts. A second application may be made when the fruits begin to color. The material should be well worked into the soil. Powdered charcoal has been suggested as an addition to help improve aeration as well as to take up some of the nutrients and make them slowly available.

Weak infusions of manure and compost are preferred by some gardeners. Their practice is to water the pots each week during the summer with weak liquid manure. Another common practice is to shape a ridge of manure and compost around the rim of the pot, about 1½ to 2 inches high. Each time the tree is watered, there is a leaching of fertilizer into the soil mass from the ring of compost.

Another interesting practice is the almost daily light syringing of the foliage with rainwater. This procedure is almost universally suggested in the older horticultural literature. New meaning is added to this practice by the discovery that water may be taken up by the leaves. Further, excess salt accumulations may be leached or washed from the leaves by syringing, and this may be beneficial when the plants are grown indoors or otherwise protected from the action of rain.

Pruning

Pruning is an important practice in successful pot culture. The trees are treated as individuals to a greater degree than in more extensive growing. Pruning consists, in the main, of thinning out of branches so that the tree does not become too thick, and in heading back so that it does not become scraggly. Cuts made in thinning out should be made back to a strong lateral shoot or to a main scaffold branch. If terminals are cut, the cut should be made close to a bud so that the wound may more easily and more rapidly heal over.

Pruning must, however, consider the fruiting habit of the particular class of fruit being grown. That is, peach and nectarine fruits are borne on one-year-old wood, so that cutting back the terminal growth removes blossom buds and is actually a thinning operation. On the other hand, the cherry, plum, apple, and pear bear largely on spurs on wood two years old and older. Heading back, accordingly, does not reduce the crop on these trees to the same extent.

Pruning must also consider the shape of tree that is desired. Pyramidal and bush forms are both useful. If a pyramidal tree is desired, there will be more heading back of side branches, and the favoring of a strong central growth. On the other hand, if a bush tree is desired, there will be a more general heading back of branches over the entire tree. Varieties that tend to spread may be helped by shortening branches that spread excessively, and varieties that tend to grow strongly upright may be helped by checking the leaders more severely.

The pinching of shoots during the growing season is a form of pruning. If properly done, it will reduce the amount of wood pruning and will result in a more shapely and attractive plant during the growing season. For example, a strong leader may be checked several times during the season by pinching out the soft terminals when the shoots are about 6 inches long. It is considered a mistake to permit the growth to reach 9 or 10 inches in length and then to cut back to 3 or 4 inches.

Repotting

Repotting, which is largely a root-pruning operation, should be done yearly in 10- or 12-inch pots, and every other year in larger pots. If not done, the plants will tend to outgrow the pot and get out of hand. Further, root pruning which attends repotting reduces the intake of nitrogen from the soil, and tends to promote fruit-bud formation. In the repotting operation the soil is loosened to a depth of 4 or 5 inches, 2 or 3 inches from the rim. When placed in a slanting position, the tree and ball of earth will come out easily.

About a third of the soil should be removed from the ball, and the large roots trimmed back about a third. A small hand weeder, a wire hook, or a hand rake make good tools with which to comb away the soil from the ball. The new soil should be rammed firmly into place as when the tree was originally planted, and the pot filled so that the top roots of the tree are covered.

If, in the case of 16- or 18-inch pots, the tree is not repotted, but merely top-dressed, the soil should be removed to a depth of about 4 inches, even though the process may destroy some surface roots. In fact, rather than to bury the surface roots under new soil when the pot is filled, they should be cut off. New roots will soon occupy the new soil mass.

Fruit Thinning

Thinning of the fruit is necessary to prevent overloading. Five or 6 fruits are enough for a small peach tree the first year it bears, and 10 are maximum. With a small-fruited kind, the number may, of course, be larger. A general rule for peach trees is to permit 3 to 6 additional peach fruits for each year of fruiting. Apricot fruits being smaller, 15 or 20 may be allowed the first fruiting year, and 6 additional each year thereafter. Plum trees may carry 20 to 30 fruits in the first fruiting year, and 8 to 12 additional each year thereafter. Apple and pear trees may bear 6 to 12 apples the first fruiting year, depending upon the size of the fruit, and may bear 3 to 6 fruits additional each year thereafter.

Glass Houses

From this much it may be gathered that some sort of glass and more or less heat are helpful additions to the most successful pot culture of fruit trees. It is, of course, essential if tropical fruits are grown. Yet, unless some particular fancy warrants the hard forcing of temperate-region fruit trees during the coldest days of the winter, very satisfactory results can be had with a cool house, which requires little or no heat.

In either case, the house should be not too hot, not too shady, sufficiently ventilated, and with plenty of head room and working room. Independent houses ranging north and south are best, but ¾-span and lean-to types with northern exposure may also be used if they have sufficient head room. Doors should be wide, and if much handling is done, tracks and a small flange-wheeled hand truck will be found useful.

Birds can be a great nuisance, especially with cherries and plums. Ventilators, wall flaps, and doors should be coarsely screened. And when the trees are placed outdoors, some sort of protection may be necessary. This is much more so in Europe than in America. A screened area, using pipe for the framework and coarse (¾-inch galvanized netting for coverage, is ideal so far as protection is concerned. For individual trees, squares of mosquito netting thrown over the tree and secured around the trunk are good. While protection of this kind may sound like considerable trouble, it is really a very simple matter and very much worthwhile.

Several arrangements are suggested for within the house, by Josh Brace. If it is a 14-foot house, trees may be placed in beds 5 feet wide on each side of a 4-foot center path. The trees in each bed are arranged 4 feet apart in two rows, with the front row 2 feet from the end of the house and the back row 4 feet from the end of the house, thus staggering the trees.

An 18-foot house, 5 feet wide at the eaves and 10 feet at the apex, is considered especially suited to pot culture. With a 4-foot center path, there will be two 7-foot beds, permitting three rows of trees in each bed, 4 feet apart in the row. The front and back rows should then be 2 feet from the end of the house, and the middle row in each bed should be 4 feet, thus staggering the plants. To accommodate to the head room, the rows nearest the center path may be the taller-growing pyramid forms, the second or middle rows may be half-standards, and the back rows may be bush trees. If the house is larger, it is possible to have a center bed and two side beds.

It is not necessary to place the different fruits in different houses, but it is easier to handle them if they are grouped according to kinds, that is, the peaches in one section of the house, cherries in another, and so on.

Where a glass house is used, the insect and disease problems are those of the greenhouse. The common red spider and whitefly are the worst pests, and are controlled with the materials commonly employed in the greenhouse for this purpose. Where no glass house is used, the pest problems are those of the garden and the orchard.

When trees are forced, the pots are brought inside soon after the middle of December and are placed at 45 to 55 degrees F. until after the trees have blossomed and the fruit has set. Thereafter the temperature is increased to 60 or 65 and even to 70 degrees as the fruit increases in size. After the pits of stone fruits have hardened, the temperature may be raised to 75 to 85 degrees.

The writer has grown peaches successfully in a greenhouse both at 50 to 60 degrees F. and at 80 to 90 degrees F. At the higher temperature the fruits develop much more rapidly than at the lower temperature, especially during the interval from full blossom to pit hardening, thus shortening the interval between blossoming and fruit maturity. Higher temperatures after the pits have hardened do not speed maturity especially, but do result in a higher quality and more highly colored fruit.

It is important that the night temperature be at least five degrees lower than the day temperature, and a differential of 10 to 15 degrees is better.

After the trees have fruited, the pots may be plunged outdoors and treated as any outdoor tree. Care should be taken not to overlook the fact that because of confined roots, the trees will need water off and on at regular intervals. And mouse damage can be severe if straw or some similar material is used as a protective mulch around the pots.

A résumé of year-round operations for potted trees in a cool house is suggested below. For a warm house for forcing, the schedule would be similar, excepting that the trees would be brought indoors during late December, January, or February:

March—bring the trees into the house; water once a week; temperature maintained at about 40 degrees F., not below 35 degrees F.

April—trees blossom, are pollinated, and set fruit; water as needed; temperature about 45 to 50 degrees F.

May—fruits are one-third grown; thin fruits; pinch shoots in order to shape the tree; surface feed or build a ring of compost 2 inches high around rim of pot; water as needed.

June—move stone fruits outdoors as soon as they have fruited, and move all apple and pear trees outdoors; plunge the pots; water as needed.

July—move other stone fruits outdoors as soon as they have fruited; plunge pots; water as needed.

August—move stone fruits outdoors as soon as they have fruited; plunge pots; water as needed.

September—water as needed.

October—pot new trees and repot old trees.

November, December—cover pots with straw, and protect from cold and rodents.

Varieties

At the Thomas Rivers and Sons establishment at Sawbridgeworth, England, where potted trees have been grown for several generations, certain varieties have been selected for their adaptability to such treatment. The peach, nectarine, apricot, cherry, and plum are especially suited; apples and pears are satisfactory; and such plants as the mulberry, fig, grape, and various other species, may also be employed.

The successful varieties are divided into those that seem well adapted to hard forcing in a warm house and those that seem better adapted to slower forcing in a cooler house. Warm-house peach varieties are Thomas Rivers, Alexander, and Waterloo. A cool-house peach is Duke of York. Warm-house nectarines are Lord Napier and Stanwick. A cool-house nectarine is Downton. Apricots are considered better if not forced in a warm house, successful cool-house varieties being Moorpark, St. Ambrose, and Royal. All varieties of European plum are said to do well as potted plants, especially the Gage types, of which Reine Claude and Jefferson are useful for forcing in a warm house. Varieties better suited to a cool house are President, Pond's Seedling, Early Rivers, Reine Claude du Bavay (Bavay), Golden Drop, Oullins, Ickworth, Monarch, Grand Duke, and Bryanston.

Sweet cherries do well, also; and among these the Early Rivers, Elton, Governor Wood, and Black Tartarian are considered good for forcing in a warm house, and the Governor Wood, Black Tartarian, Napoleon (Royal Anne), Schmidt, Mezel, Emperor Francis, Florence, and Hedelfingen are considered useful for the cool house.

Among apples, the most suited varieties reported are Coxe's Orange Pippin, Bismarck, King, Mother, and Alexander. Among pears, the suggested varieties are Doyenné du Comice, Superfin (Beurré Superfin), Louise Bonne de Jersey, Glou Morceau, Pitmasten Duchess, and Durandeau.

Experiences in America with American varieties of these fruits have shown thoroughly satisfactory results with practically all the varieties that have been tried. This generalized statement may perhaps be due to the fact that potted tree culture has not been extensive enough to indicate the refinement that more widespread and critical culture might show.

Furthermore, the severity of American winters, and the cost of heating, reduces the incentive to hard-force such bulky plants as fruit trees. Accordingly, complications or dormancy have not generally been introduced, but they might become very real if hard forcing was to be used. For example, the Elberta peach requires a relatively long period of chilling before dormancy is broken, whereas the Babcock peach requires a relatively short period. If varieties were to be brought indoors as early as mid-December, for rapid forcing, it is possible that those that have a shorter and more easily broken period of dormancy might succeed, whereas those that have a longer chilling requirement might fail. The fact that peaches and nectarines, which have a relatively short period of dormancy, are used in England for forcing in a hothouse, whereas the apple and pear, which have a relatively long period of dormancy, are grown in a cool house, gives support to this theory.

It might be expected from this that the varieties of fruits that have proved successful in warm sections, such as the Babcock peach, might be especially suited to early and hard forcing, inasmuch as this variety was developed to meet the unusually warm winter temperatures of southern California, where sufficient chilling does not occur to break the rest period of the more common commercial varieties of peaches. On the other hand, varieties that have been developed in northern and colder sections, such as the Red Astrachan apple, might be better suited to slow forcing and a cool house.

At all events, most attention in America has been given to growing in cool houses, and in such houses a wide range of varieties has been successfully used. In the cool greenhouses at the New York State Agricultural Experiment Station at Geneva, New York, the following varieties have been grown for nine years with more or less satisfaction, and none that have been tried have been complete failures:

Apples on EM I rootstocks: Baldwin, Carlton, Delicious, Gravenstein, Jonathan, Kendall, King, Lodi, McIntosh, Macoun, Medina, Milton, Northern Spy, Ogden, Orleans, Rome, R. I. Greening, Sweet McIntosh, Winesap, Yellow Newtown.

Apricots on peach rootstocks: Doty, Oullins, Stoke, Toyahvale.

Cherries (Duke) on mahaleb rootstocks: Reine Hortense, Royal Duke.

Cherries (sour) on mahaleb rootstocks: English Morello, Montmorency.

Cherries (sweet) on mahaleb rootstocks: Bing, Black Tartarian, Elton, Emperor Francis, Giant, Kirtland, Lyons, Schmidt, Windsor, Yellow Spanish.

Nectarines on peach rootstocks: Hunter, Lippiatt Late Orange, Rivers Orange, Surecrop.

Peaches on peach rootstocks: Belle, Champion, Elberta, Greensboro, Golden Jubilee, Halehaven, Marigold, Oriole, Paloro, Rochester, St. John, South Haven, Valiant, Vedette, Veteran.

Plums (European) on Myrobalan rootstocks: Albion, Golden Drop, Grand Duke, Hall, Imperial Épineuse, Italian Prune, Reine Claude, Shropshire, Stanley, Tragedy, Victoria, Yellow Egg.

Plums (Japanese) on Myrobalan rootstocks: Beauty, Burbank, Formosa, Santa Rosa.

Pears on Orange quince rootstocks: Alex Lucas, Bartlett, Beurré d'Amanlis, Beurré d'Anjou, Clapp Favorite, Duchess d'Angoulême, Flemish Beauty, Glou Morceau, Louise Bonne de Jersey, Marguerite Marillot, Marechall, Passé Crassane, Phelps, Pitmaston, Pulteney, Seckel, Tyson, Vicar of Winkfield, Willard, Winter Nelis.

The above list carries with it no special varietal recommendation. It merely indicates the wide range of varieties that can be successfully grown in pots. The final choice will rest with individual likes and dislikes. In general, however, the choice should be made to assure varieties of quality and of interest. Productivity and shape of tree should be secondary. Following is a list of a few varieties the writer has especially esteemed:

Apples: Melba, Milton, Gravenstein, Coxe's Orange Pippin, Delicious, and Macoun.

Apricots: Tilton.

Cherries (Duke): Reine Hortense.

Cherries (sour): Montmorency.

Cherries (sweet): Early Rivers, Lyons, Black Tartarian, Emperor Francis, Schmidt, and Giant.

Nectarines: Rivers Orange and Surecrop.

Peaches: Oriole, Delicious, Rochester, Pallas, and Champion.

Plums (European): Oullins, Jefferson, De Montfort, Mirabelle, Imperial Épineuse and Sannois.

Plums (Japanese): Beauty and Formosa.

Pears: Elizabeth, Tyson, Bartlett, Dana Hovey, and Beurré Bosc.

Pollination

Facilities for pollination should not be overlooked. If the plants are hard-forced they may be in blossom before bees and pollinating insects are flying or if the house is tightly screened with close-mesh screening, they may be excluded from the house. Under favorable conditions of a cool house that permits the entry of bees and insects and where there are a number of varieties present, satisfactory pollination may result. But to be sure of the set it is a simple task to brush the open blossoms of one variety with a twig or spray of open blossoms of another, or otherwise to provide for cross-pollination. At Longwood Gardens in Pennsylvania, nectarines are successfully pollinated by agitation with a light spray of water (Fig. 154).

Sour cherries, nectarines, and peaches (with the exception of the Mikado and J. H. Hale varieties) are fertile to their own pollen and need no cross-pollination. Apples, pears, plums, sweet cherries, and Duke cherries are for the most part self-sterile and must be cross-fertilized to ensure fruit.

FRUIT TREES UNDER GLASS, NOT IN CONTAINERS

Fruit trees may be grown successfully under glass, not in containers, but planted directly in the soil, as is done for hothouse grapes. Nectarines and apricots have been grown in this way at Longwood Gardens, Kennett Square, Pennsylvania, for many years, and with great success. Some specimens of nectarine trees trained on wire mesh in palmette form have been grown in this way for forty years or more. They have attained a height of twelve or more feet, and a similar spread.

Very little pruning is done during the growing season. Care is taken not to cut back the extension growth or otherwise to force strong shoot growth. Desired lateral shoots are carefully spaced and tied to the wire mesh. Excess buds are rubbed off, and excess lateral shoots are similarly thinned out by cutting back to the branch from which they arose. No heading back is done, which might induce lateral shoot development. The result is a tree of spurlike growths, nicely and evenly spaced, that completely cover the supporting frame (Fig. 153).

In the dormant season, pruning consists of thinning out spurs and shoots. Extension growth is cut back to outside laterals to keep the tree in hand, but is never headed back.

Fig. 153. Forty-year-old nectarine tree grown under glass as a palmette supported on wire mesh, in full bloom. *Courtesy of Longwood Gardens, Kennett Square, Pennsylvania*

Since the nectarine and apricot are self-fruitful, no cross-pollination is necessary. Vibration of the flowers is sufficient to effect pollination, fertilization, and fruit set. A simple procedure to ensure self-pollination has been to spray the trees with light water spray while they are in full bloom (Fig. 154).

Fertilizers are applied four times during the season. The first application is of nitrogen, made in early spring just as growth starts. A complete fertilizer is applied shortly after bloom, and again in midsummer.

Fig. 154. Pollination of nectarine tree under glass by agitation of the blossoms with a water spray. The nectarine is self-fruitful. *Courtesy of Longwood Gardens*

The fourth application is of wood ashes, made in midsummer. The amounts applied are just enough to cover the entire soil area with a light surface coating. Regular crops of high-quality fruit are secured by these methods.

CITRUS AND RELATED SPECIES

Citrus and related species have long been popular when grown in pots and tubs, as already noted. Many forms are naturally small plants, but the size can be still further controlled by the use of dwarfing rootstocks. The result is plants which may range from 2 to 4 feet in height. Specific rootstocks, mostly strains or clones of the Trifoliate orange (*Poncirus trifoliata*) should be used for specific varieties, as observed in the chapter dealing with dwarfing rootstocks for citrus. When this is done, the resulting trees are somewhat predictable as to size, behavior, and performance. They have proved especially suited to portable pots and tubs. In this way they can be grown where they might not otherwise succeed, since they can be moved to favorable locations in the garden, on the patio or terrace, or in the home. The large-fruited Ponderosa lemon is a favorite, spectacular plant in cold northern climates, grown indoors in a sunny window. There are a great many varieties and interesting ornamental forms of citrus and related species that can be handled similarly, with much satisfaction.

There are nurseries in California, Florida, and to a lesser extent elsewhere, that specialize in dwarf citrus plants and that have been responsible for the development of considerable interest along the Pacific Coast north into Canada. The fruit does not develop the best edible quality under the cool summer temperatures that prevail in some of the northern regions, but it is colorful and often persists the year around. The flowers are fragrant and the foliage is evergreen, so that the plants are attractive during all four seasons of the year. Finally, citrus plants seem adapted to pot and tub culture. In fact, Thomas Rivers, in 1858, pictured the virtues of the orange as grown in similar and somewhat adverse conditions in England and on the Continent: "As an ornamental tree in the greenhouse and conservatory, it is an old friend; and perhaps no tree in the known world has suffered, and does suffer, such vicissitudes, yet living and seeming to thrive under them. It *glories* in a tropical climate, and yet *lives* and grows after being poked into those cellar-like vaults used for its winter quarters on the Continent; it gives flowers in abundance under such treatment, and would even give its fruit—albeit uneatable—if permitted. Nearly the same kind of cultivation has been followed for many, many years in England; it has rarely had heat sufficient to keep the tree in full vigor, and its roots in pots or tubs must have suffered severely from having been placed out-of-doors in summer on our cool damp soil, and in winter on a stone floor still more cold. If roots could make their complaints audible, what moanings should we hear in our orangeries all the winter!"

Varieties

A wide assortment of varieties is available. Among those that have proved satisfactory are:

Oranges: Hamlin, Pineapple, Robertson Navel, Washington Navel, Ruby, Summernavel, Shamoudi, Temple, Trovita, Valencia, Seedless Valencia, Tarocco.

Tangerines—Mandarin oranges: Kinnow, King, Kara, Dancy, Dweet, Owari Satsuma, Clementine (Algerian) Mandarin.

Lemons: Eureka, Lisbon, Meyer, Ponderosa, Villafranca.

Grapefruit: Duncan, McCarty, Marsh Seedless, Ruby Pink, Triumph.

Limes: Bearss Seedless, Mexican, Rangpur, Tahiti.

Tangelos: Minneola, Sampson.

Kumquats: Nagami, Meiwa.

Ornamental forms that are especially valuable as potted plants are:

Chinotto and Myrtifolia oranges (*Citrus Aurantium* var. *myrtifolia*).
Citron (*C. medica*).
Fingered citron (*C. medica* var. *sarcodactylis*).
Eustis and Lakeland limequats (*C. aurantifolia* x *Fortunella* sp.).
Meyer and Ponderosa lemon.
Rangpur lime.
Otaheite orange (*C. otaitense*).
Calamondin (*C. mitis*).
Chinese box orange (*Severina* sp.).
Nagami, Marumi, and Meiwa kumquats (*Fortunella* sp.).
Triphasia trifolia or limeberry.
Trifoliate orange (*Poncirus trifoliata*).

Among these, the most recommended are the Meyer lemon, the Ponderosa lemon, the Otaheite orange, the Nagami kumquat, and the Chinotto orange.

Containers

Dwarf citrus plants can be grown for a while in 2-gallon and even 1-gallon cans. Containers that are 16 to 18 inches in diameter are satisfactory for small plants for several years and are not too heavy to move about. Larger, square redwood boxes are more commonly used for large plants, and are most attractive. These should be about 18 to 24 inches deep (Fig. 155).

Soil, Fertilizer, Water, and General Care

When purchased from a nursery that specializes in dwarf citrus trees, the plant may be received in a container holding a mixture of two-thirds redwood sawdust and one-third very fine sand. The unit is light and easily handled, but such a light root-medium quickly dries out and requires frequent watering. The plant is best transplanted to a larger container to which a soil mixture is added of about half to one-third peat moss, vermiculite, or aged compost, and half to two-thirds good garden soil. The root ball can be slipped from the original container intact.

As in any planting of this kind, bottom drainage should be provided from the box, and a layer of gravel or crushed rock placed in the bottom before the soil mixture is added. The plant should be set so that the root ball is covered but so that there is a space at the top for watering. A good practice is to mulch the surface with peat, gravel, shavings, or crushed brick.

Citrus prefers to be a little on the dry side. With the light soil mixture that has been suggested, drainage is rapid and there is not much chance of overwatering. Young plants may need water once a week, or more often in a dry, warm period. When watering, the soil mass should be thoroughly wet, and then should not be watered again until the plant approaches the wilting point.

Fertilizers may be added lightly during the growing season, in the form of a prepared commercial plant food mixture. The application is made when the plant is watered. The amount will depend upon the size of the plant and the container, and is best judged by observing the color of the foliage, which should be deep green. It is better to err on the side of too little rather than too much. For small containers, one level tablespoonful a month is sufficient, while large boxes or tubs may require three or four times as much.

Fig. 155. Owari Satsuma Mandarin orange growing in a wooden container as a true dwarf. *Courtesy of Four Winds Growers*

Citrus trees require very little pruning. Any strong-growing branches may be supressed by pinching or cutting back, but the plants will soon adjust to a symmetrical shape. Any sucker growth that arises from below the crown of the plant is probably from the dwarfing rootstock, and must be removed promptly if it appears.

Plants grown indoors are tender and are easily injured by cold. Temperatures approaching freezing should be avoided. From December to February the trees should be permitted a rest period with temperatures near 60 degrees F. or slightly below. After this rest, the temperature should be raised, to bring the plants into blossoming and fruiting.

28

BONSAI, OR JAPANESE
DWARFED PLANTS

A N INTERESTING use of dwarf plants has come from the Orient in what is perhaps more of art than of scientific horticulture, though obviously a combination of both. Various plants, principally woody plants, sometimes several hundred years old, are grown as dwarfs in shallow containers, and trained and shaped to suit artistic tastes, employing detailed and elaborate horticultural skills and techniques (Fig. 156).

The method has become very popular among the Japanese, and has acquired the Japanese term *bonsai*. The word is said to be derived from *bon*, a shallow pan or tray, and *sai*, to plant—meaning planted in a shallow tray or pan. The word is pronounced "bone-sigh," and is used variously. Thus, the Japanese speak of "bonsai culture," of a single "cherry bonsai," of "several bonsai," and of "bonsai," meaning the entire field of ornamental dwarf-plant culture. Ordinary potted plants are called *hachiuye* (literally "pot planted"), and are not to be confused with bonsai.

Bonsai in Europe and America

From time to time considerable interest has been shown in bonsai in both Europe and America. But this interest, intense though it may have been, has been largely confined to a few individuals or groups, and has been sporadic at best. In fact, bonsai culture in the Occidental world

519

may be thought of in terms of a hobby. It has never reached anything like the proportions found in the Orient, principally Japan. The reasons for this failure of bonsai to "catch on" as yet in Europe and America in any large fashion will be fairly obvious from the discussion of bonsai in the Orient in the following pages. Great patience, exactness, and close daily attention are required, coupled with horticultural skill and an artistic touch. These qualifications are all too infrequently encountered in any large segment of Occidental populations.

Further, bonsai are most easily handled in an equable climate, which might be called maritime in contrast to a continental climate of extremes. For example, east of the Rocky Mountains and north of the Mason-Dixon line in the United States, the severe cold of winter and the frequent high temperatures of summer are difficult to cope with. Further, American heated homes in winter are notorious for being hot and dry. This is not conducive to bonsai culture.

Yet there are locations in America that are well suited to bonsai. For example, the Pacific Coast, particularly the Pacific Northwest, is ideally suited. And where greenhouse facilities are available, excellent conditions can be provided.

Fig. 156. **Pine tree grown as a bonsai.**

Further, there are a number of excellent books on bonsai, some of which are listed in the Bibliography. And experts are available who have been highly successful both in practicing the art and in teaching it.

Accordingly, in spite of some of the limitations that have been mentioned, the opportunity exists for individuals and groups to pursue the art of bonsai culture. There is much satisfaction and enjoyment to be derived from it.

THE HISTORICAL BACKGROUND

Because of the vogue for bonsai in Japan, the culture of dwarf plants for ornamental purposes has come to be associated almost exclusively with that country. Historically, however, dwarf ornamental plants were first cultivated by the Chinese, from whom the Japanese secured their interest and information, subsequently modifying and developing them to suit Japanese desires.

Bonsai in China

In China, during the Sixth Dynasty (first and second centuries C.E.), pines and junipers were trained as dwarf trees in the palaces of the royal families, to represent and symbolize the reverence and majesty of age. The juniper and pine are themselves thought of as symbolic of great age, so that by employing dwarf culture it became possible to bring these symbols into the palaces. During the T'ang Dynasty (eleventh to thirteenth centuries C.E.) dwarf plants reached the peak of development. Among the more popular forms were *Pinus* (pine), *Juniperus* (juniper), *Prunus mume* (Japanese apricot), *Punica granatum* (pomegranate), and *Malus halliana* (Hall's crab apple).

In modern time several centers in China are famous for pruning and training dwarf plants; namely, the Soochow and Yangchow region in Kiangsu, the Kaifeng and Loyang region in Honan, the Feng Tai and Peiping regions in Hopei, and the Peh-Pei and Chengtu regions in Szechwan. The plants most used are *Pinus* (pine), *Malus* (apple), *Syringa* (lilac), *Lagerstroemia* (Crape myrtle), Gardenia, *Cercis* (Red bud), *Prunus mume* (Japanese apricot), flowering forms of peach (*Prunus persica*), Azalea and Rhododendron.

There are two principal methods used by the Chinese in developing dwarf plants; (1) by special cultural practices, and (2) by propagating on dwarfing rootstocks. By far the most are produced by the first method, employing small pots containing specially blended soil, and in which the plants are carefully pruned and trained to attain the desired artistic or poetic design. Many of these designs are named and are well known by those names throughout China, and these names are always in the basic rhythm of four. For example, one of these is

> "Chien-chan
> Chu-Lien"

which is translated "Ten-Thousand Foot Falls of Pearls." Another is

> "Kwei-Fei
> Tseu-Kiu,"

which means "Kwei-Fei, an ancient queen of the T'ang Dynasty radiant with wine, in artistic pose."

All these specially trained plants, which are dwarfed by culture, pruning, and training, are to be distinguished from other naturally growing or dwarfish potted plants.

The second method of dwarfing plants, namely, by the use of dwarfing rootstocks, is confined largely to fruit trees (Fig. 157). These, too, are grown in pots, though when used in this way they are employed as ornamentals. They are also pruned and trained to special forms, and their culture demands repotting each year in order to replenish the soil. Peaches and plums are grafted onto a dwarf peach known as Shu Shih Tao, which is short-internoded form of *Prunus persica*. Apples and crab apples are worked onto a dwarf form of *Malus halliana* (Hall's crab apple); pears on *Chaenomeles lagenaria* (flowering quince); and persimmons on *Diospyrus lotus*. Oranges and mandarins are worked onto the dwarf *Citrus limetta*; kumquat, on *Poncirus trifoliata*; and the Buddha-fingered citron, on *Citrus sunki*.

Bonsai in Japan

In Japan, there are evidences in writings and in pictures of dwarfed trees in ceramic containers among the upper classes during the Kamakura period (1180–1333). The following centuries witnessed steady expansion of interest in trained trees in pots, though most specimens were relatively large and were contained in substantial pots. Vogues and styles changed during the years. The Japanese apricot (*Prunus mume*), *mume*, or *ume*, is said to have been a favorite of the Samurai around the year 1700. It was, however, during the Tempo period (1830–1844) that bonsai culture became of general interest to the masses, aided by freer interchange in travel and commerce with the mainland of China. In the hands of the Japanese, bonsai culture has reached its greatest perfection. Tokyo is the great center.

It is, then, the popularity of ornamental dwarf plants in Japan, together with the refinement and development of the art there, rather than because of any claim to priority or to having originated ornamental dwarf plant culture, that makes bonsai so particularly Japanese.

BONSAI IN GENERAL

To grasp the significance of bonsai to the Japanese, their artistic nature must be emphasized. A Japanese author, Shinobu Nozaki, has likened bonsai to a short form of poetry esteemed among the Japanese for centuries; it is called the haiku, and comprises only seventeen syllables. One of these, a composition of Besho, said to be the greatest haiku writer of Japan, is translated:

> "On a bare tree bough
> A crow came to sit
> On an autumn eve!"

To the Japanese, quoting from Nozaki, "this poem gives a perfect picture of a lonesome autumn evening, where a single crow, flying to its nest, stops on its way to sit in a tall tree, whose bare and ragged outline, now holding the gaunt black figure of the forlorn bird, is silhouetted in silvery gray against the sunset sky. The solitude of an autumn eve cannot be better and more graphically pictured than this."

Fig. 157. **Tolman apple tree in bonsai form. The tree is 17 inches tall, 16 years of age, and is bearing 21 fruits.**

In the same way, a bonsai is expected to portray in miniature some particular theme, in keeping with the Oriental aim to find beauty in the simple and unassuming. For example, the black pine of Japan, found along the northern coast, is esteemed for its majestic size, rugged outline, and masculine beauty. A successful black pine bonsai, then, "brings before the eye of the viewer the loveliness of the pine woods on a remote island or wave-washed ocean beach, and carries away his mind in poetic meditation of the imagined scene."

Forms of Art

As in any form of art, there are works of varying greatness. Of the million or more in Japan, perhaps a thousand are said to be famous as masterpieces. The age of the plant, the length of time since transplanting, and the artistic shape and beauty are the standards of quality. A fine bonsai, just as with some fine painting, may sell for a fabulous sum. Similarly, there are named collections of bonsai. The Brooklyn Botanic Garden, Brooklyn, New York, maintains a valuable collection, and has become a center of bonsai interest in America. At the other end of the scale, the low grades may sell for a trifle. The more popular and most widely used grades are intermediate.

Bonsai are usually employed in artistic arrangements of simplicity. For example, a single artistic scroll may be placed on the wall in back of a table upon which a bonsai rests. On one side there may be a small carved image and on the other side an interesting naturally shaped piece of rock. On various special occasions, particularly at New Year's, bonsai are used for their decorative value.

BONSAI CULTURE

Where the Plants Are Procured

Plants for training as bonsai are either procured in the wild and brought under cultivation or propagated especially for this purpose from seed, from layers, from cuttings, or by grafting or budding. On rocky cliffs, in alpine climates, in old gardens, and in any other situations where plants may be naturally dwarfed by their environment, they are sought out and carefully transplanted to the nursery. There they may be painstakingly cared for until the root system is adjusted to the new environment, after which they are trimmed and shaped as desired, and potted. To say that a bonsai is several hundred years old does not, therefore, necessarily mean that the plant has been grown as a bonsai for that many years, but more likely that it has been found in the wild as a dwarfish plant and has been estimated to be that age. It may have been in culture only a relatively few years. Plants fifteen to twenty years under cultivation are, however, not uncommon, and those that have been kept alive in pots for sixty to ninety years can be found.

Propagation

When propagated especially for bonsai culture by the various horticultural means previously enumerated, no special technique is involved other than that usually employed in the propagation of plants. After the plant is propagated, however, it is subjected to a technique that will maintain it as a dwarf plant, and in the form desired. In this, special consideration is given to soil, to watering, to fertilizer applications, to clipping and pruning, and to bending with metal strips, wire, or thread.

The Soil

The soil must be of a type that drains well but is not overly fertile. To realize this condition, great pains are taken in the preparation of the different soil materials and in their blending and mixing. Every enthusiast has his own special materials and his own special mixtures. Among the materials mentioned in Japanese literature are decomposed leaves of deciduous trees; black, humus-containing forest soils; humus-containing soil from rice fields and low places; mountain sand derived from weathered granite and not too fine in particle size, usually secured near the source of a stream rather than where the water moves more slowly so that it deposits smaller particles; soil from a flower bed or garden that is mixed with a slow-acting fertilizer; chopped sphagnum moss; and clay from riverbeds.

Most mixtures contain a large proportion of humus and sand, and avoid silt and clay particles. A general or all-purpose mixture is composed of one part loam, one part leaf mold, and one part sand. For the mountain pine a mixture of seven parts of riverbed sand and three parts manured garden soil is recommended, and for fir, four parts of manured garden soil, three parts of decayed vegetable soil, and three parts of river sand. Yet the mixture depends to considerable degree upon the particular species being grown. For example, the pear, which is recognized as enjoying a relatively heavy soil, is grown in five parts manured soil and three parts river sand, to which is added two parts clay; and the humus-loving rhododendron is grown in finely cut-up bog moss or three parts of bog moss, four parts of soft sand, and three parts of manured garden soil.

Further, the soil mixture is carefully sieved and often separated into a half-dozen different size particles measuring from $3/8$ to $1/32$ of an inch. A convenient separation is particles about the size of a pea, those about the size of a grain of wheat, and those about the size of bird seed or millet seed. Any finer particles are not included. When the plant is planted, the container is first soaked in water, and a screen or portion of broken pot is placed over the drainage hole in the pot or pan so that it will not become plugged with soil. A layer of the larger particles (pea size) is then carefully spread on the bottom of the container to a depth of about an inch, followed by a layer of medium-size particles (wheat grains) to a depth of about half an inch and upon which the roots of the plant are spread. The pot is then filled the remainder of the way with the smallest-size (millet seed) particles, the soil being carefully spread and firmed about the roots with a flat wooden stick or spoon. Sometimes as many as six layers of different sizes are used, particularly for large containers. The detailed attention given to soil preparation and mixing indicates its relative importance in bonsai culture.

Fertilizers

Fertilizers are sparingly used and are usually of a slow-acting kind of low nitrogen content. A favorite material is an oil cake made from rape seed, which is sometimes mixed with a small amount of bone meal. Application is made either in the form of a powder placed in two or three small heaps on the surface and near the sides of the pot containing the plant, or in a liquid form from a water extract of the material.

Little or no fertilizer application is made during either midsummer or midwinter. During spring, however, and again in fall, the plants are fertilized at weekly intervals, the spring applications being usually of the more quickly available liquid type, whereas for the fall applications the powdered form is preferred.

Pine trees and other slow-growing types require very little fertilizer, and may go for a year or more untreated, depending upon the color of the foliage and the growth of the plant. On the other hand, flowering and fruit-bearing plants are more generously treated. As little as ¼ teaspoonful of a 6 per cent nitrogen fertilizer may be sufficient for a small tree in a 5-inch container, whereas 3 teaspoonfuls may be required for a larger plant in a 12-inch container.

Watering

Because of the shallowness of the container and the small soil mass, bonsai must be carefully attended to keep them from drying out. Fruit-tree bonsai, after they have become old enough to fruit, are watered by soaking the bottom of the pot in water. Most plants, however, are watered at the surface of the soil, and alpine plants are, further, syringed at night. In fact, frequent syringing of the leaves is practiced for most plants by many experts. This practice tends to wash out any excess salt accumulation on the leaves and also provides moisture to the plant. It is now known that plants are able to absorb both water and nutrients through their leaves.

During the heat of the summer plants may need to be watered as frequently as twice a day, whereas in the winter once in three days may be sufficient, depending upon the dryness of the air. In spring and summer, watering once a day is usually enough. Gardeners watch the color of the surface soil, and do not permit it to change to the dull green or whitish color that accompanies drying out. Further, the pot itself should not be allowed to dry out, especial care being given to this possibility during winter months.

It is recommended in Japan that bonsai be kept outdoors, perhaps in a specially built stand, as much as possible. During the heat of summer they should be protected from excessive heat by shading with lath or screen, or brought indoors. Similarly, in winter they must be protected from excessive cold. An unheated greenhouse is considered a good location, and burying of the pot outdoors in winter is also a successful practice for hardy plants where winters are not severe.

Transplanting

Bonsai must be transplanted to fresh soil from time to time because of the exhaustion of the soil. Further, the root pruning that accompanies such an operation tends to keep the plant in hand. Evergreen plants are transplanted in early spring or in late summer. Deciduous plants are transplanted in early spring or in late fall. The operation should be completed before new growth starts. The longer a plant can be maintained without transplanting, the more valuable it is considered.

Containers

The selection of the container and the artistic placing of the plant are considered quite important. In fact, pots costing several dollars may be used, though most are in the neighborhood of a few cents. The name of the maker is stamped on pots, and they are selected with this in mind, just as with porcelain dinnerware.

Unglazed pots are used only for the first few years, while establishing a plant brought in from the wild or while a young plant is being trained. Glazed pots of both Japanese and Chinese manufacture are used for the permanent bonsai, the Chinese being often very colorful—yellow, blue, purple, and crimson. The colors of Japanese pots are generally white, ochre, black, blue, and green. Whitish- or brownish-colored pots are used for evergreens, while for deciduous trees gray and blue as well as white and brown are used. Plants with red flowers are put in white pots; those with white flowers, in red pots; and those with yellow flowers, in blue pots.

Sizes of pots range in diameter from 2 inches to 25 inches, averaging mostly about 16 inches, and in depth from 1 inch to 10 inches, averaging from 3 to 5 inches. Shapes most used are round, elliptical, and rectangular.

The pot, then, is carefully selected for both its utility and its artistic quality in relation to the plant to be placed in it. The tree is never set in the center of the pot. In a rectangular pot it is placed in back of the center and to one side. Hanging trees are placed in relatively deep pots in order to retain balance. Good specimens are planted alone. Less perfect specimens are often planted with other trees.

Training and Shaping

The training of a plant is a task calling for great patience and skill. A worker may spend an entire day training a single specimen. Work begins in spring with the new growth. The main trunk may be bent so as to give it a hanging effect, or some other interesting shape. For this purpose, metal plates and strong copper wire are used. Sixteen-gauge wire is used for large branches. The wire is heated slightly, wrapped spirally around the trunk, bent into the desired shape, and stiffened by cooling with water. Hand vises, iron supports, and cotton thread are additional parts of the equipment.

The main branches are next shaped in similar manner, followed by the smaller twigs and branches. Twelve-gauge wire may be sufficient for small branches. The young shoots are clipped back as desired, and the entire plant is trimmed for symmetry and attractiveness.

Since the plants are very slow-growing, there is little likelihood of constriction by wire or braces before a year or two. They may then be removed, and if the plant has attained the desired shape, they are removed. The careful technique of bending and fastening the wires and of shaping the plants is perhaps the most important phase of bonsai training.

FORMS OF BONSAI

It is of interest to note in passing that the vogue for certain species and forms has changed from time to time in Japan. From about 1829 to 1853 the Japanese apricot, the flowering cherry, oranges, and pines were preferred. From 1854 to 1867, pine and camellia were most grown. From 1868 to 1897 the red pine, black pine, and roses were given most attention. About 1902, the black pine, cedar, cypress, juniper, fir, pomegranate, maple, mulberry, Japanese apricot, and tree peony were in vogue. Since 1914, the trend has been toward black pine, cedar, juniper, fir, Japanese apricot, pomegranate, roses, wisteria, and azalea.

There are any number of interesting forms and patterns in which plants can be grown as bonsai. And, just as with fruit trees trained to special forms in Europe, the limit is defined solely by the imagination and perseverance of men. Thus, there may be such forms as pyramidal, columnar, wide conical, winding, erect, slant, cascade, raft, rock-planting, single, multiple, group, windswept, exposed-root, twisted-trunk, octopus, driftwood, and so on. Yet the vast majority of bonsai can be classified roughly according to (1) size, (2) potting arrangement, and (3) basic style or shape. Further, each of these can be subdivided into four groups.

Size Grouping

Based on size, the four groupings are (1) large, (2) medium, (3) small, and (4) miniature.

Large bonsai are 26 to 40 inches in height, and are usually grown in good-sized pots or tubs. There may be some question regarding the designation of these large forms as bonsai; but they are definitely not merely plants in tubs or large pots. They are trained as bonsai. Because of their size they are most often used outdoors.

Medium-sized bonsai are 12 to 26 inches in height, and are the most popular size in the Orient.

Small bonsai are 7 to 12 inches in height. The Japanese call them by a term that is translated "bonsai that can be carried in the hand." They are well suited to a windowsill or small table indoors, and therefore are especially esteemed in modern American homes and apartments.

The miniature bonsai are termed "baby bonsai" by the Japanese, and are under 7 inches in height. Many of these are grown as tiny plants in 1-inch pots, and are called "finger-tip bonsai."

Style by Number of Plants in a Group

The simplest style is that of a single plant, which may nevertheless be branched into several trunks.

The second style is that comprising two or more plants in a group. There are special names for the various numbers, such as "two trunks," "three trunks," and so on.

The third style consists of several trunks developed from a single connected root. Pine and maple are frequently used, and a forest effect is created. A related form is the "stump with shoots," developed by cutting off a tree just above the roots and training the shoots that arise. The fig, cherry, and pomegranate are especially suited. The "raft" form is developed from a horizontal stem or shoot, from which suckers are similarly induced to rise.

The fourth style is developed from one or more trees planted on a rough stone and suggestive of a tree on a rocky crag.

Style by Shape and Kind of Trunk

The simplest of these is the upright trunk.

The second style employs a slanting trunk developed as though growing on the side of a mountain slope.

The third style is the cascade, or "overhanging a cliff," in which the plant is trained downward over the edge of the container as though hanging over a cliff. Chrysanthemums are popular in this form, but pine, citron, and Japanese quince are also used.

The fourth form is the "gnarled or twisted trunk." It is developed to depict the force of weather, such as "windswept," "split trunk," "torn," and "decayed." A popular form is the "rock" form, produced by planting a tree carefully in or on a rock and training the root to conform to the shape of the rock.

PLANTS EMPLOYED

Although several hundred species of plants have been used as bonsai at one time or another, the more popular forms number in the neighborhood of fifty to seventy-five. They range all the way from conifers to broad-leaved evergreens, deciduous trees, flowering trees and shrubs, bamboos, reeds, and perennial herbaceous forms. A representative list of plants commonly used as bonsai follows:

Evergreen Trees

Cedar—*Cedrus atlantica* var. *glauca*
Cryptomeria—*Cryptomeria japonica*

Cypress—*Chamaecyparis obtusa* (Hinoki restinospera, in many varieties); *C. pisifera squarrosa*

Juniper—*Juniper chinensis sargentii* (Sargent juniper)

Pine—*Pinus densiflora* (Japanese red pine); *P. pumila* (Japanese stone pine); *P. parviflora* (Japanese white pine); *P. thunbergii* (Japanese black pine)

Spruce—*Picea jezoensis* (Yeddo spruce)

Yew—*Taxus cuspidata nana* (Dwarf Japanese yew)

Deciduous Trees

Beech—*Fagus japonica* (Japanese beech)

Gingko—*Gingko biloba* (Maiden-hair tree)

Liquidambar—*Liquidambar styraciflua* (Sweet gum)

Maple—*Acer buergerianum* (Trident maple); *A. palmatum* (Japanese maple, small-leaved forms).

Zelkova—*Zelkova serrata* (Gray-bark elm)

Flowering Trees and Shrubs

Apple—*Malus domestica; M. halliana; M. micromalus; M. sargentii; M. toringo*

Apricot—*Prunus mume* (Japanese apricot or Japanese umi, many forms)

Azalea—*Azalea* spp., many forms

Cherry—*Prunus incisa; P. glandulosa; P. serrulata* (flowering cherry); *P. tomentosa* (Nanking cherry)

Citrus—*Fortunella japonica*

Hawthorn—*Crataegus cuneata; C. oxyacantha*

Holly—*Ilex serrata*

Pear—*Pyrus serotina*

Pomegranate—*Pumica granatum nana* (Dwarf pomegranate)

Persimmon—*Diospyros lotus*

Pyracantha—*Pyracantha coccinea* (Scarlet firethorn); *P. koidzumi* var. Low Dense

Quince—*Chaenomeles japonica* (Japanese quince)

Grasses and Herbs

Acorns—*Acorus gramineus* (Sweet flag)
Bamboo—*Bambusa multiplex* var. Chinese God; *Phyllostachys aurea*
  (Golden bamboo)
Horsetail—*Equisetum hyemale* (Scouring rush)
Reeds—*Phragmites communis* (Common reed)
Sedges—*Cyperus alternifolia* var. *gracilis* (Slender umbrella plant)

29

DWARFED FRUIT TREES
AS ORNAMENTALS

F R U I T trees are measured for the most part for the value of the fruit they produce, yet if some of them had no fruit to offer they would still be useful as ornamental plants. Too frequently their dual role is overlooked. For one who wishes to take the time, there is ever-increasing satisfaction in the intimate knowledge of various fruit trees in their different moods and in different seasons of the year. In blossom, in form and shape of tree, in foliage characters, in individuality and attractiveness of fruit, in winter dress, they present personalities that are distinctive and interesting and well worth knowing—a hobby in itself.

DECIDUOUS FRUITS

The Fruit Tree as a Flowering Tree

By proper choice of varieties of fruit trees, the beauty of the flower may be blended with the utility of the fruit. Fruit trees may be used individually as specimen plantings, in groups against a background, or along a border or fence row, just as any flowering tree is used. More attention might well be paid to varietal differences in color of blossom, season of bloom, and size, beauty, and profusion of bloom. The dwarf fruit tree by virtue of its small stature, abundance of bloom, and early age of flowering and fruiting adds to the usefulness of fruit trees in these respects. Though not approaching the beauty of the flowering crabs or the Japanese cherries, many forms can nevertheless be found that have value for their flowers.

The pear, the Japanese plums, the European plums, and the sweet and sour cherries make interesting masses of white. The peach is commonly admired for the delicate pink color of its blossoms. Not all varieties of peaches are attractively large-flowered. Many varieties, such as Elberta and Crosby, have small salmon-pink blossoms. Others, such as Greensboro, Alton, and Carman, have the desired large, showy blossoms; and there is at least one variety of peach, Summer Snow, which has white blossoms, and another, Blood Leaf, which has a noticeable reddish tinge to the flowers.

The apple flower combines the pink of the unopened flower with the whiteness of full bloom. Some, such as Oldenburg, Red Astrachan, and Coxe's Orange Pippin, and the various crab types, are free-flowering. Others, such as Northern Spy and Yellow Newtown, are less so. Some flowers are large and showy, as are Oldenburg and R. I. Greening; others are smaller and more delicate, such as Jonathan and Coxe's Orange Pippin. Delicious carries a decided pink to the unopened flowers which makes them very attractive.

The blooming season of fruit trees covers a wide range of three to four weeks, depending upon the season. In the latitude of western New York, the season is heralded in late April or early May by the almond, apricot, native plums, sweet cherry, and Japanese plum. Following in quick succession are the peach, European plum, sour cherry, and pear. The apple and medlar conclude the season in late May, with such early-blooming varieties as Oldenburg leading the apple procession and such late-blooming varieties as Rome, Macoun, and Northern Spy bringing it to a close. It may be noted in passing that contrary to much belief, early-ripening varieties do not necessarily bloom earlier than late-ripening varieties. A more detailed account of the time of blossoming of the more important varieties of tree fruits will be found in the discussion in Chapter 25 on pollination.

The fragrance of apple blossoms is a byword. Peach blossoms, cherry blossoms, and plum blossoms also have fragrance. The blossoms of the pear, on the other hand, are not overly fragrant and are actually unpleasant in some varieties at full bloom.

Dwarf apple trees on the EM IX rootstock may be expected to carry some blossom the first year set, and may be veritable snowballs of bloom by the third or fourth year set. Apple trees on this rootstock are very dwarf, no taller than a man, and if low-headed at 12 to 14 inches, are more like low bushes of bloom than like trees. On the EM IV and EM VII rootstocks they may become as large as small peach trees and will bloom the second or third year set; on the EM I, II, and V, they are about the size of peach trees, and will bloom the second or third year set; and on the EM XIII and XVI they are about the size of full-grown sour cherry trees, and will bloom the third or fourth year set. The pear on the quince rootstock blooms at two to three years of age and develops to about as large as peach trees. The sour cherry on the mahaleb rootstock reaches about the size of full-grown peach trees and should bloom the third year set. Japanese varieties of plum will attain a size similar to a peach tree, and bloom the second or third year set. The European varieties of plum on dwarfing rootstocks will reach about the same size as a peach tree, and bloom the third year set. The peach will bloom the third year set.

Those varieties that tend to biennial bearing, such as Baldwin, are inclined to biennial blossoming as dwarf trees; and those varieties tending to annual bearing, such as McIntosh and Coxe's Orange Pippin, tend to annual flowering as dwarf trees, as well.

Flowering Peaches

The burst of flowers on various ornamental red (Early Double Flowering Red), pink (Early Double Flowering Pink), and white flowering peaches is one of the thrills of early spring. Unfortunately, the fruits are small, hard, and bitter. Unfortunately, also, the trees are vigorous-growing, and are adapted only to southern regions, such as Florida, the Gulf States, the Southwest, and southern California. However, there is a beautiful double-flowered, evergreen dwarfed variety (Chinese Dwarf Mandarin) that suggests further developments in combining dwarfness with double flowering. Some forms of dwarf flowering peach are sufficiently hardy in Michigan. As pot plants for early spring forcing they have become quite popular.

Plant breeders have succeeded to some degree in incorporating improved fruit characters into new hybrid forms. Four varieties produced in California are the Daily News Three Star, the Daily News Four Star, Altair, and Saturn. The first-named variety is a deep-pink, double-flowered peach with large, late-ripening fruits. The second is a large-flowered, pink double form with large white-fleshed, early-ripening freestone fruits. Altair combines large double light-pink flowers and large highly colored, white-fleshed freestone fruits ripening in midseason. Saturn has large double rose-pink blossoms, and large yellow-fleshed freestone fruits ripening in midseason. The fruit of none of these varieties is high in quality, but it is edible.

Flordahome is a pink-flowered variety introduced in 1960 from Florida. It is a second-generation seedling of Chico 11 (PI 146130) and *Prunus davidiana* (C-26712). Chico 11 is a seedling selected from the Shau Thai peach (PI 65821), introduced from China as seed. A special feature of Flordahome is its low chilling requirements (approximately 400 hours of 45 degrees F.), which adapts it to regions of mild winters. This quality also makes it easy to force.

Flowers of Flordahome are 2 inches in diameter with 32 to 36 petals per flower. The fruit is very early, freestone, oval, nearly 2 inches in diameter, lightly blushed with red, with soft white flesh of fair quality.

The Form and Shape of Trees

The various sizes and shapes of tree the different varieties provide are worth at least a passing glance. At once comes to mind the standard pair of plum trees, Abundance and Burbank, seen in so many farmyards, placed there by enterprising nurserymen of forty or fifty years ago—the one variety, Abundance, tall and upright, and the other, Burbank, short and spreading. The Seckel pear tree is naturally a shapely Christmas-tree-like pyramid. The Kieffer pear tree, unless severely pruned, is a coarse, scraggly grower. The Montmorency cherry tree is a vigorous-growing roundish tree; the English Morello cherry is a small almost weeping type. The McIntosh apple tree grows to very large size but with symmetrical roundish shape; the R. I. Greening is broad and spreading; the Northern Spy is upright spreading; and the Cortland, Gallia, and Rome Beauty trees are smallish and somewhat willowy and drooping.

Foliage Characters

Although seldom considered from the standpoint of color and general nature of the foliage, fruit trees show considerable differences in this regard as well. Some varieties, such as the Baldwin and R. I. Greening apples, have large thick, rich dark-green leaves. Others, such as the Jonathan, Coxe's Orange, and Winesap apples, have smaller, finer leaves that give a lighter, more feathery appearance. The common cultivated varieties of Oriental pear have relatively large, roundish leaves as contrasted with the smaller, more oval leaves of the typical European type. The foliage of peach varieties that have yellow-fleshed fruit carries a slight yellowish tinge, whereas those varieties with white-fleshed fruits are noticeably clearer green. There is one peach variety, Blood Leaf, that has reddish-green leaves. Among plums, the Pissardii types have distinctly reddish leaves. The Japanese types have the elongate, thin, peachlike leaf; and the European types have oval, thicker, darker-green leaves.

Fall coloring, too, is of interest. The Seckel pear becomes russet brown, and the leaves persist. The Kieffer pear often becomes a mass of brilliant, flaming red, and many sweet cherries take on a beautiful clear golden yellow garb before suddenly dropping their leaves.

Individuality and Attractiveness of the Fruit

Quite aside from the utility of the fruit, there is real pleasure and satisfaction in familiarity with the characters of size, shape, color, aroma, flesh texture, flavor, and all those numerous small details of individuality that stamp a variety almost as a personage. The dwarf fruit tree adds to the interest and the possibilities by permitting a greater number of varieties to be grown in a small area, by favoring fruiting at an early age, and by favoring the maturity of varieties that might not mature on standard rootstocks under similar conditions.

There are no limits to the possibilities in this direction, for there are literally thousands of varieties of fruits, and there are new acquaintances to be made each season as new varieties are introduced. By way of suggestion, the matter of size ranges among apples from the small-fruited Hyslop, Dolgo, American Beauty, and Transcendent crab apples to the large-fruited Wolf River. The small-fruited uncommon variety, Lady, with attractive red blush on one cheek, suggests in flavor the old-fashioned Christmas candies. The Coxe's Orange Pippin, a favorite dessert apple of England, brings deep yellow flesh and an unusual and rich flavor. The Delicious variety carries trim long oval shape and a crown of "five points," while the Senator gives flattened or oblate shape. The Melba brings white flesh and high quality in early season; the glistening, juicy Macoun provides mild, refreshing flavor in December and January. Golden Delicious introduces smooth waxy yellow color and excellent quality. Northern Spy contrives to bring together attractive size, shape, color, and flavor that approach perfection.

The pear also offers a great range of interest. There is the smallish sweet-meated Seckel; the larger bronzed, buttery Beurré Bosc; the large, coarse Duchess d'Angoulême; the red-cheeked, early-ripening Elizabeth; and the golden Dana Hovey.

Peaches present the honey-type Pallas with sweetish "honeyed" flavor; the large-fruited, firm J. II. Hale; the white-fleshed Champion; the yellow-fleshed Golden Jubilee; the early clingstone Mikado; and the late freestone Crosby, Salwey, and Smock.

Plums offer perhaps the widest range of interest of any of the fruits, running the gamut of colors, shapes, flavors, and sizes. There are the clear yellow, almost translucent Japanese variety, Shiro; the red-fleshed Satsuma, Wickson, and Santa Rosa; the small, meaty American Mirabelle; and the spicy Shropshire Damson. Pearl and Jefferson bring attractive yellow color and high quality, Pacific brings large size and blue color, and Hungarian brings long, datelike fruits.

The cherries, too, provide some degree of variety. Among sweet cherries there are the soft-fleshed Black Tartarian and the firm-fleshed Lambert; the black Schmidt and the white-fruited Napoleon; the early-ripening Coe and the late-ripening Giant; and the roundish Governor Wood and the long-pointed Elton. Sour cherries provide Early Richmond and Montmorency with light-colored juice, and English Morello and Chase with dark-colored juice. The Duke types bring characters midway between the sweet and sour types.

The Fruit Tree in Winter

In winter, too, fruit trees show individuality and charm. Outlined against the white landscape, silhouetted against a leaden sky or draped with snow, they make a pretty sight. And not only are there differences between distinct classes of fruit, there are also differences between varieties within the same class. The R. I. Greening apple tree bends and twists a few long sturdy gray branches outward in a spreading pattern, while the Early McIntosh sends a number of brownish branches upward in more erect fashion. The McIntosh tends to a symmetrical, naturally graceful and rounded top. The Cortland develops a network of finer, more willowy branches. The Yellow Transparent bark is a yellowish-brown; that of Ben Davis is reddish-brown; and that of Northern Spy is grayish- or greenish-brown.

Pear trees are among the most picturesque of fruit trees in winter. Dark in color, irregular or even angular in shape, they develop an interesting framework that shows to best advantage in winter. On a cold, clear midwinter night when a full moon casts shadows upon a clean white blanket of snow, a pear tree is a most fascinating object. The quince, too, with its bushlike shape, dark color, and much-branched habit is an interesting midwinter plant.

The cherry tree offers a smooth, reddish-brown bark. The peach has an open shape and light-brown or yellowish-brown bark. The Japanese plum is peachlike. The European plum has smooth, soft-looking dark gray bark. The apricot twists and bends its branches gracefully upward.

CITRUS AND RELATED FORMS

Where citrus and citrus relatives can be grown, they are most valuable as ornamental plants, quite aside from the value of the fruit itself. Most citrus have beautiful dark-green, glossy evergreen foliage. In addition there are interesting variegated forms. The blossoms are not very showy, but they are attractive and most fragrant, besides being almost everblooming in some varieties. Finally, the fruit is exceptionally showy and attractive, providing a great range of size, shape, and color. It may hang on the tree for a long time and overlap the bloom (Fig. 158).

A suggested selection has been made, by W. P. Bitters, of citrus for ornamental planting in California, which is here summarized:

Sour orange (Citrus aurantium)—grown for hundreds of years in the Mediterranean region, principally as an ornamental and shade tree, with abundant bright red-orange fruit and lush, dense, dark-green foliage.

Bouquet sour orange—identical with Bouquet des Fleurs, commonly called Bouquet, not to be confused with the Bergamot orange; rarely over 8 or 10 feet, grown as an ornamental shrub and hedge in California and Florida, extensively grown in southern France for perfume manufacture; fruit medium in size, sour, usable for ade and marmalade only.

Chinotto sour orange—one of the best of the naturally occurring dwarf forms; it bears heavy crops of attractive bright deep-orange-colored fruit, 1½ inches in diameter, which persist throughout the season, useful for preserves and juice.

Myrtifolia sour orange—small, symmetrical, somewhat columnar dwarf tree or shrub with thornless branches and very small, closely set leaves; a prolific bloomer, but sets poorly; useful for hedges and corner accents.

Salicifolia sour orange—a willowy-leaved selection; fruit not edible.

Citron (Citrus medica)—one of the oldest and best known of citrus fruits, tender to winter cold, best grown as individual trees; blossoms throughout most of the year; fruits large (5 to 7 inches in length), long-keeping, used mostly for peel.

Fingered citron—known as "Buddha's Hand"; floral end of fruit split into a number of fingerlike sections.

Limequat (C. aurantifolia x Fortunella sp.)—small, dwarfish plant, more hardy than the lime, thorny; fruit poor; Eustis and Lakeland varieties best.

Meyer lemon—often called Dwarf lemon; semidwarf, thornless, with dense lemonlike foliage; fruit light orange, juicy, lemon-flavored, available throughout the year.

Fig. 158. **Dwarf Meyer lemon used at corner of house.** *Courtesy of W. P. Bitters*

Ponderosa lemon—often called Wonder lemon or American Wonder lemon; small tree to 8 or 10 feet; fruit 5 or 6 inches in diameter, lemon-color, available throughout the year.

Rangpur lime—an acid lime, sometimes called a red lime, sometimes confused with Otaheite orange; a bushy dwarf tree, but may reach 15 or 20 feet; fruit oblate, up to 2½ inches in diameter, yellow-orange to reddish-orange, flesh and juice deep orange.

Otaheite orange—resembles an acidless lime; dwarf tree, nearly thornless, spreading; flowers tinted with purple; fruit orange to reddish-orange, 1½ inches in diameter, with odor of a lime; esteemed as potted plant.

Calamondin (Citrus mitis)—belongs to tangerine or loose-skinned group; tree tends to grow tall and columnar, also useful as pot variety on dwarf rootstocks or on its own roots, cold-resistant; fruits heavily year round, called "Sechi Chieh" (four seasons) by Chinese, ½ inch in diameter, deep orange, spectacular.

Cleopatra mandarin—also called spice tangerine and Ponki; small, somewhat pendant, dense tree, attractive with numerous small bright orange-red fruit, year round.

Skekwasha (Citrus depressa)—similar to calamondin and Cleopatra mandarin; fruit of better eating quality.

Fingerlime (Microcitrus sp.)—often referred to as Australian wild lime; handsome tall tree or small shrub, spiny; leaves very small, new growth long, pendant, purplish; flowers small, pink; fruit 3 inches long, ¾ inch in diameter, aromatic.

Kumquat (Fortunella sp.)—called golden orange and golden bean; dwarfish, but differing in growth habit; fruit small, 1 inch in diameter, bright orange, with edible peel, year round; useful as pot plant on dwarf rootstock; best varieties Nagami, Marumi, and Meiwa.

Wampee (Clausena lansium)—tree to 20 feet; leaves compound, like walnut; flowers in large panicles at end of branches, resembling lilac; fruit brown, less than an inch in diameter, edible, prized by the Chinese.

BIBLIOGRAPHY

Abbott, E. S. Freshen indoor decor with dwarf fruit trees. *Horticulture* 38: 418. 1960.

Abjornson, Eberhard. "Ornamental Dwarf Fruit Trees." A. T. De La Mare Company. New York. 1929.

Adriance, G. W., and F. R. Brison. "Propagation of Horticultural Plants." McGraw-Hill Co. New York. 1939.

American Pomological Society Proceedings. Boston, Mass. 1851–1909.

Argles, G. K. A review of the literature on stock-scion compatibility in fruit trees, with particular reference to pome and stone fruits. *Imp. Bur. Fruit Prod. Tech. Com.* 9:1–115. 1937.

Bailey, L. H. "The Pruning-Manual." The Macmillan Company. New York. 1923.

Baltet, C. "The Art of Grafting and Budding" (translated from the French). 6th Ed. Crosby Lockwood. London. 1910.

Barrett, H. C. Compatibility of certain pear-quince combinations. *Fruit Var. and Hort. Digest* 14:3–6. 1959.

Barry, Patrick. "The Fruit Garden." Orange-Judd Co. New York. 1851.

Batchelor, L. D., and H. J. Webber. "The Citrus Industry." Vol. II. Univ. Calif. Press. Berkeley, Calif. 1948.

Bates, Carlos G. The windbreak as a farm asset. *U.S.D.A. Farmers' Bul.* 1405. Rev. Ed. 1936.

Batjer, L. P., and Henry Schneider. Relation of pear decline to rootstock and sieve-tube necrosis. *Proc. Amer. Soc. Hort. Sci.* 76:85–97. 1960.

Beach, F. H. The home fruit garden. *Ohio Agr. Exp. Sta. Bul.* 252. 1947.

Beakbane, A. B. Intensive methods of apple and pear growing. *Jour. Royal Hort. Soc.* 72:145–154. 1946.

———. Possible mechanisms of rootstock effect. *Ann. Appl. Bot.* 44:517–521. 1956.

Beakbane, A. B., and E. C. Thompson. Anatomical studies of stems and roots of hardy fruit trees. II. The internal structure of the roots of some vigorous and some dwarfing apple rootstocks, and the correlation of structure with vigor. *Jour. Pom. and Hort. Sci.* 17:141–149. 1939.

Bitters, W. P. Dwarfing citrus rootstocks. *Calif. Citrograph* 34:516. Oct. 1949.

———. Citrus for ornamental planting in California. *Lasca Leaves* 7:29–44. 1957.

Blake, M. A. Observations upon summer pruning of the apple and peach. *Proc. Amer. Soc. Hort. Sci.* 14:14–23. 1918.

Bould, C. The mineral nutrition of fruit crops. In "Modern Commercial Fruit Growing." Ed. by T. Wallace and R. G. W. Bush. Country Life. London. 1956.

Brace, Josh. "The Culture of Fruit Trees in Pots." John Murray. London. 1904.

Bradford, F. C., and B. G. Sutton. Defective graft unions in the apple and the pear. *Mich. State Agr. Exp. Sta. Tech. Bul.* 99. 1929.

Brase, K. D. Similarity of the Clark dwarf and East Malling rootstock VIII. *Proc. Amer. Soc. Hort. Sci.* 61:95–98. 1953.

———. Observations on four released Malling-Merton rootstocks. *Farm Research.* March-May. 8. 1963.

Brase, K. D., and R. D. Way. Rootstocks and methods used for dwarfing fruit trees. N. Y. *State Agr. Exp. Sta. Bul.* 783. 1959.

Bretonneau, M. C. New stocks for grafting scions of fruit. *Rev. Hort.* 1:210–212 (3d Ser.). 1847.

Brooklyn Botanic Garden. "Pruning Handbook" 8:No. 2. Brooklyn, N.Y. 1952.

———. Handbook on Dwarfed Potted Trees—The Bonsai of Japan. Plants and Gardens 9:No. 3. 1956.

———. Trained and sculptured trees. Plants and Gardens 17:No. 2. 1961.

Brossier, M. J. Les problèmes posés par la sélection du cognassier. *Rpt. Cong. Pom. Orléans* 81–88. 1959.

Brown, J. W. Chemical studies on the physiology of apples. *Ann. Bot.* 40:129–147. 1926.

Bruinsma, J. Chemical control of crop growth and development. *Neth. Jour. Agric. Sci.* 10:409–427. 1962.

Buchloch, G. Zur Physiologie der Unterträglichkeit von Birnen: Quitten-Veredlungen. *XVth Int. Hort. Congress.* 43. 1958.

———. The lignification in stock-scion junctions and its relation to compatibility. In "Phenolics in Plants in Health and Disease." Pergamon Press. London. 1960.

———. Verwachsung und Verwachsungsstörungen als Ausdruck des Affinitätsgrades bei Pfropfungen von Birnenvarietaten auf Cydonia oblonga. *Beiträge zur Biologie der Pflanzen* 37:183–240. 1962.

Bukovac, M. J., S. H. Wittwer, and H. B. Tukey. Effect of stock-scion interrelationships on the transport of P^{32} and Ca^{45} in the apple. *Jour. Hort. Sci.* 33:145–152. 1958.

Bunyard, E. D. The history of the Paradise stocks. *Jour. Pom.* 1:166–176. 1920.

Bunyard, George, and Owen Thomas. "The Fruit Garden." Country Life. London. 1904.

Bush, Raymond. "Tree Fruit Growing." Penguin Books. Middlesex, England. 1943.

———. "Frost and the Fruit Grower." Cassell and Co. London. 1945.

Butterfield, H. M. Dwarfing ornamental trees and shrubs (mimeographed). Univ. Calif. (undated).

Calvino, Mario. Un porta-innestra per nunificate l'olivo. *Proc. IXth Int. Hort. Congress* 252–254. 1930.

Campbell, A. I. Apomictic seedling rootstocks for apples: Progress Report II. *Ann. Rpt. Long Ashton Res. Sta.* (1959):50 56. 1960.

Carlson, R. F., and H. B. Tukey. Cultural practices in propagating dwarfing rootstocks in Michigan. *Mich. Agr. Exp. Sta. Quart. Bul.* 37:492–497. 1955.

——— Fourteen-year orchard performance of several apple varieties on East Malling rootstocks in Michigan (Second Report) *Proc. Amer. Soc. Hort. Sci.* 74:47–53. 1959.

Chang, W. T. Studies in incompatibility between stock and scion, with special reference to certain deciduous fruit trees. *Jour. Pom. and Hort. Sci.* 15:267–325. 1938.

Chidamian, Claude. "Bonsai Miniature Trees." D. Van Nostrand Co. Princeton, N.J. 1955.

Christopher, E. P. "The Pruning Manual." The Macmillan Company. New York. 1957.

Cline, R. A. Studies on the interrelationships between mineral nutrition and the physiological and morphological dwarfing response of *Malus domestica* (Borkh.). Ph.D. thesis, Purdue University. West Lafayette, Ind. 1960.

Colby, H. L. Stock-scion chemistry and the fruiting relationships in apple trees. *Plant Phys.* 10:483–498. 1935.

Coxe, William. "A View of the Cultivation of Fruit Trees." Philadelphia, Pa. 1817.

Croux Fils. Catalogue général. Châtenay-Malabry. France. 1934–1935.

Daniel, L. The morphology and physiology of grafts (translated title). *Rev. Gén. Bot.* 6:5–21; 60–75. 1894.

Dawson, R. F. Accumulation of nicotine in reciprocal grafts of tomato and tobacco. *Amer. Jour. Bot.* 29:66–71. 1942.

Delbard, Georges. "Les beaux fruits de France." Georges Delbard. Paris. 1947.

Delplace, E. "Manuel d'arboriculture fruitière." J. Lamarre. Paris. 1933.

Dillon, F. C. and D. F. Dillon. The story of the Four Winds tree dwarf citrus. Mission San Jose, Fremont, Calif. 1959.

Du Breuil, M. A. Culture des arbres et arbrisseaux. Paris. 1876.

Duhamel du Monceau. "Traites des arbres fruitier." Paris. 1768.

Dunkin, Henry. "The Pruning of Hardy Fruit Trees." J. M. Dent and Sons. London. 1934.

Dwarf Fruit Tree Association. News Letter. East Lansing, Mich. 1958.

Eames, H. L., and L. G. Cox. A remarkable tree-fall and an unusual type of graft-union failure. *Amer. Jour. Bot.* 32:331–335. 1945.

Ferrari, G. P. "Hesperides; sive de malorum aureorum cultura et usu libri quatuor." Hermanni Scheus. Rome. 1646.

Floor, J., and A. K. Zweede. "Handbook for the identification of apple rootstocks." Lab. voor. Tuinboowpl. Wageningen. (N.A.K. Publication.) 1937.

Florida Agricultural Experiment Station. Pot Culture of Citrus Fruits. Gainesville, Florida. 1949.

Garner, R. J. The recognition of some apple and plum rootstocks in the nursery. *Ann. Rpt. East Malling Res. Sta.* (1945):130. 1946.

———. "The Grafter's Handbook." Oxford University Press. New York. 1949.

———. The nursery behavior of the Malling-Merton and Malling XXV apple rootstocks. *Ann. Rpt. East Malling Res. Sta.* (1952):55–63. 1953.

Garner, R. J., and H. D. Hammond. Studies in incompatibility of stock and scion. II. The relation between time of budding and stock-scion compatibility. *Ann. Rpt. East Malling Res, Sta.* (1937):154–57. 1938.

Gibault, Georges. Notice sur un tableau du musée du Louvre et l'origine des espaliers. *Jour. Soc. Nat. Hort. France.* (February, 1905) 113–118. 1905.

Gleisberg, W. Die Obstunterlagenselektion. *Der Züchter* 2:151–170. 1930.

Glenn, E. M. Plum rootstock trials at East Malling. *Jour. Hort. Sci.* 36:28–38. 1961.

Goldschmidt, V. H., and A. V. Delap. The spindle bush method of growing apple and pear trees. In "The Fruit Year Book 1950." *Royal Hort. Soc.* London. 1950.

de Haas, P. G. Studien über die "Freimachung" an 27 jährigen Birnen- und Apfelbuschbäumen. *Gartenbauwissenschaft* 10:610–650. 1937.

Harris, H. J., and J. J. Wood. Dwarf apple trees on Vancouver Island. *Canada Dept. of Agr., Saanichton Branch, Pub.* 171. 1959.

Hartman, H. T., and D. E. Kester. "Plant Propagation." Prentice-Hall. Englewood Cliffs, N.J. 1959.

Hatton, R. G. Paradise apple stocks. *Jour. Royal Hort. Soc.* 42:361–399. 1917.

———. Paradise apple stocks: Their fruit and blossom described. *Jour. Royal Hort. Soc.* 44:89–94. 1919.

———. A first report on quince stocks for pears. *Jour. Royal Hort. Soc.* 45:269–277. 1920.

———. Stocks for stone fruits. *Jour. Pom* 2: 1–37. 1921.

———. The behavior of certain pears on various quince roots. *Jour. Pom. and Hort. Sci.* 7:216–233. 1928.

———. Plum rootstock studies. *Jour. Pom.* 14:97–136. 1936.

Hatton, R. G., J. Amos, and A. W. Witt. Plum rootstocks. *Jour. Pom.* 7:63–99. 1928–1929.

Hatton, R. G., and N. H. Grubb. Field observations on the coincidence of leaf scorch upon the apple. *Jour. Pom. and Hort. Sci.* 4:65–77. 1924.

Hearman, J. The Northern Spy as a rootstock when compared with other standardized European rootstocks. *Jour. Pom. and Hort. Sci.* 14:246–275. 1936.

Hedrick, U. P. "The Plums of New York" J. B. Lyon. Albany, N.Y. 1911.

———. "The Cherries of New York." J. B. Lyon. Albany, N.Y. 1915.

———. "The Peaches of New York." J. B. Lyon. Albany, N.Y. 1917.

———. "The Pears of New York." J. B. Lyon. Albany, N.Y. 1921.

———. Dwarf apples. *N.Y. Agr. Exp. Sta. Bul.* 406. 1915.

———. Stocks for plums. *N.Y. Agr. Exp. Sta. Bul.* 498. 1923.

Herrero, J. Studies of compatibility graft combinations with special reference to hardy fruit trees. *Jour. Hort. Sci.* 26:186–237. 1951.

Hoblyn, T. N. Manurial trials with apple trees at East Malling 1920–1939. *Jour. Pom. and Hort. Sci.* 18:325–343. 1940–1941.

Horticulturist. Albany, N.Y. 1846–1875.

Hovey, C. M. Bush apple trees. *Mag. Hort.* 31:33–36. 1865.

Howe, G. H. Growth and yield of apple trees pruned in various ways. *N.Y. Agr. Exp. Sta. Bull.* 500. 1923.

Hudson, James. Fruit trees in pots. In "The Fruit Garden" by George Bunyard and Owen Thomas. Country Life. London. 1904.

Hull, George F. "Bonsai for Americans." Doubleday & Company, Garden City, New York. 1964.

Jensen, Martin. "Shelter Effect: Investigations Into the Aerodynamics of Shelter and Its Effects on Climate and Crops." The Danish Technical Press. Copenhagen. 1954.

John Innes Institute. Ann. Rpt. (1960):15–17. 1961.

Kenworthy, A. L. Nutrient element composition of leaves from fruit trees. *Proc. Amer. Soc. Hort. Sci.* 55:41–46. 1950.

Knight, R. C., J. Amos, R. G. Hatton, and A. W. Witt. The vegetative propagation of fruit tree rootstocks. *Rpt. East Malling Res. Sta.* 14, 15 for 1926–1927. II. Supplement: 11–30. 1928.

Knight, T. A. Account of some experiments in the descent of sap in trees. *Royal Soc. London. Phil. Trans.* 277–289. 1803.

———. On the formation of the bark of trees. *Royal Soc. London. Phil. Trans.* 103–113. 1807.

———. Physiological observations upon the effects of partial decortication, or ringing of the stems or branches of fruit trees. *Trans. Hort. Soc. London.* 4:159–162. 1822.

Kobayashi, Norio. "Bonsai-Miniature Potted Trees." Japan Travel Bureau. Tokyo. 1951.

Kramer, P. J., and T. T. Kuzlowski. "Physiology of Trees." McGraw-Hill Book Co. New York. 1960.

Krenke, N. P. "Wundkompensation Transplantation und Chimären bei Pflanzen." Julius Sprenger. Berlin. 1933.

Langley, B. Of the management of fruits trees after planting. In "Pomona." London. 1729.

Lecolier, Paul. Culture rationnelle des arbres fruitiers en pots. *Jour. Soc. Nat. Hort. France.* 254–274. February, 1905.

Lorette, Louis. "The Lorette System of Pruning." Translated by W. R. Dykes. London. 1925.

———. "La taille Lorette." 6th Ed. Bibliothèque de la Revue Jardinage. Versailles. 1926.

Lucas, Fr. "Die Lehre vom Baumschnitt für die deutschen Gartenbearbeitet." Eugene Ulmer. Stuttgart. 1909.

Lucas, I. B. "Dwarf Fruit Trees." A. T. De La Mare Co. New York. 1946.

Luckwill, L. C., and A. I. Campbell. The use of apomictic seedling rootstocks for apples: Progress report. *Ann. Rpt. Long Ashton Res. Sta.* (1953):47–53. 1954.

McKenzie, D. W. Rootstock-scion interaction in apples with special reference to root anatomy. *Jour. Hort. Sci.* 36:40–47. 1961.

McKenzie, W. F. "Fruit Culture for the Amateur." Garden Publication. London. 1947.

Maclean, Gordon A. "The Pillar System of Intensive Apple Production." The Abbey Press. Abingdon, Berkshire, England. 1948.

Magazine of Horticulture. Boston, Mass. 1835–1868.

Maney, T. J. Dwarfing apple trees by the use of an intermediate dwarf section in the trunk of the tree. *Proc. Ia. State Hort. Soc.* 78:127–135. 1943.

Maurer, E. "Die Unterlagen der Obstgehölze." Paul Parey. Berlin. 1939.

Ministry of Agr. and Fisheries, Bul. 2. "Tree Fruits." London. 1935.

———, Bul. 119. "Plums and Cherries." London. 1948.

———, Bul. 133. "Apples and Pears." London. 1948.

Ministry of Agr., Fisheries and Food, Bul. 135. "Fruit Tree Raising: Rootstocks and Propagation." London. 1956.

Mosse, B. A study of bark-wood relationship in apple stems. *Rpt. East Malling Res. Sta.* (1951) 39:70–75. 1952.

————. Further observations on the effects of ring-grafting peaches with an incompatible rootstock variety. *Jour. Hort. Sci.* 35:275–281. 1960.

————. Graft-incompatibility in fruit trees. Tech. Communication 28. Commonwealth Bureau of Horticulture and Plantation Crops. East Malling, Maidstone, Kent, England. 1962.

Nelson, S. H., and H. B. Tukey. Effects of controlled root temperatures on the growth of East Malling rootstocks in water culture. *Jour. Hort. Sci.* 31:55–63. 1956.

Northwestern Dwarf Fruit Tree Association (News Letter). Wenatchee, Wash. 1959.

Nozaki, Shinobu. "Dwarf Trees (Bonsai)." The Sanseido Company. Tokyo, Japan. 1940.

Preston, A. P. Some new apple rootstocks. *Rpt. 13th Int. Hort. Congress* 267–281. 1953.

————. Apple rootstock studies: the M IX crosses. *Ann. Rpt. East Malling Res. Sta.* (1953):89–94. 1954.

————. Apple rootstock studies: thirty-five years' results with Lane's Prince Albert on clonal rootstocks. *Jour. Hort. Sci.* 33:29–38. 1958.

————. Apple rootstocks studies: thirty-five years' results with Cox's Orange Pippin on clonal rootstocks. *Jour. Hort. Sci.* 33:194–201. 1958.

————. Thirty-five years' results with Worcester Pearmain on clonal rootstocks. *Jour. Hort. Sci.* 34:2–8. 1959.

————. Effect of tree density in an exposed apple orchard. *Ann. Rpt. East Malling Res. Sta.* (1959):52–56. 1960.

————. Pruning trials with dessert apples. *Ann. Rpt. East Malling Res. Sta.* (1959):122–127. 1960.

————. The behavior of fourteen clonal apple rootstocks in an exposed hilltop orchard. *Ann. Rpt. East Malling Res. Sta.* (1961):63–66. 1962.

Proebsting, E. L. Further observations on structural defects of the graft union. *Bot. Gaz.* 86:82–92. 1928.

de la Quintinye, J. "Instructions pour le jardins fruitiers et potages." Paris. 1690.

Rea, John. "Flora: Seu, de florum cultura." London. 1665.

Rémy, P. "Les porte-greffes du pommier." *Jour. Fruit Reg.* April 5, 1960.

Ritter, C. M., and L. D. Tukey. Growth and fruiting of various apple varieties in response to several rootstocks. *Penn. Agr. Exp. Sta. Bul.* 649. 1959.

Rivers, Thomas. "The Orchard House; or the Culture of Fruit Trees in Pots Under Glass." London. 1858.

————. "The Miniature Fruit Garden." 1st American Ed. New York. 1866.

Rivière, Auguste. "Traité d'arboriculture fruitière." Villefranche. 1928.

Roberts, A. N. Growth-controlled stocks in the orchard management picture. *Proc. Wash. State Hort. Soc.* 53:202–206. 1957.

Rogers, W. S. Advances in rootstock research. *Jour. Royal Agr. Soc. Eng.* 118:64–75. 1957.

Royal Horticultural Society and Geoffrey Cumberlege. "The Fruit Garden Displayed." Oxford University Press. London. 1951.

Samuels, E. W., and A. G. Dickson. The mechanism of controlled growth of dwarf apple trees. *Jour. Arnold Arboretum* 37:307–313. 1956.

Sax, Karl. Dwarf trees. *Arnoldia* 10:73–79. 1950.

———. The control of tree growth by phloem blocks. *Jour. Arnold Arboretum* 35:251–258. 1954.

Sax, Karl, and A. G. Dickson. Phloem polarity in bark regeneration. *Jour. Arnold Arboretum* 37:173–179. 1956.

Schneider, H. Anatomy of bark of bud union, trunk, and roots of quick decline affected orange trees on sour orange rootstocks. *Hilgardia* 22 (16):567–581. 1954.

Scientific Horticulture. A glossary of terms used in pruning fruit trees. 11:67–74. 1952–1954.

Sharpe, R. H. Flordahome—a double pink ornamental peach. *Fla. Agr. Exp. Sta. Circ.* S-125. 1960.

Shaw, J. K. The propagation and identification of clonal rootstocks for the apple. *Mass. Agr. Exp. Sta. Bul.* 418. 1944.

Shewell-Cooper, W. E. "Up-to-Date Fruit Growing." The English Universities Press. London. 1938.

Southwick, Lawrence. "Dwarf Fruit Trees." The Macmillan Company. New York. 1948.

Sprenger. A. M. (A table of identification.) *Cultura* 216–224. 1923.

———. Standardizierung von Obstunterlagen. *Gartenbauwissenschaft* 1:93–99. 1929.

de Stigter, H.C.M. Studies on the nature of the incompatibility in a cucurbitaceous graft. Publication 147, *Landbouwhogeschool te Wageningen* 56:1–56. 1956.

———. Translocation of C^{14} photosynthates in the graft muskmelon/*Cucurbita ficifolia. Acta Botanica Neerlandica* 10:466–473. 1961.

Thompson, C. M. "The Pruning of Apples and Pears by Renewal Methods." Faber and Faber. London. 1949.

Tufts, W. P. Pruning bearing deciduous fruit trees. *Calif. Agr. Exp. Sta. Bul.* 386. 1925.

Tukey, H. B. Stock and scion terminology. *Proc. Amer. Soc. Hort. Sci.* 35:378–382. 1938.

———. Dwarf and semi-dwarf apple trees for commercial planting. *Rpt. Conn. Pom. Soc.* 52:76–83. 1942.

————. Time interval between full bloom and fruit maturity for several varieties of apples, pears, peaches, and cherries. *Proc. Amer. Soc. Hort. Sci.* 40:133–140. 1942.

Tukey, H. B. (Ed.). "Plant Regulators in Agriculture." John Wiley and Sons. New York. 1954.

Tukey, H. B., and K. D. Brase. Granulated peat moss in field propagation of apple and quince stocks. *Proc. Amer. Soc. Hort. Sci.* 27:100–108. 1931.

————. Influence of the cion and of an intermediate stem-piece upon the character and development of roots of young apple trees. N.Y. *State Agr. Exp. Sta. Tech. Bul.* 218. 1933.

————. What yield of rooted shoots may be expected from mother plantations of Malling apple and quince rootstocks. *Proc. Amer. Soc. Hort. Sci.* 33:338–345. 1936.

————. Random notes on fruit tree rootstocks and plant propagation. N.Y. *State Agr. Exp. Sta. Bul.* 649. 1934.

————. Random notes on fruit tree rootstocks and plant propagation. II. N.Y. *State Agr. Exp. Sta. Bul.* 657. 1935.

————. Random notes on fruit tree rootstocks and plant propagation. III. N.Y. *State Agr. Exp. Sta. Bul.* 682. 1938.

————. Behavior of Malling apple rootstocks in soils of high, medium, and low moisture content. *Proc. Amer. Soc. Hort. Sci.* 37:305–310. 1940.

————. Three-year performance of sixteen varieties of apples on Malling IX rootstocks. *Proc. Amer. Soc. Hort. Sci.* 38:321–327. 1940.

————. The dwarfing effect of an intermediate stem-piece of Malling IX apple. *Proc. Amer. Soc. Hort. Sci.* 42:350–364. 1943.

————. Differences in congeniality of two sources of McIntoch apple budwood propagated on rootstock USDA 227. *Proc. Amer. Soc.· Hort. Sci.* 45:190–194. 1944.

Tukey, H. B., and R. F. Carlson. Five-year performance of several apple varieties on Malling apple rootstocks in Michigan. *Proc. Amer. Soc. Hort. Sci.* 54:137–143. 1949.

Tukey, R. B., R. L. Klackle, and J. A. McClintock. Observations on the uncongeniality between some scion varieties and Virginia Crab stocks. *Proc. Amer. Soc. Hort. Sci.* 64:151–155. 1954.

Tukey, R. B., and E. L. Schoff. Mulching as an agricultural technique. *Eastern Fruit Grower* 23:5. 1959.

Tukey, H. B., and H. B. Tukey, Jr. Practical implications of nutrient losses from plant foliage by leaching. *Proc. Amer. Soc. Hort. Sci.* 74:671–676. 1959.

Tydeman, H. M. Descriptions of some quince rootstocks. *Ann. Rpt. East Malling Res. Sta.* (1947):59–64. 1948.

————. A description and classification of the Malling-Merton and Malling

XXV apple rootstocks. *Ann. Rpt. East Malling Res. Sta.* (1952):55–63. 1953.

———. A description of certain M IX crosses. *Ann. Rpt. East Malling Res. Sta.* (1953):86–88. 1954.

———. Description of the Malling apple rootstocks. *Ann. Rpt. East Malling Res. Sta.* (1954):64–66. 1955.

———. A description and classification of certain plum stocks. *Ann. Rpt. East Malling Res. Sta.* (1956):75–80. 1957.

United States Department of Agriculture. Pruning hardy plants. *Farmers' Bul.* 1870. 1957.

Upshall, W. H. Throw away those ladders. *American Fruit Grower* 12. November, 1959.

———. Costs can be cut with dwarf apple trees. *Rural New-Yorker* 111: No. 5963. 28. 1961.

Vavilov, N. I. Wild progenitors of the fruit trees of Turkestan and the Caucasus and the problems of the origin of fruit trees. *Proc. IX Int. Hort. Congress* 271–286. London. 1930.

Waugh, F. A. "Dwarf Fruit Trees." Orange-Judd Company. New York. 1906.

Weber, H. J. Rootstock relations as indicating the degree of congeniality. *Proc. Amer. Soc. Hort. Sci.* 23:30–36. 1926.

Weber, H. J., and L. D. Batchelor "The Citrus Industry." Vol. I. Univ. Calif. Press. Berkeley. 1943.

Weiss, G. M., and D. V. Fisher. Growing apple trees on dwarfing rootstocks. *Canada Dept. of Agr., Summerland Branch Sta. Publication.* March, 1960.

Whitehead, Stanley B. "Fruit From Trained Trees." J. M. Dent and Sons. London. 1954.

Witt, A. W., and R. J. Garner. Rootstock trials. *Ann Rpt. East Malling Res. Sta.* II—Supplement 1928–1930:22–31. 1931.

Woodruff, N. P., and A. W. Zingg. Wind tunnel studies of shelterbelt models. *Jour. Forestry* 51:173–178. 1953.

Wright, D. Macer. "Dwarf Fruit Trees." Faber and Faber. London. 1953.

———. "Dwarf Pyramid Fruit Culture." Faber and Faber. London. 1959.

Yashiroda, Kan. "Bonsai Japanese Miniature Trees." Charles T. Brantford Co. Newton, Mass. 1960.

Yeager, A. F. Trunkless apple trees. *Proc. Amer. Soc. Hort. Sci.* 33:39–40. 1936.

Yoshima, Yuji, and G. M. Halford. "Miniature Trees and Landscapes." Charles E. Tuttle Co. Rutland, Vt. 1957.

Young, Floyd D. Frost and the prevention of frost damage. *U.S.D.A. Farmers' Bul.* 1588. 1929.

Zeiger, Donald, and H. B. Tukey. An historical review of the Malling apple rootstocks in America. *Mich. State Agr. Exp. Sta. Cir. Bul.* 226. 1960.

INDEX

Library of Congress Cataloging in Publication Data
(For library cataloging purposes only)

Tukey, Harold Bradford, 1896–1971.
  Dwarfed fruit trees for orchard, garden, and home.

  Bibliography: p.
  Includes index.
  1. Dwarf fruit trees. I. Title.
SB357.5.T8 1978   634   77–12289
ISBN 0–8014–1126–2